Leonidas Deligiannidis, George Dimitoglou and Hamid R. Arabnia (Eds.)
Artificial Intelligence

Intelligent Computing

Series Editors
Leonidas Deligiannidis, George Dimitoglou, Hamid R. Arabnia

Volume 1

Artificial Intelligence

Machine Learning, Convolutional Neural Networks and Large Language Models

Edited by
Leonidas Deligiannidis, George Dimitoglou
and Hamid R. Arabnia

DE GRUYTER

Editors

Dr. Leonidas Deligiannidis
School of Computing and Data Science
Wentworth Institute of Technology
550 Huntington Ave
Boston, MA 02115
USA
deligiannidisl@wit.edu

Dr. Hamid R. Arabnia
Computer Science
University of Georgia
415 GSRC
Athens, GA 30602-7404
USA
hra@uga.edu

Dr. George Dimitoglou
Department of Computer Science & IT
Hood College
401 Rosemont Ave
Frederick, MD 21701
USA
dimitoglou@gmail.com

ISBN 978-3-11-134400-3
e-ISBN (PDF) 978-3-11-134412-6
e-ISBN (EPUB) 978-3-11-134417-1
ISSN 2943-4432

Library of Congress Control Number: 2024934022

Bibliographic information published by the Deutsche Nationalbibliothek
The Deutsche Nationalbibliothek lists this publication in the Deutsche Nationalbibliografie; detailed bibliographic data are available on the internet at http://dnb.dnb.de.

Cover image: Naeblys/iStock/Getty Images Plus
Typesetting: Integra Software Services Pvt. Ltd.

www.degruyter.com

Preface

It gives me great pleasure to introduce this collection of papers to the readers of the book series "**Intelligent Computing**" (De Gruyter). This book is entitled *Artificial Intelligence: Machine Learning, Convolutional Neural Networks and Large Language Models.*

From the computer science perspective, the core of Artificial Intelligence (AI) includes Machine Learning. In recent years the growth in utilizing AI applications has been exponential. One reason for this exponential growth has been the advancement in Machine Learning; many give credit for this advancement to Deep Learning (via Convolution Neural Networks, CNN) and new applications in Large Language Models (LLMs). This book covers the emerging trends in AI, Machine Learning, CNNs, and LLMs. Machine Learning methods heavily rely on large datasets. Although the topic of Data Science is not explicitly addressed in this book, many algorithms and methodologies that appear in this book utilize Data Science methodologies.

This book is composed mainly of selected papers that were accepted for the 2022 and 2023 International Conferences on Computational Science and Computational Intelligence (CSCI: December, Las Vegas, USA and the 2022 and 2023 International Conferences on Artificial Intelligence (CSCE/ICAI: July, Las Vegas, USA). Selected authors were given the opportunity to submit the extended versions of their conference papers for publication consideration in this book. An important mission of CSCI and CSCE annual conferences includes *"Providing a unique platform for a diverse community of constituents composed of scholars, researchers, developers, educators, and practitioners. The Congress makes concerted effort to reach out to participants affiliated with diverse entities (such as: universities, institutions, corporations, government agencies, and research centers/labs) from all over the world. The congress also attempts to connect participants from institutions that have teaching as their main mission with those who are affiliated with institutions that have research as their main mission. The congress uses a quota system to achieve its institution and geography diversity objectives."* Since this book is composed of the extended versions of the accepted papers of CSCI and CSCE annual conferences, it is no surprise that the book has chapters from highly qualified and diverse group of authors.

We are very grateful to the many colleagues who offered their services in organizing the CSCI and CSCE conferences. Their help was instrumental in the formation of this book. The members of the editorial committees appear on the websites of CSCI and CSCE.

We express our gratitude to Steve Elliot (De Gruyter Editor) and Aleksandra Ślosarczyk (De Gruyter Editorial Project Manager). We hope that you benefit from reading this book as much as we did.

Leonidas Deligiannidis, Ph.D.
George Dimitoglou, Ph.D.
Hamid R. Arabnia, Ph.D.

https://doi.org/10.1515/9783111344126-202

Contents

Machine learning (ML)

Omobayo Ayokunle Esan, Munienge Mbodila, and
Patrick Mukeninay Madimba

Detection of lesions in breast image using median filtering and convolutional neural networks

Abstract: Many people all over the world are impacted by cancer, a serious health issue. This illness has already claimed many lives and will do so in the future. Breast cancer has recently surpassed cervical cancer as the most prevalent cancer in women. At present, mammography screening is used for early detection of any form of lumps or lesions in breast images before developing cancer. Accurate screening and detection of breast lesions is a challenging issue for many medical practitioners even with the use of mammography imaging due to the process of interpretation of mammogram results, which are often done subjectively through visual analysis consequently leading to some breast lesions going unnoticed. The skewed median filtering technique on a convolutional neural network (CNN) is used to accurately and quickly identify breast lesions in their early stages. A publicly accessible breast dataset (Wisconsin) was used for the experiments. The results demonstrate that the proposed methods outperform other techniques in terms of F1 score, precision, recall, and accuracy with 0.9661, 0.9881, 0.9783, and 98.64%, respectively in comparison with the other methods. This model can help radiologists identify breast regions where cancer is most likely to develop in the future.

Keywords: breast lesion, nose removal, accuracy, diagnosis, CNN, detection

1 Introduction

One of the leading causes of cancer-related mortality among women is breast cancer [1]. Breast cancer remains a significant global health concern, emphasizing the critical need for accurate and efficient methods to detect lesions in breast images [2].

Acknowledgments: The authors gratefully acknowledge the financial support and resource made available by the Walter Sisulu University, South Africa.

Omobayo Ayokunle Esan, Department of InformationTechnology Systems, Walter Sisuslu University, South Africa, e-mail: oesan@wsu.ac.za
Munienge Mbodila, Department of InformationTechnology Systems, Walter Sisuslu University, South Africa, e-mail: mmbodila@wsu.ac.za
Patrick Mukeninay Madimba, Mukemad Medical Health Practice Inc., South Africa, e-mail: mukemad2@gmail.com

https://doi.org/10.1515/9783111344126-001

Screening finds disease indicators, like breast cancer. The purpose of human breast screening is to identify the disease at an early stage. A variety of screening methods, including thermography, magnetic resonance imaging (MRI), breast exams, and mammography, are available; mammography screening is the popular method that most medical practitioners normally use to detect tumors in breast image [2, 3].

Currently, human experts manually interpret mammography images [2, 4]. Manual mammography image visual inspection is done based on human judgment and is subject to human error [2, 5]. Furthermore, visualization of the mammography image manually requires medical practitioners to spend much time in detection of any suspicious patterns in the image and this practice can be laborious and overwhelming.

The report generated from World Health Organization (WHO) in 2021 indicated that in the last three years, over 7.8 million women had been diagnosed with breast cancer, which makes the disease to be common globally [6]. Every year, 45.5% of new cases of cancer are reported worldwide [1]. Early detection has been shown to increase a patient's chances of survival. A type of cell that invades surrounding tissues in the body and does not grow normally is called a cancerous tumor. The two categories of breast cancer tumors are benign and malignant [7]. Due to higher diagnosis rates, a variety of medical imaging techniques, including MRI, X-ray, CT, ultrasound, and endoscopy, are used in the medical field for many complex medical image analyses. Faster screening time with scalability, consistency, and precision are necessary in the current mammography imaging diagnostic system [8].

At present, the diagnostic imaging methods commonly used by medical practioners to physically examine different organs [2, 3] is ultrasound. However, depending on the operator, the quality and interpretation of images may differ [2–4]. There is a high error rate in breast ultrasound examination because of the appearance of noise and damage in the breast images, which makes it difficult to determine whether the breast is cancerous or not [2, 9].

In this era of advancing technology, the fusion of traditional image processing techniques with cutting-edge deep learning approaches has shown promise in enhancing breast lesion detection. To increase radiologists' ability to accurately identify lesions in breast images, this study uses median filtering in conjunction with a convolutional neural network (CNN) technique to address the noise gap and inaccuracy in breast image detection. To enhance and identify lesions impacted by noise, this work develops an intelligent framework based on the merging of the CNN and median filter techniques.

This paper explores the integration of median filtering, a well-established image enhancement technique, with CNNs, a state-of-the-art deep learning architecture, for the precise and timely identification of breast lesions in medical images. By combining the strengths of these two methodologies, this study aims to address the challenges associated with breast cancer diagnosis, ultimately contributing to early detection and improved patient outcomes. Hence, the main research question raised is as follows:

1.1 Research question and contributions

The research poses the question: *"How can medial filtering be integrated into deep learning to enhance the detection of unobserved breast lesions in mammography imaging that can lead to cancer?"* The research contributes in the following ways:

- **Removal of noise:** This can enhance the accuracy of the mammography imaging in detecting unnoticed lesions in the breast image.
- **Improved accuracy:** Reduction in the false errors in the current breast mammography for the detection of lesions in breast images can improve the detection of unobserved lesion by medical practitioners.
- **Reduction of computational time:** Reducing computational time on graphic processing unit (GPU), thereby addressing a critical aspect of real-time detection systems and enabling more responsive treatment.
- **Detailed experimental evaluations:** The proposed technique is evaluated with different metrics and benchmarked with related state-of-the-art breast detection techniques, using publicly available data.

This paper is divided into the following sections, which are listed in the following order: Section 2 gives a summary of related techniques that are currently in use as well as the theoretical underpinnings of the suggested technique. CNNs and median filtering are explained in detail in Section 3. Section 4 discusses a number of techniques evaluations and experiments. Section 5 contains the concluding remarks.

2 Background

The detection of breast lesions is a vital aspect of breast cancer diagnosis and early intervention. Breast cancer is one of the most prevalent malignancies globally and a leading cause of cancer-related deaths among women. To improve the accuracy and efficiency of lesion detection in breast images, various image-processing techniques and machine-learning approaches have been explored [2]. Among these, median filtering has been widely used to reduce image noise and enhance relevant features, while CNNs have emerged as powerful tools for image classification and object detection. This background explores various techniques and their challenges in breast lesion detection, with a focus on the potential synergy between these techniques in improving the accuracy of breast lesion detection. By leveraging their combined strengths, this study aims to contribute to the development of more robust and reliable techniques for breast cancer diagnosis and patient care.

2.1 Related techniques

Researchers have conducted a number of studies to identify breast abnormalities before they become cancerous. Table 1 provides an overview of the currently available methods for detecting breast cancer, including information on the problem addressed, the methods used, the results obtained, and their limitations.

One can see that the existing literature regarding noise and unobserved lesion detection in the breast within mammography imaging systems has apparent limitations. Previous research has contributed immensely but not enough work has considered noise disturbances to reveal unobserved lesions in the mammography image screening that may lead to breast cancer. This research greatly strengthens solutions to these weaknesses with the development of the integration of median filtering and CNN techniques to address the noise and misinterpretation of lesions in the breast image due to unobserved lesions during mammogram screening. The theoretical background used in this research is explained in Section 2.2.

2.2 Selected theoretical techniques

Median filtering

The nonlinear filter processing technique based on statistics is called median filtering, as described in [2, 22]. As in eq. (1), the noise value in the image is substituted by the average value of its neighbors (mask), where $f(x,y)$, $g(x,y)$ is the original image and the output image W is a two-dimensional mask, which can be linear, square, circular, etc.

$$g(x,y) = med\{f(x-i,\ y-j),\ i,\ j \in W \tag{1}$$

One of its advantages is the ability to average the filter to effectively remove unwanted noise from images while remaining simple to apply. The next section explains the underpinning theory of CNN that is used to extract and detect lesions in images.

Convolutional neural network

CNN is a special type of artificial neural network that filters input through a convolutional layer to produce useful information for the network [23]. The following sections explain in more detail the different iteration layers and activation functions that CNNs commonly use.

Table 1: Some of the related works on breast cancer detection methods.

Citations	Problem addressed	Method used	Result obtained	Limitations
[10]	Reducing misclassification problems during breast cancer prediction.	A bagging technique with gradient boosted tree (GBT) was used.	The experimental result shows that integrating the bagging technique with GBT can effectively reduce misclassification and improve breast cancer prediction.	The approach could not detect microcalcification masses in the breast.
	Early detection of breast cancer.	The decision tree technique was used for classification.	The experimental study conducted gives an accuracy of 94.563% with an incorrect classification of 5.4363%.	The approach utilizes only a few datasets for implementation.
[11]	Lack of existence of a precise and reliable system for the diagnosis of benign or malignant tumor.	Logistic regression and group method data handling (GMDH) neural networks were used.	The results of the simulation demonstrate that the suggested method produces an accuracy of 99.4% for WBCD, 99.6% for WDBC, and 96.3% for the WPBC dataset that was used.	The approach did not consider feature extraction.
[12]	Discovering hidden relationships in data.	The likelihood that women will develop breast cancer was calculated using the Naive Bayes method.	Experimental results show that the classifier provides an improved accuracy of 98% with low computational efforts and very high speed.	Only a few attributes of the dataset were used for the implementations.
[13]	Extraction of tumor classification features and detection of suspicious regions with low background contrast that may be tumors.	Support vector machines (SVM), segmentation, and mammography advancements were all used.	Experimental findings show that this method had an 88.75% sensitivity rate when tested on 75 mammography images from the Mini-MIAS database.	The approach was used with only a few datasets and does not justify the model's effectiveness when used with a large dataset.

(continued)

Table 1 (continued)

Citations	Problem addressed	Method used	Result obtained	Limitations
[14]	Unnatural and uncontrollable cell development in the breast.	Machine learning (ML) method is used.	The outcomes demonstrate that the huge data set is required in order to increase accuracy when using the ML algorithm.	
[15]	Inaccurate diagnosis of breast cancer due to varied image quality and interpretation.	Deep learning-based anomaly (sliced-Wasserstein auto-encoder) is used.	Experimental results showed that sliced - Wasserstein auto-encoder model outperformed the other AE-based models.	The approach is computationally expensive.
[16]	Accurate detection of breast cancer cells.	Quantum convolutional neural networks (QCNNs).	According to experimental findings, this method can train a ten-qubit array to learn from the label of the input dataset and generate results with a low error rate.	The approach is computationally expensive.
[17]	Classification of new breast cancer images for newly generated images.	Image preprocessing, hybrid median filtering (HMF) for eliminating noise, and image contrast are enhanced using quadrant dynamic histogram equalization (QDHE), ROI segmentation using USE-Net deep learning model, and classification using random forest (RF) with extreme boosting (XGB).	The model's output had a 97.8% accuracy, a 98.45% sensitivity, a 95.24% specificity, a 98.96% precision, and an F1-score of 98.7%.	Requires more time to be implemented. Hence, might not be suitable for real-time practice.

[18]	Misinterpretation of mammography results in an unnecessary biopsy of the false-positive results, lowering the patient's odds of survival.	A new deep learning (DL) model based on a combination of transfer-learning (TL) and long short-term memory (LSTM).	The classification of mammographic data using the proposed model produced overall accuracy, sensitivity, specificity, precision, and area under curve (AUC) values of 99.23%, 98.8%, 99.1%, 96%, and 0.998, respectively.	Requires the knowledge of experts' domain.
[19]	Difficulties in characterizing calcification mammography in an automatic and robust way.	The calcification was characterized by descriptors obtained from deep learning and handcrafted descriptors.	According to experimental results, the filtered depth features have the best classification accuracy (89.32%) and sensitivity (86.89%) of any feature set.	Computationally expensive.
[20]	Inaccurate detection of microcalcification and radiologist's visual interpretation of mammograms.	CAD system based on mask R-CNN with multi-task learning to detect breast cancer and segment mammogram images.	The PCA + Mask R-CNN gives a specificity of 0.9 and an AUC of 0.9.	Requires knowledge of the expert domain.
[21]	Breast cancer early detection with the highest accuracy.	A decision tree-based data mining technique.	The decision tree's accuracy in the first trial was 100%, while it was 97.9% in the follow-up inquiry.	The issue of noise that affects detection accuracy is not addressed.

Layer: convolutional

This layer performs the convolution operation using convolution filters, also known as kernels, which have a defined size and cover the entire input data. One-pass filtering guides the recognition of patterns from previous layers, as shown in eq. (2), as the filter moves through the input matrix:

$$(f_k)_{i,j} = (W_k * X)_{i,j} + b_k \tag{2}$$

where $(f_k)_{(i,j)}$ is the convolution image representing X as the input image, W_k as the weight, and b_k as the bias.

Layer: activation

ReLU (Rectified Linear Unit) applies an unsaturated activation function, setting the negative values of the activation map to zero as in eq. (3), to remove them from the map. The nonlinear properties of the decision function and the entire network are improved by:

$$f(x) = \max(0, x) \tag{3}$$

ReLU uses eq. (3) without affecting the receptive field of the convolutional layer.

Layer: pooling

The pooling layer in CNN is used for downsampling. In addition to implementing rotational and translational transform invariance as in eq. (4), grouping also reduces dimensionality:

$$y[i] = \sum_{k=1}^{k} x[i + r.k]w[k] \tag{4}$$

where, $x[i]$ is the 2D input image, $w[k]$ is the filter of length k, r is the sampling step, and $y[i]$ is the output of the convolution image. The methodology used in the detection of lesion (abnormalities) in breast images is described in detail in the following section.

3 Method

The development of deep learning computer vision frameworks has witnessed remarkable progress over the past decade, revolutionizing the way we perceive and interact with visual data. These frameworks have played a pivotal role in advancing the field of computer vision by providing researchers and developers with the tools and resources to create sophisticated neural networks for tasks like image classification, object detection, semantic segmentation, and more [23, 24]. Their open-source nature and user-friendly interfaces have democratized deep learning, enabling a broader community of scientists, engineers, and enthusiasts to harness the power of CNNs and other deep learning architectures [2, 23, 24, 26]. This ongoing evolution of deep learning computer vision frameworks has not only accelerated the development of innovative applications in fields like autonomous vehicles, healthcare, and augmented reality but also promises further breakthroughs in our ability to understand and interpret visual information in the digital age.

This section mainly discusses the advancement of deep learning combined with medical image processing to detect breast lesions. The proposed system architecture is divided into three main stages: and image acquisition stage, image preprocessing stage, and detection stage, as shown in Figure 1.

3.1 Phase 1: collected data

Breast images used in this study were obtained from the publicly available Wisconsin dataset [2, 24]. The images are sent to an image preprocessing phase for further processing.

3.2 Phase 2: image preprocessing and feature extraction stage

To achieve computational efficiency, image preprocessing has become a standard procedure in image processing [2]. The top layer illustrates this, as shown in Figure 1.

Noise removal: median filtering

The acquired breast images are filtered to remove noise and artifacts from the image. In this research, the median filtering technique is used as described in eq. (1) to discard artefacts from the breast image and help in discrimination of the boundary between the required object and background [2].

Image background subtraction

The enhanced image output is fed into the feature extraction stage in this research in order to segment the breast. Here, the image differencing technique [2, 25] is used to extract the pertinent features from the breast image, as shown in eq. (5).

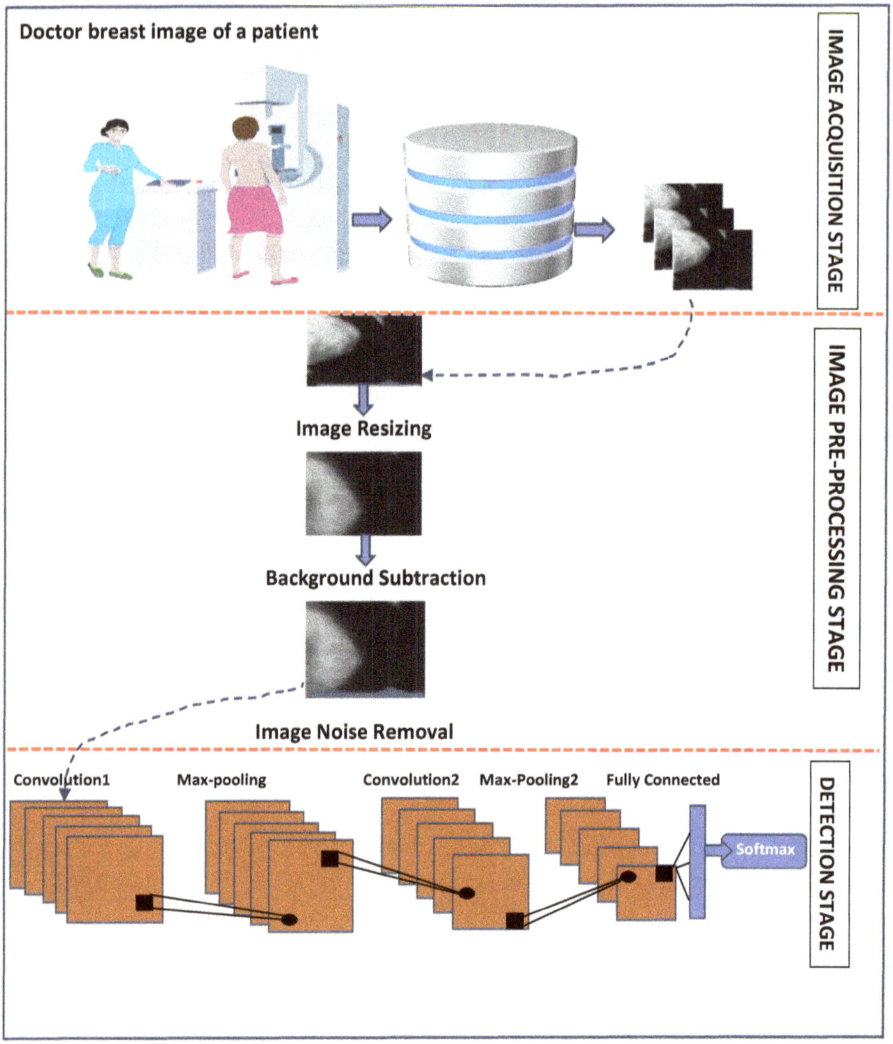

Figure 1: System architecture of detection of breast lesion using median filtering and CNN technique.

$$Foreground = |I(x,y,t) - I(x,y,t-1)| \tag{5}$$

where, $I(x,y,t)$ represents the current image, and $I(x,y,t-1)$ represents the frame from the prior image. The foreground picture that is fed into the detection stage is preserved during this operation.

3.3 Phase 3: convolutional neural networks

The CNN receives the output of the extracted features, as shown in Figure 1. Three pooling layers and three convolutional layers are used in this case. Table 2 illustrates the filter sizes of the 256-, 128-, and 64-channel convolutional layers used. The ReLU activation function activates the maximum pooling layer consisting of two steps after each convolutional layer and the filter size is 2 × 2. To determine the location and semantics of breast cancer, deep features are extracted from the image breast using the final fully connected layers. To classify and detect pixels at the pixel level, the encoded data must be linkable to the original pixels. In the visualization used to conduct this study, Table 2 shows the list of parameters for the fully connected network layer.

Table 2: Parameters of the network used for the implementation.

Layer	Parameter	Activation
Conv -^{1}1	(32, 3, 3)	ReLU
Pool_max	(2, 2)	–
Conv -^{1}2	(32, 3, 3)	ReLU
Pooling 2	(2, 2)	–
Deconv -^{1}1	(32, 4, 4)	ReLU
Deconv -^{1}1	(32, 4, 4)	ReLU
Conv -^{1}2	(32, 2, 2)	
Softmax		

As shown in eq. (3), the ReLU activation function is used to increase the nonlinear characteristics of the decision function in the neural network. The Softmax function is then used to categorize the image's patterning as normal or abnormal. The batch size, which is used to train each image frame, the epoch parameter selection, which controls how the model adapts to the detection of lesions, the kernel, which is used by convolution to extract features, and the learning rate, which controls how much adjustment would be applied to the model for loss gradient, are the parameters that are embedded in this model during implementation.

3.4 Model training, validation, and testing

The dataset is divided into training, validation, and testing sets as shown in Figure 2 using cross-validation technique; the model is tested on validation dataset after being trained. The training, testing, and validation datasets are partitioned to 80%, 10%, and 10%, respectively. Overfitting and bias are avoided by repeating this procedure.

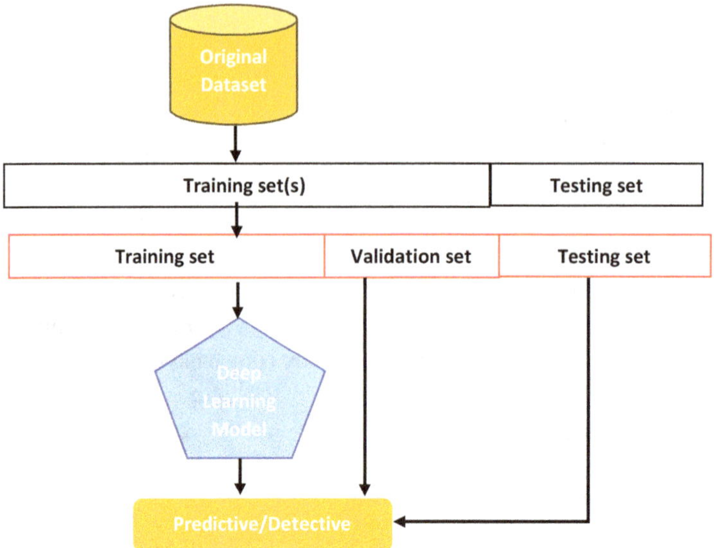

Figure 2: Flowchart of the dataset cross-validation.

3.5 Building model by median filtering and CNN methods with algorithmic analysis

The implementation pseudocode for the suggested model is displayed in Figure 3. The resulting images are median-filtered to remove any noise or unwanted artifacts that could affect the model's performance. The output of the enhanced breast image is then passed to background subtraction to separate the foreground from the background image. The anterior chest image is then sent to CNN for detection.

3.6 Performance evaluation mechanism

Through qualitative and quantitative measurements, this section introduces the performance evaluation of the suggested technique. In the quantitative test, the quality comparison between the filtered frame and the noisy images are shown.

The Mean Square Error (MSE) and Root Mean Square Error(RMSE) are computed as in eqs. (6) and (7), respectively [2, 26]:

$$MSE = \frac{1}{MN}\sum_{i=1}^{MN}\left(y_i - y_i^a\right)^2 \tag{6}$$

Algorithm 1: Breast Cancer detection using median filtering and CNN

```
Input: Breast Image I
Initialize histogram H as the starting point of k
    for i = 1 to M do
     for j = 1 to N do
      for k = -n/2 to n/2 do
       remove X_{i+k, j-N/2-1} from H
       add X_{i+k, j+N/2} from H
       update k
     endfor
   Y_{i,j} ← median(h, k)
    endfor
endfor
// pseudocode for background subtraction
Input: breast Image I, B
Output: foreground for every input image
Initialize B as image background [B(x, y, t)]
I ← input frame [I(x, y, t)]
    if (I, B) > threshold
      return forground
      else
 background
// convolutional Neural Network
input: Breast Image {Y}
output: model parameter θ
initialize θ from Gaussian distribution (μ = 0, σ = 0.001)
    for 1:n                                   //n is the number of training image
     for l = 1:3                              // layers to fine tune
      for(l=0;1<L; 1++){
       for(m=0;m<M; m++){
       for(n=0;n<N; n++){
         Sum = bias[l]
          for(k = 0; k < K; k + +){
           for(sl = 0; sl < SL; sl + +){
            for(s2 = 0; s2 < S2; s2 + +){
          Sum+ = weight[k][l][s1][s2] × input[k][m + sl][n + s2];
    }
      Output[l][m][n] = activation_func(sum);
end
```

Figure 3: Pseudocode for detection of lesion in breast image using proposed method.

$$RMSE = \sqrt{\frac{1}{MN}\sum_{i=1}^{MN}\left(y_i - y_i^a\right)^2} \tag{7}$$

where, y_i and y_i^a denote the pixel values of the noisy and original breast image respectively, and $M \times N$ is the breast image size.

Confusion matrix: This is a tabular representation of breast pattern instances that are correctly detected and those that are incorrectly detected, as illustrated in Table 3.

Table 3: Confusion matrix adapted from [27].

	Predicted/detected		
Actual		Normal	Abnormal
	Normal	w	x
	Abnormal	y	z

where w represents True Positive (TP), x represents True Negative (TN), y represents False Positive (FP), and z is False Negative (FN). These terms are further explained in eqs. (8)–(11).

Precision: This is the division of normal instances considered to be normal with summation of normal instance considered to be normal and abnormal instance considered to be normal by the model [28]. This is represented in eq. (8):

$$Precision = \frac{w}{w+y} \tag{8}$$

Recall: This is the division of normal instances considered to be normal with summation of normal instance considered to be normal and abnormal instance considered to be abnormal by the model [29], which can be calculated as in eq. (9):

$$Recall = \frac{w}{w+z} \tag{9}$$

F1-score: This is the ratio of precision multiplied by recall to summation of precision and recall then multiplied by 2, as shown in eq. (10):

$$f1 - score = 2 \times \frac{Precision \times Recall}{Precision + Recall} \tag{10}$$

Accuracy: This is the summation of normal instances considered to be normal and abnormal considered to be abnormal to the sum of instances in the dataset by the model, which is computed as in eq. (11):

$$Acc. = \frac{w+z}{w+z+x+y} \tag{11}$$

The metrics in eqs. (6)–(11) are used for the implementation and validation of the results in this research. The next section shows the experimental evaluation and results of the proposed selected method for detection of lesion in breast images.

4 Experimental evaluations and results

4.1 Data description and experimental settings

Python 3.6 is the implementation tool used in this study. As the platform to deploy systems with the necessary Python libraries and the Spyder 3 IDE platform to perform Python programming, we chose Jupyter (Anaconda 3) Python. CNN model hardware requires high GPU configuration for high performance to train the largest possible network scale.

4.2 Experimental results and analysis

This section evaluates the performance of the proposed method compared to other currently used detection methods on a publicly available breast dataset (Wisconsin) [2, 24].

Experiment 1: observations with abnormality in breast image frames

Evaluating the effectiveness of the proposed technique on noisy images is the main goal of this section. Figure 4(a) is the breast image containing 10% noise. As can be observed, the added noise in the Figure 4(a) is lighter compared to Figure 4(d) with 70% noise level [2].

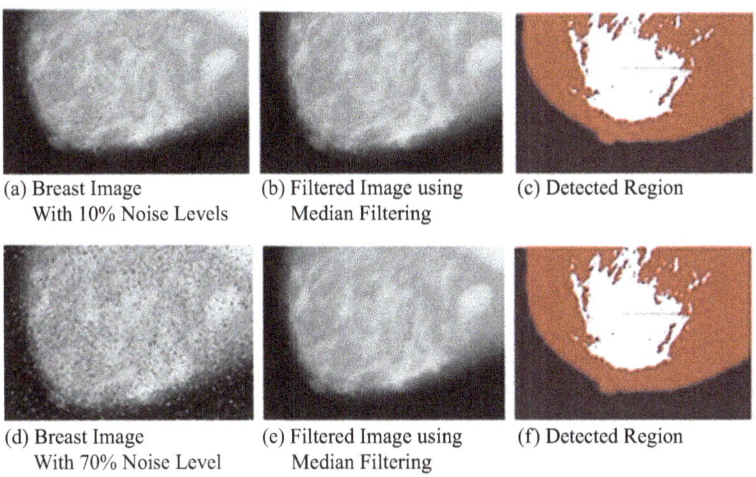

Figure 4: Quantitative evaluation of image with noise and the enhanced Image using median filtering and CNNs.

Breast images are shown with noise in Figures 4(a) and (d); Figures 4(b) and (e) are the results of the median filtering technique applied to noisy breast images. The detection zones are shown in Figures 4(c) and (f). According to Table 4, one can observe how well the proposed model performs at noise levels ranging from 10% to 70%.

Table 4: Comparing the noisy image and the filtered image based on the noise levels.

	Noisy image		Median-filtered	
Noise level (%)	**MAP**	**RMSE**	**MAP**	**RMSE**
10	17.33	41.92	18.69	42.22
20	18.81	42.45	19.29	43.21
30	19.49	44.51	20.94	45.94
40	20.11	46.61	22.24	48.84
50	25.28	48.88	26.31	51.11
60	27.94	52.22	29.07	54.46
70	28.15	56.73	29.98	57.88

The MSE and RMSE quantitative evaluations between noisy and enhanced images are shown in Figures 5 and 6, respectively.

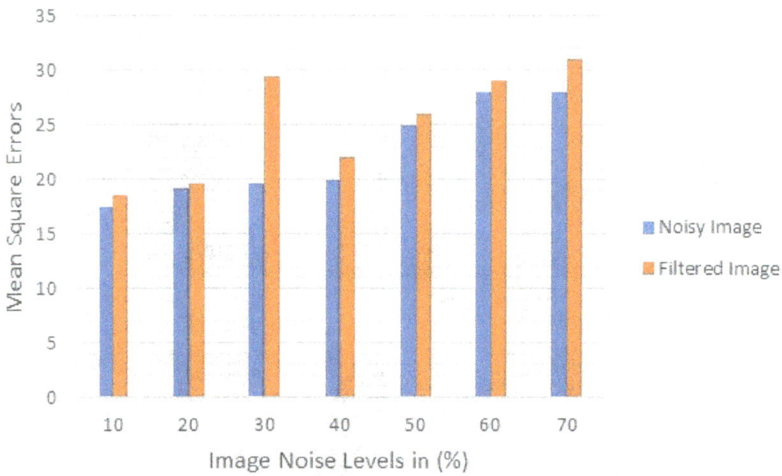

Figure 5: The MSE comparative values between the filtered and noisy image.

The trend of the median-filtered results is shown in Figures 5 and 6, which significantly reduces noise compared to noisy images. This proves that the results will be higher when the MSE and RMSE scores are lower. As mentioned in Section 3.4, to evaluate the effectiveness of parameter optimization on the proposed model for detecting unob-

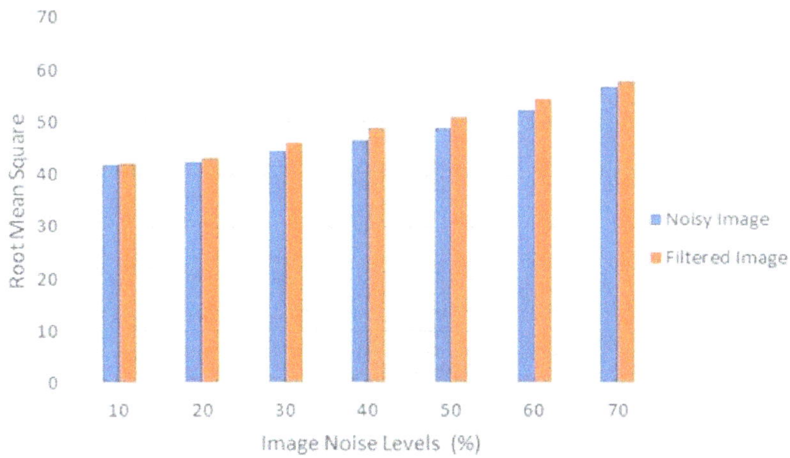

Figure 6: The RMSE comparative values between the filtered and noisy image.

served lesions in breast images, a CNN is trained for detection by varying some hyperparameters, including batch size, epoch, kernel size, filter, and learning rate. The results of these changes are presented in Table 5.

Table 5: Training parameters of the proposed model.

Batch size	Epoch	Kernel-size	Filter	Learning rate	Training loss	Validation loss	Training accuracy	Validation accuracy
256	10	5.0	64.0	0.001	0.0913	0.0921	0.7686	0.7883
128	15	3.0	32.0	0.001	0.0802	0.0783	0.7689	0.8711
32	20	3.0	32.0	0.001	0.0611	0.0721	0.7736	0.8579
64	25	3.0	64.0	0.001	0.0632	0.0604	0.7764	0.9098
256	30	3.0	32.0	0.003	0.0664	0.0559	0.9864	0.9967

Table 5 shows how the proposed model's performance accuracy varies and how the evaluation of batch size, epoch, kernel size, and training affects the testing dataset. No overfitting occurred during training, as shown by comparing model performance between the training set and the validation set. These training and validation loss values are shown in Figure 7.

The training loss value increases from 0.0913 to 0.0529, while the validation loss value increases from 0.0921 to 0.0201, as shown in Figure 7. The proposed strategy is compared with other detection models to better verify the performance of the proposed technique. This is done by running quantitative experiments with cross-validation using 90% training data and 10% testing data. In Figures 8(a)–(d), the confusion matrix is shown.

Confusion matrices for identifying lesions in breast images using three separate models and proposed techniques are presented in Sections 4.5 (a)–(d). Additionally,

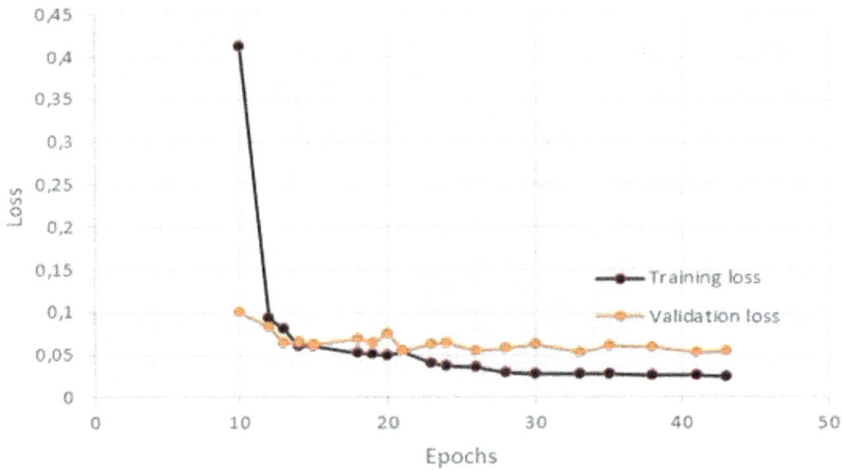

Figure 7: Loss values for detection of lesion in breast image.

Table 6 provides an overview of the performance of the existing models compared to the proposed model in terms of recall, precision, F1 score, and precision.

Table 6 provides a recall of 0.8368 for Naive Bayes; 0.8442 for GoogleLeNet CNN, 0.8748 for Resnet-VGG, and 0.9783 for the proposed model. The proposed model's accuracy for Naive Bayes, GoogleLeNet CNN, and Resnet-VGGand, in that order, is 0.8793, 0.8318, 0.8436, and 0.9781. The proposed model has an accuracy of 98.64%, while the accuracy of Naive Bayes is 88.89%, that of GoogleLeNet CNN is 85.64%, and that of Resnet-VGG is 84.79%. Notably, the high quality noise reduction of the median filter and CNN used during detection explains the high detection accuracy of the proposed technique on the publicly available breast dataset (Wisconsin) compared to other detection models.

Table 6: Summary of performance metrics in conventional techniques.

Models	Recall	Precision	FPR	F1-score	Accuracy (%)
Naïve Bayes	0.8368	0.8793	0.4202	0.8388	88.89
GoogleLeNet CNNs	0.8442	0.8388	0.5676	0.8245	85.64
Resnet-VGG	0.8748	0.8436	0.4481	0.8039	84.79
Proposed model	0.9783	0.9881	0.0734	0.9661	98.64

Experiment 2: evaluation of computational time

Since one of the objectives of this research is to reduce the computational time in order to address a critical aspect of real-time suspicious lesion detection systems and create

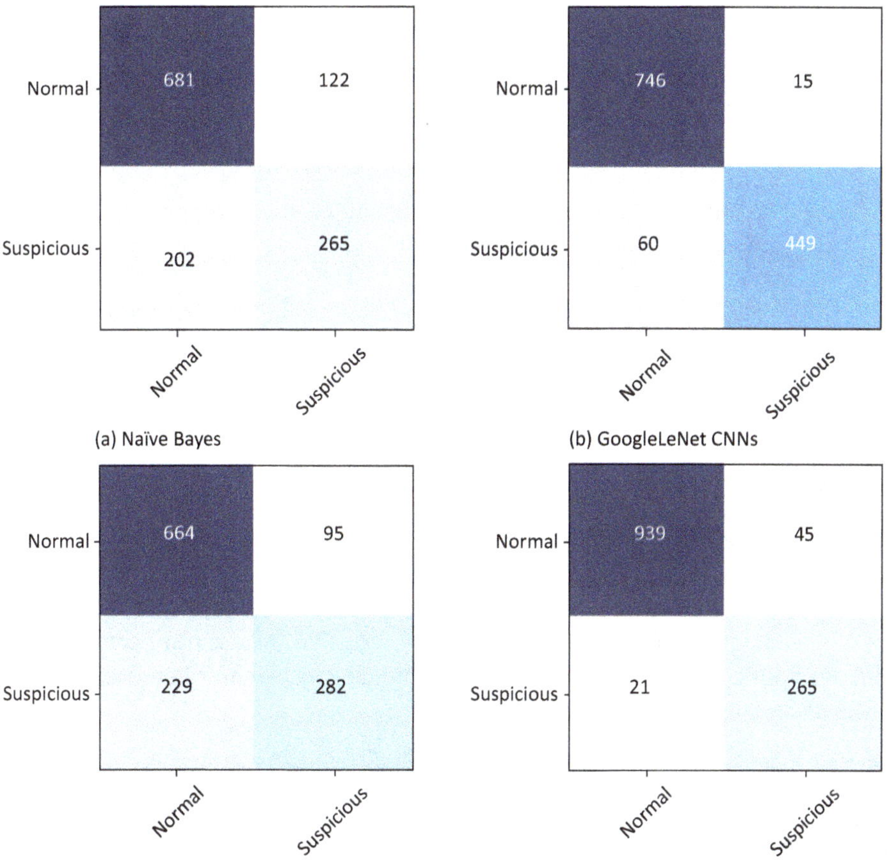

(a) Naïve Bayes (b) GoogleLeNet CNNs (c) Resnet-VGG (d) Proposed Model

Figure 8: Confusion matrix for detection of lesion in breast image using Naïve Bayes, GoogleLeNet CNNs, Resnet-VGG, and the proposed model.

more responsive treatment, the goal of this experiment is to evaluate the computation time of four models using the Wisconsin dataset. Figure 9 shows the time spent on Naive Bayes, GoogleLeNet CNNs, Resnet-VGG, and the proposed model on a graphics processing unit (GPU) to learn how to identify lesion patterns in a breast image dataset.

From the Figure 9, it is clear that the processing time of the proposed model on GPU outperforms other models used in the experiment, as shown in Figure 9. Thus, this result shows that radiologists can consider using the proposed model as an alternative model for lesion detection in breast images.

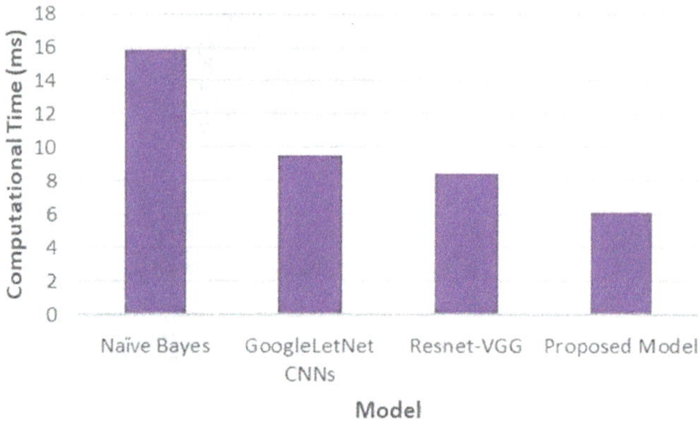

Figure 9: Comparison of model processing time on GPU.

4.3 Evaluation of performance

We also examine the performance of the proposed model in terms of detected anomalies, the techniques used, precision, recall, F1-score, and the accuracy compared to other methods currently used for detection techniques. appear anomalous on the Wisconsin dataset, as shown in Table 7.

Table 7: Performance comparison of proposed technique with existing methods.

Ref.	Detected anomalies	Technique	Precision	Recall	F1-score	Accuracy
[30]	Breast cancer	Logistic regression (LR) Naïve Bayes (NB) k-Nearest neighbor (k-NN)	LR:0.33 NB: 0.17 K-NN: 0.33	LR:0.33 NB: 0.06 K-NN: 0.66	LR:0.73 NB: 0.64 K-NN: 0.66	LR:0.75 NB: 0.69 K-NN: 0.74
[31]	Breast cancer	Internet of Things (IoT) and ML	98	97	96	98
[32]	Breast cancer	Random Forest (RF), Gradient Boosting (GB), Decision Tree (DT), Naïve Bayes (NB)	RF: 93.33 GB: 95.16 DT: 89.55 NB: 93.44	RF: 95.65 GB: 93.65 DT: 95.23 NB: 90.47	RF: 95.93 GB: 94.40 DT:92.30 NB:91.93	RF:97.07 GB:95.90 DT:94.15 NB:94.15
[33]	Breast cancer	Deep learning Xception	97.60	97.60	97.58	–

Table 7 (continued)

Ref.	Detected anomalies	Technique	Precision	Recall	F1-score	Accuracy
[34]	Breast cancer	k-Nearest Neighbor (k-NN)	0.960	1.00	0.98	0.97
[35]	Benign and malignant tumors in breast images	Supervised machine learning such as Random Forest (RF), Logistic Regression (LR), Naïve Bayes (NB), k-nearest Neighbor (k-NN)	RF: 0.97 LR: 0.98 k-NN: 0.94 NB: 0.93	–	RF: 0.97 LR: 0.98 k-NN: 0.96 NB: 0.95	RF: 97 LR: 98 k-NN: 96 NB: 95
Proposed technique	Lesion and noise in breast image	Median filtering and CNN	0.9881	0.9783	0.9661	98.64

Compared with other techniques developed using the Wisconsin dataset, Table 7 demonstrates that the proposed technique performs better in the breast cancer detection study, with the highest accuracy of 98.64%. This model is suitable for real-time applications due to its accuracy.

5 Conclusion

This study demonstrated how to identify lesions in a breast image using CNNs and median filtering. CNNs are used to extract significant features and learn from the image features, while the median filter is used to eliminate noise from the breast image. The 2000-image Wisconsin dataset, which is accessible to the public, is used to assess the suggested technique and perform k-fold cross-validation. A 98.64% recognition accuracy was attained. This outcome shows that the suggested technique performs satisfactorily in identifying lesions in the breast image early on before they develop into cancer. Additionally, this study provides an insightful comparison with several other advanced breast cancer detection methods currently in use. The results show that the proposed method outperforms state-of-the-art methods. This is unmistakable evidence that the work of radiologists can be improved by applying the proposed technique, allowing the system to identify breast cancer in women earlier.

References

[1] R. K. Yadav, P. Singh, and P. Kashtriya, "Diagnosis of Breast Cancer Using Machine Learning Techniques – A Survey," *International Conference on Machine Learning and Data Engineering*, vol. 218, no. 2023, pp. 1434–1443, 2023.

[2] O. A. Esan, M. Mbodila, and P. M. Madimba, "Detection of Suspicious Clusters in Women's Breast Image Using Convolutional Neural Networks," *Proceedings of the 2023 Conference on Health Informatics and Medical Systems (part of CSCE 2023 Congress), IEEE CPS*, 2023.

[3] J. Elmore, K. Armstrong, C. Lehman, and S. Fletcher, "Screening for Breast Cancer," *JAMA*, vol. 293, no. 10, pp. 1–26, 2005, doi: 10.1001/jama.293.10.1245.

[4] S. Siviengphanom, Z. Gandomkar, S. Lewis, and P. Brennan, "Global Radiomic Features from Mammography for Predicting Difficult-To-Interpret Normal Cases," *Journal of Digital Imaging*, vol. 36, no. 4, pp. 1541–1552, 2023, doi: 10.1007/s10278-023-00836-7.

[5] E. Ekpo, M. Alakhras, and P. Brennan, "Errors in Mammography Cannot be Solved Through Technology Alone," *Asian Pacific Journal of Cancer Prevention*, vol. 26, no. 2, pp. 291–301, 2018, doi: 10.22034/APJCP.2018.19.2.291.

[6] M. Arnold *et al.*, "Current and Future Burden of Breast Cancer: Global Statistics for 2020 and 2040," *Breast*, 2022, doi: 10.1016/j.breast.2022.08.010.

[7] Y. Feng *et al.*, "Breast Cancer Development and Progression: Risk Factors, Cancer Stem Cells, Signaling Pathways, Genomics, and Molecular Pathogenesis," *Genes and Diseases*, vol. 5, no. 2, pp. 77–106, 2018 May 12, PMID: 30258937; PMCID: PMC6147049, 2018, doi: 10.1016/j.gendis.2018.05.001.

[8] S. Hussain *et al.*, "Modern Diagnostic Imaging Technique Applications and Risk Factors in the Medical Field: A Review," *BioMed Research International*, vol. 5164970, 2022, doi: 10.1155/2022/5164970.

[9] D. Thigpen, A. Kappler, and R. Brem, "The Role of Ultrasound in Screening Dense Breasts – A Review of the Literature and Practical Solutions for Implementation," *Diagnostics (Basel)*, vol. 8, no. 1, 2018, doi: 10.3390/diagnostics8010020.

[10] T. Desyani, Y. Kasmayanti, A. Saifudin, and Yulianti, "Baggung Techniques to Reduce Misclassification of Breast Cancer Prediction Based on Gradient Bosted Tree (GBT) Algorithm," *Journal of Physics Conference Series*, vol. 1477, no. 032010, pp. 1–6, 2020.

[11] Z. Khandezamin, M. Naderan, and M. J. Rashti, "Detection and Classification of Breast Cancer Using Logistic Regression Feature Selection and GMBH Classifier," *Journal of Biomedical Information*, vol. 11, no. 2020, pp. 1–15, 2020.

[12] S. Kharya, S. Agrawal, and S. Soni, "Naive Bayes Classifiers: A Probabilistic Detection Model for Breast Cancer," *International Journal of Computer Applications*, pp. 26–31, 2014.

[13] Y. I. A. Rejani, and D. S. T. Selvi, "Early Detection of Breast Cancer Using SVM Classifier Technique," *International Journal on Computer Science and Engineering*, vol. 1, no. 3, pp. 127–130, 2009.

[14] R. K. Yadav, P. Singh, and P. Kashtriva, "Breast Cancer Using Machine Learning Techniques – A Survey," *International Conference on Machine Learning and Data Engineering*, vol. 218, pp. 1434–1443, 2023.

[15] C. Yun *et al.*, "A Study on the Effectiveness of Deep Learning-Based Anomaly Detection Methods for Breast Ultrasonography," *Sensors (Basel)*, vol. 23, no. 5, 2023, doi: 10.3390/s23052864.

[16] A. Bisarya *et al.*, "Breast Cancer Detection Using Quantum Convolutional Neural Networks: A Demonstration on a Quantum Computer," *medRxiv*, 2020, doi: https://doi.org/10.1101/2020.06.21.20136655.

[17] S. M. Almutairi, S. Manimurugan, M. M. Aborokbah, C. Narmatha, S. Ganesan, and P. Karthikeyan, "An Efficient USE-Net Deep Learning Model for Cancer Detection," *International Journal of Intelligent Systems*, vol. 2023, no. 8509453, pp. 1–14, 2023.

[18] A. Saber, A. G. Hussien, W. A. Awad, A. Mahmoud, and A. Allakany, "Adapting the Pre-trained Convolutional Neural Networks to Improve the Anomaly Detection and Classifcation in Mammographic Images," *Scientific Reports*, vol. 13, no. 14877, pp. 1–17, 2023.

[19] H. Cai *et al.*, "Breast Microcalcification Diagnosis Using Deep Convolutional Neural Network from Digital Mammograms," *Computational and Mathematical Methods in Medicine*, vol. 2019, no. 2717454, pp. 1–10, 2019, doi: https://doi.org/10.1155/2019/2717454.

[20] M. Kelkha, and Y. K. Tamandani, "Breast Cancer Detection Using Deep Multilayer Neural Networks," *Journal of Epigenetics*, vol. 3, no. 1, pp. 27–34, 2022, doi: 10.22111/jep.2022.41712.1041.

[21] O. Tarawneh, M. Otair, M. Husni, H. Y. Abuaddous, M. Tarawneh, and M. A. Almomani, "Breast Cancer Classification Using Decision Tree Algorithms," *(IJACSA) International Journal of Advanced Computer Science and Applications*, vol. 13, no. 4, pp. 676–680, 2022.

[22] S. Sreejith, and J. Nayak, "Study of Hybrid Median Filter for the Removal of Various Noises in Digital Image," *First International Conference on Advances in Physical Sciences and Materials*, vol. 1796, no. 2020, pp. 1–7, 2020, doi: 10.1088/1742-6596/1706/1/012079.

[23] S. J. Shri, and S. Jothilakshmi, "Anomaly Detection in Video Events Using Deep Learning," *International Journal of Innvative Technology and Exploring Engineering (IJITEE)*, vol. 8, no. 9, pp. 1313–1316, 2019.

[24] X. Hu, S. Hu, X. Zhang, H. Zhang, and L. Luo, Anomaly Detection Based on Local Nearest Neighbor Distance Descriptor in Crowded Scenes, The Scientific World Journal, vol. 2014. no. 632575, pp. 1–12, 2014.

[25] S. Haifeng, and X. Chao, "Moving Object Detection Based on Background Subtraction of Block Updates," *2013 6th International Conference on Intelligent Networks and Intelligent Systems (ICINIS)*, pp. 1–10, 2013, ISBN:978-1-4799-2809-5.

[26] O. A. Esan, and I. O. Osunmakinde, "Towards Intelligent Vision Surveillance for Police Information Systems," *Lecture Notes in Network and Systems (COCS'22)*, vol. 503, pp. 136–148, 2022.

[27] O. A. Esan, and I. O. Osunmakinde, "A Computer Vision Model for Detecting Suspicious Behaviour from Multiple Cameras in Crime Hotspots Using Convolutional Neural Networks," *González-Briones, A., et al. Highlights in Practical Applications of Agents, Multi-Agent Systems, and Complex Systems Simulation. The PAAMS Collection*, vol. 1678, 2022, doi: https://doi.org/10.1007/978-3-031-18697-4_16.

[28] O. A. Esan, D. O. Esan, M. Mbodila, F. A. Elegbeleye, and K. Koranteng, "Surveillance Detection of Anomalous Activities with Optimized Deep Learning Technique in Crowded Scenes," *Bulletin of Electrical Engineering and Informatics*, vol. 12, no. 3, pp. 1674–1683, 2023.

[29] D. Esan, P. A. Owolawi, and C. Tu, "Anomalous Detection in Noisy Image Frames Using Cooperative Median Filtering and KNN," *IAENG International Journal of Computer Science*, vol. 49, no. 1, pp. 1–9, 2022.

[30] V. P. C. Magboo, and M. S. A. Magboo, "Machine Learning Classifiers on Breast Cancer Recurrences," *25th International Conference on Knowledge-Based and Intelligent Information & Engineering Systemsd*, 2020.

[31] V. N. Gopal, F. Al-Turjman, R. Kuma, L. Anand, and M. Rajesh, "Feature Selection and Classification in Breast Cancer Prediction Using IoT and Machine Learning," *Measurements*, vol. 109442, no. 178, pp. 1–8, 2021.

[32] Y.-T. Bau, T. Sasidaran, and C.-L. Goh, "Improved Machine Learning Algorithms for Breast Cancer Prediction," *Journal of System and Management Sciences*, vol. 12, no. 3, pp. 251–266, 2022.

[33] B. S. Abanasser, M. R. J. AL-Hiealy, I. S. Zaqout, and S. S. Abu-Naser, "Breast Cancer Detection and Classification Using Deep Learning Xception Algorithm," *(IJACSA) International Journal of Advanced Computer Science and Applications*, vol. 13, no. 7, pp. 223–228, 2022.

[34] H. Chen, N. Wang, X. Du, K. Mei, Y. Zhou, and G. Cai, "Classification Prediction of Breast Cancer Based on Machine Learning," *Computational Intelligence and Neuroscience*, vol. 23, no. 6530719, pp. 26–31, 2023.

[35] S. Nag, and J. Nag, "A Comparative Analysis of Machine Learning Approaches for Prediction of Breast Cancer," *Journal of Emerging Investigators*, vol. 3, pp. 1–9, 2021.

Martin Roa-Villescas, Jin-Guo Liu, Patrick W. A. Wijnings, Sander Stuijk, and Henk Corporaal

Pushing the boundaries of probabilistic inference through message contraction optimization

Abstract: A key aspect of intelligent systems is their capacity to reason under uncertainty. This task involves calculating probabilities of relevant variables while considering any available information, a process commonly referred to as probabilistic inference. When working with discrete variables, the primary operations in probabilistic inference algorithms involve adding and multiplying multidimensional arrays with labeled dimensions, known as factors. The algorithmic complexity is dictated by the highest dimensional factor involved in any calculation; a concept referred to as the induced tree width. Despite advances in state-of-the-art techniques focused on reducing this metric, many real-world problems remain too complex to solve through existing probabilistic inference algorithms. In this work, we introduce a new method for adding and multiplying factors, which leads to marked improvements in inference performance, particularly for more complex models. Furthermore, this method serves as the core of a novel optimization framework introduced in this work, which employs metaprogramming to further enhance the runtime performance of probabilistic inference algorithms. Our method complements current leading-edge techniques aimed at reducing the induced tree width, thereby extending the range of models that can be effectively solved using exact inference. To validate the performance of our approach, we compare it against two other open-source libraries designed for probabilistic inference. Our method demonstrates an average speedup of 23 times on the UAI 2014 benchmark set. For the 10 most complex problems of this set, the average speedup increases to 64 times, highlighting the scalability of our method.

Keywords: probabilistic inference, message-passing algorithms, probabilistic graphical models, computational efficiency

---

Martin Roa-Villescas, Eindhoven University of Technology, Eindhoven, The Netherlands,
e-mail: m.roa.villescas@tue.nl
Jin-Guo Liu, Hong Kong University of Science and Technology, Guangzhou, China,
e-mail: jinguoliu@hkust-gz.edu.cn
Patrick W. A. Wijnings, Eindhoven University of Technology, Eindhoven, The Netherlands,
e-mail: p.w.a.wijnings@tue.nl
Sander Stuijk, Eindhoven University of Technology, Eindhoven, The Netherlands,
e-mail: s.stuijk@tue.nl
Henk Corporaal, Eindhoven University of Technology, Eindhoven, The Netherlands,
e-mail: h.corporaal@tue.nl

https://doi.org/10.1515/9783111344126-002

1 Introduction

Probabilistic graphical models (PGMs) are a fundamental component of modern Bayesian machine learning. Their significance stems from their ability to model uncertainty in complex systems using the fundamental laws of probability theory. PGMs, unlike other more data-driven approaches such as neural networks, have the capability to incorporate prior knowledge into their underlying causal model. These models offer a concise and intuitive approach for encoding large joint probability distributions that grow exponentially with the number of variables. Additionally, they form the foundation for various algorithms that enable efficient probabilistic inference, which is the focus of this paper.

While there are efficient algorithms for certain classes of PGMs (such as trees and singly connected networks), performing exact probabilistic inference on general graphs with discrete variables is known to be an NP-hard problem [1, 2]. Furthermore, it has been demonstrated that even approximate versions of these algorithms are NP-hard [3, 4]. As a result, designing efficient algorithms that can handle the complexity of probabilistic inference is a major challenge in applying PGMs to real-world problems.

A successful approach to confront the aforementioned challenge is the *variable elimination algorithm* (VEA) [5, p. 298]. The VEA exploits local dependencies in a PGM's structure to enable efficient inference. It does this by computing expressions once, caching the results, and thereby mitigating (although not entirely resolving) the exponential blowup. These computations can be represented as messages being propagated through the PGM that characterize the system; thus, the VEA and its variants are commonly known as message-passing algorithms. In numerous applications, however, it is often necessary to calculate the marginal probabilities of several variables in a PGM. For instance, in medical diagnostics, it is typically necessary to estimate the probabilities of multiple potential diseases. The *junction tree algorithm* (JTA), developed by Lauritzen and Spiegelhalter [6] and refined by Jensen [7], is an alternative implementation of the VEA that offers considerable computational benefits in the context of multiple query variables.

The complexity of probabilistic inference using message-passing algorithms is governed by the size of the largest data structure encountered in any of its intermediate computations, a measure known as the *induced tree width*. These computations involve sum-product operations, represented as messages traversing the graph. Although state-of-the-art methods have made significant progress in minimizing this metric, thus extending the feasibility of exact inference, many real-world problems remain intractable. This paper introduces a novel method for computing sum-product messages that offers a substantial improvement in inference performance, particularly for increasingly complex models. It establishes an optimal contraction sequence for input factors to minimize memory usage and execution time, and then recasts each pairwise contraction as batched *general matrix multiply* (GEMM) operations to harness their computational efficiency. Our approach complements the current state-of-the-art methods designed to minimize the induced tree width, thereby further expanding the spectrum of tractability for exact inference in increasingly complex models.

The key contributions of our work are as follows:

1. A novel method for the evaluation of sum-product messages in the context of message-passing algorithms, leading to a remarkable improvement in scalability for models of increasing complexity (Section 3).
2. An open-source implementation of the JTA, licensed under the MIT open-source license and accessible on GitHub at `https://github.com/mroavi/JunctionTrees.jl`. This implementation not only capitalizes on the computational advantages of our proposed method but also introduces a novel metaprogramming-based framework used to optimize runtime overhead by shifting as many computations as possible to the compilation phase. This strategy is particularly advantageous for operations that must be executed repeatedly at runtime, which appear often in probabilistic inference algorithms (Section 4).
3. A comprehensive experimental evaluation against the two existing open-source libraries for probabilistic inference using the UAI 2014 inference competition benchmark problems. The experiments were conducted on two different platforms in terms of computational resources: a high-end workstation and a Raspberry Pi 4. Our method demonstrates a significant speedup across a variety of benchmark problem sets, pushing the boundaries of exact inference by making it tractable for more complex models (Section 5).

This work is an extended version of [8]. It expands upon the original paper by (1) presenting experimental results of our message contraction optimization method on two distinct platforms: a high-end workstation and a Raspberry Pi, with the aim of validating that the results observed on the more capable platform are applicable to resource-constrained devices, and (2) introducing a novel metaprogramming-based framework for probabilistic inference, which leverages the generally larger compute power during compile-time to alleviate runtime limitations, often faced by embedded devices. Moreover, the use of metaprogramming for this purpose facilitates the integration of optimization passes into existing programs, akin to how compilers operate.

The rest of this paper is organized as follows. We begin in Section 2 with a review of probabilistic inference in PGMs using the junction tree algorithm. Section 3 introduces a method for computing sum-product messages that leads to remarkable scalability for models of increasing complexity. Section 4 outlines our novel metaprogramming-based framework for probabilistic inference. Section 5 presents an experimental evaluation of our method within the context of the JTA. A discussion of the results follows in Section 6. Finally, Section 7 concludes this work.

2 Inference in junction trees

In this section, we describe the *junction tree algorithm* (JTA), arguably the most popular method for performing exact inference on PGMs. For a more elaborate presentation, readers are directed to [9]. The JTA was initially developed by Lauritzen and Spiegelhalter [6] and later refined by Jensen *et al.* [7]. This refined version is known as the *Hugin* architecture and serves as the basis for a software tool of the same name. Contemporaneously, inspired by the work of Lauritzen and Spiegelhalter, Shenoy and Shaffer introduced the *Shenoy-Shaffer* architecture, an abstract framework for computing marginals, applicable to any domain that satisfies a series of axioms [10]. Today, both the Hugin and Shenoy-Shaffer architectures are among the most renowned.

Before delving into the JTA, we clarify the notation employed in this work. Italic uppercase letters (A, B, C) represent *random variables*, while their corresponding italic lowercase letters (a, b, c) stand for the variables' values. The assignment of a value a to a variable A is expressed as $A = a$. Boldface uppercase letters $(\boldsymbol{X}, \boldsymbol{Y}, \boldsymbol{Z})$ denote sets of variables, and boldface lowercase letters $(\boldsymbol{x}, \boldsymbol{y}, \boldsymbol{z})$ represent their associated values. An assignment of values $\boldsymbol{x}$ to a set of variables $\boldsymbol{X}$ is notated as $\boldsymbol{X} = \boldsymbol{x}$.

Figure 1 presents an overview of the algorithm. The JTA is a method for performing inference in *probabilistic graphical models* (PGMs). PGMs encapsulate the mathematical modeling of reasoning in the presence of uncertainty. *Bayesian networks* and *Markov random fields* are popular types of PGMs. Consider the Bayesian network shown in Figure 2. It consists of a *directed acyclic graph* $G = (\boldsymbol{V}, \mathcal{E})$ and a probability distribution $P(\boldsymbol{V})$, where $\boldsymbol{V}$ is the set of variables and $\mathcal{E}$ is the set of edges connecting the variables. We assume all variables to be discrete. Each variable $V \in \boldsymbol{V}$ is quantified with a *conditional probability distribution* $P(V \mid \mathrm{pa}(V))$, where $\mathrm{pa}(V)$ are the parents of V. These conditional probability distributions, together with the graph G, induce a *joint probability distribution* over $P(\boldsymbol{V})$, given by

$$P(\boldsymbol{V}) = \prod_{V \in \boldsymbol{V}} P(V \mid \mathrm{pa}(V)). \tag{1}$$

We define a *factor* $\phi_{\boldsymbol{V}}$ over a set of variables $\boldsymbol{V}$ as a function that maps each instantiation $\boldsymbol{V} = \boldsymbol{v}$ into a nonnegative number. Note that a probability distribution is a special case of a factor. The *product* of two factors $\phi_{\boldsymbol{X}}$ and $\phi_{\boldsymbol{Y}}$ is another factor $\phi_{\boldsymbol{Z}}$, where $\boldsymbol{Z} = \boldsymbol{X} \cup \boldsymbol{Y}$ and $\phi_{\boldsymbol{Z}}(\mathbf{z}) = \phi_{\boldsymbol{X}}(\boldsymbol{x})\phi_{\boldsymbol{Y}}(\boldsymbol{y})$ for the instantiations $\boldsymbol{x}$ and $\boldsymbol{y}$ that are consistent with the instantiation $\mathbf{z}$. The *marginalization* of a factor $\phi_{\boldsymbol{Y}}$ into $\boldsymbol{X} \subseteq \boldsymbol{Y}$ is a new factor $\phi_{\boldsymbol{X}}$, where each $\phi_{\boldsymbol{X}}(\boldsymbol{x})$ is computed by summing the values of $\phi_{\boldsymbol{Y}}(\boldsymbol{y})$ for all $\boldsymbol{y}$ that are consistent with $\boldsymbol{x}$.

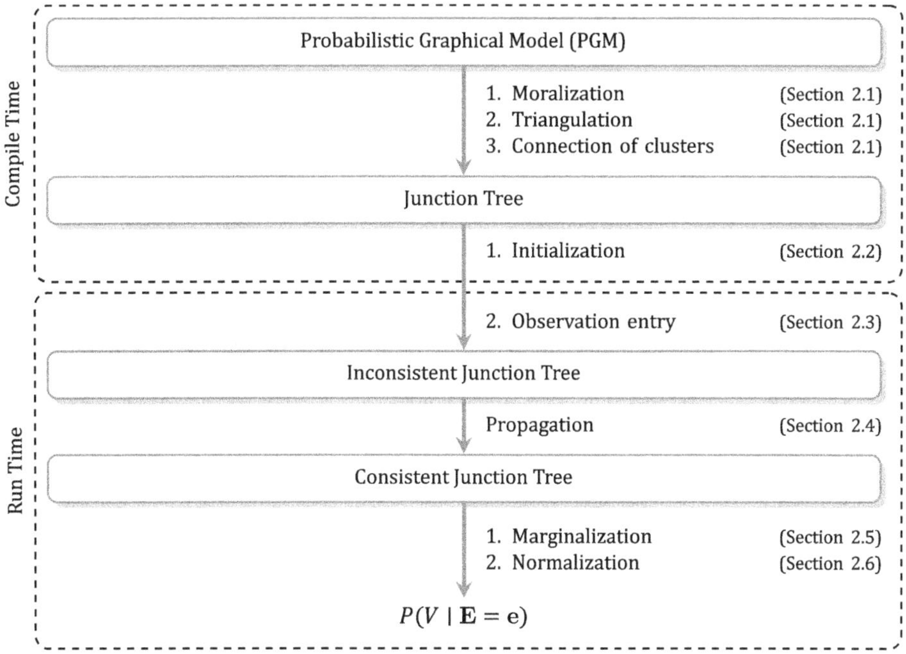

Figure 1: Overview of the Junction Tree Algorithm (JTA).

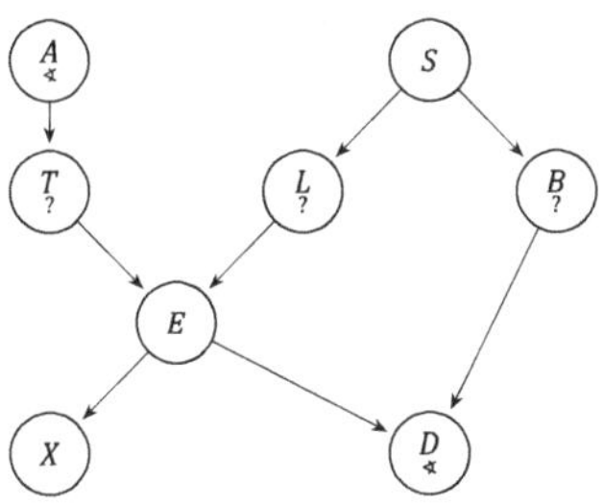

Random variable	Meaning
A	Recent trip to Asia
T	Patient has tuberculosis
S	Patient is a smoker
L	Patient has lung cancer
B	Patient has bronchitis
E	Patient has T and/or L
X	Chest X-Ray is positive
D	Patient has dyspnea

Figure 2: The ASIA network: a simple Bayesian network example from the medical domain [6], illustrating probabilistic dependencies between random variables such as diseases, symptoms, risk factors, and test results. Arrows represent conditional dependencies between variables. Evidence variables are marked by a spherical angle symbol (∢), while query variables are indicated with a question mark (?).

2.1 Junction tree construction

The JTA performs probabilistic inference on a secondary structure, known as a *junction tree*. A junction tree is constructed from a PGM using three graphical transformations: *moralization*, *triangulation*, and *connection of clusters*.

Moralization converts a directed acyclic graph into an undirected graph by dropping the directions of the edges and connecting the parents of each node. Figure 3 shows the corresponding moral graph of the Bayesian network in Figure 2.

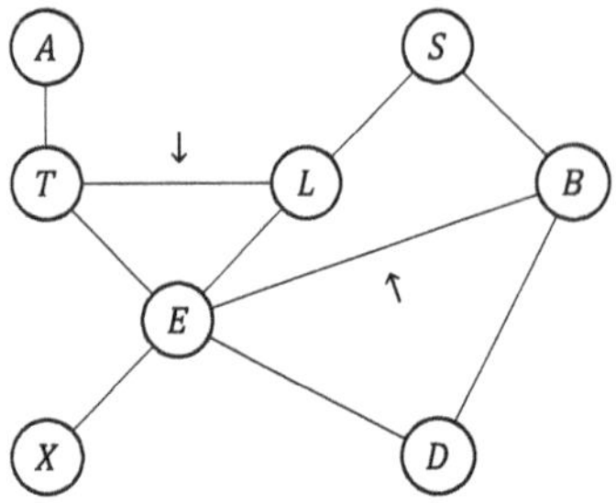

Figure 3: The moral graph of the Bayesian network in Figure 2. The arrows point to the edges introduced by this transformation.

Triangulation of an undirected graph is carried out by connecting two nonadjacent nodes in every cycle of a length greater than three. Figure 4 shows a triangulated graph of the moral graph in Figure 3. In general, a given undirected graph can have multiple possible triangulations. An optimal triangulation minimizes the sum of the *state space sizes* of the *maximal cliques* in the triangulated graph. The state space size of a clique is determined by the product of the number of possible values for each variable within that clique. In an undirected graph, a clique refers to a fully connected subgraph, and a maximal clique is a clique that is not part of any larger clique. The maximal cliques in our example are denoted with shaded boundaries in Figure 4. It is worth noting that finding an optimal triangulation is an NP-complete problem [11].

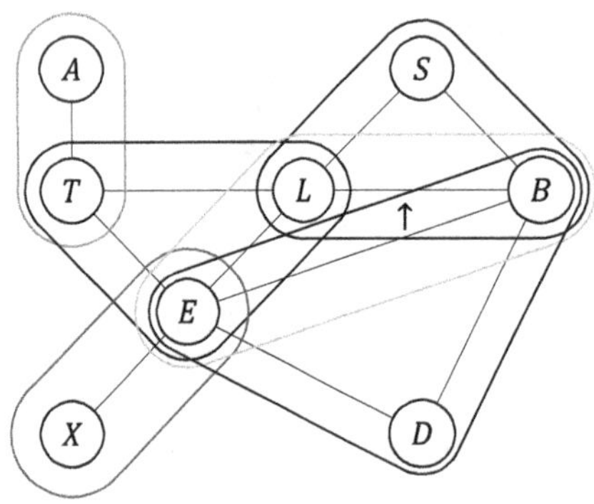

Figure 4: A triangulated graph of the undirected graph presented in Figure 3, with the arrow indicating the newly introduced edge. The shaded boundaries indicate the maximal cliques.

The maximal cliques of the triangulated graph correspond to the nodes of the junction tree. We call these *clusters*. Clusters are connected in a tree structure such that the *running intersection property* is satisfied: for any two clusters $\boldsymbol{X}$ and $\boldsymbol{Y}$ in the tree, all clusters on the path between $\boldsymbol{X}$ and $\boldsymbol{Y}$ contain $\boldsymbol{X} \cap \boldsymbol{Y}$. Edges are labeled with the intersection of their adjacent clusters. Such labels are called *sepsets*. Jensen and Jensen [12]

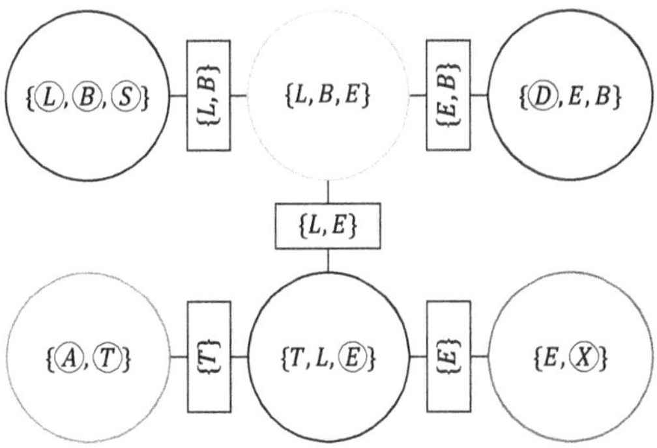

Figure 5: A junction tree constructed from the triangulated graph in Figure 4. Clusters are depicted as large circles and sepsets as rectangles. The clusters correspond to the maximal cliques in Figure 4. The encircled variables indicate which conditional probability distributions in (1) were multiplied into which cluster factors of the junction tree as part of the initialization stage.

present an optimal method to construct a junction tree from a triangulated graph. Figure 5 shows the result of applying this method to the triangulated graph in Figure 4.

2.2 Initialization

Each cluster $\boldsymbol{X}$ in the junction tree is associated with a *cluster factor* $\psi_{\boldsymbol{X}}$. Initially, all these cluster factors are set to unity. Subsequently, each conditional probability distribution $P(V \mid \mathrm{pa}(V))$ in (1) is multiplied into a cluster factor $\boldsymbol{X}$ that contains $\boldsymbol{V}$ and its parents, $\mathrm{pa}(V)$:

$$\psi_{\boldsymbol{X}} \leftarrow \psi_{\boldsymbol{X}} \cdot P(V \mid pa(V)). \tag{2}$$

The encircled variables in Figure 5 indicate which conditional distributions in (1) were multiplied into which cluster factors of our running example.

2.3 Observation entry

Observations take the form of $\boldsymbol{E} = \boldsymbol{e}$, where $\boldsymbol{e}$ is the instantiation of the set of evidence variables $\boldsymbol{E}$. These are incorporated into the junction tree by finding a cluster factor $\psi_{\boldsymbol{X}}$ for each evidence variable in $\boldsymbol{E}$ that contains this variable, and setting all its entries that are not consistent with the evidence to zero. Integrating observations into a cluster factor changes its elements locally. This creates an inconsistency, which requires propagation of the observed information to other clusters containing the same evidence variables.

2.4 Propagation

Propagation refers to a series of synchronized local manipulations between clusters that guarantee consistency throughout the entire junction tree. These manipulations are called *messages*. The propagation of messages begins by designating an arbitrary cluster as the *root*, which gives direction to the edges. Messages then flow between clusters in two recursive phases: an *inward* and an *outward* phase. In the inward phase, each cluster passes a message to its parent. In the outward phase, each cluster passes a message to each of its children. A cluster passes a message to a neighbor only after it has received messages from all its *other* neighbors. A message from cluster $\boldsymbol{X}$ to an adjacent cluster $\boldsymbol{Y}$ is a factor $\phi_{X \to Y}$, defined by

$$\phi_{X \to Y} = \sum_{X \setminus Y} \psi_X \prod_{N \in \mathcal{N}_X \setminus Y} \phi_{N \to X}, \tag{3}$$

where ψ_X is the cluster factor of $\boldsymbol{X}$ and $\mathcal{N}_X$ is the set of neighbors of $\boldsymbol{X}$. Figure 6 shows an admissible schedule for the propagation of messages of our running example.

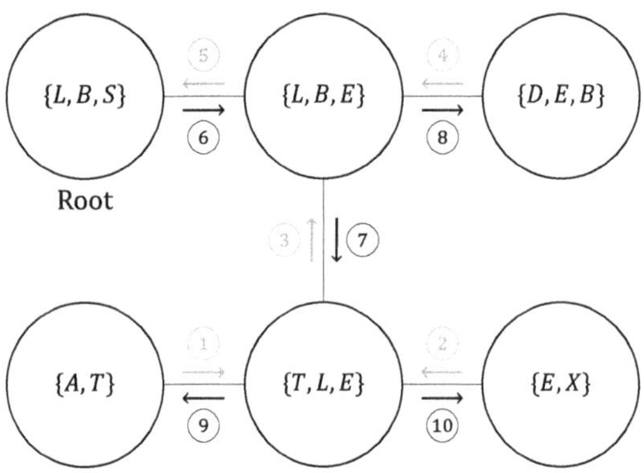

Figure 6: Admissible schedule for the propagation of messages in the junction tree algorithm. Gray messages correspond to the inward pass while black messages correspond to the outward pass.

2.5 Marginalization

After the propagation phase, each edge holds two messages; one in each direction. The joint marginal probabilities for each sepset are given by the product of the two messages passing through the corresponding edge, i.e.

$$\boldsymbol{P}(\boldsymbol{S}_{XY}, \boldsymbol{E} = \boldsymbol{e}) = \phi_{X \to Y} \cdot \phi_{Y \to X}, \tag{4}$$

where $\boldsymbol{X}$ and $\boldsymbol{Y}$ are adjacent clusters, and $\boldsymbol{S}_{XY}$ is the sepset between clusters $\boldsymbol{X}$ and $\boldsymbol{Y}$. Similarly, the joint marginal probabilities for each cluster are given by the product of the cluster's incoming messages and its factor, i.e.,

$$\boldsymbol{P}(\boldsymbol{X}, \boldsymbol{E} = \boldsymbol{e}) = \psi_{\boldsymbol{X}} \prod_{\boldsymbol{N} \in \mathcal{N}_{\boldsymbol{X}}} \phi_{\boldsymbol{N} \to \boldsymbol{X}}. \tag{5}$$

The marginal probability $\boldsymbol{P}(V, \boldsymbol{E} = \boldsymbol{e})$ for each variable of interest V is then computed from the joint marginal of a sepset or cluster containing V by marginalizing all other variables:

$$\boldsymbol{P}(V, \boldsymbol{E} = \boldsymbol{e}) = \sum_{\boldsymbol{X}' \setminus V} \boldsymbol{P}(\boldsymbol{X}', \boldsymbol{E} = \boldsymbol{e}), \tag{6}$$

where $\boldsymbol{X}'$ is a sepset or cluster that contains V.

2.6 Normalization

The last step is to compute $P(V | \boldsymbol{E} = \boldsymbol{e})$ for each variable of interest V. We do so by normalizing $P(V, \boldsymbol{E} = \boldsymbol{e})$

$$P(V|\boldsymbol{E} = \boldsymbol{e}) = \frac{P(V, E = e)}{\sum_V P(V, \boldsymbol{E} = \boldsymbol{e})}. \tag{7}$$

3 Message contraction optimization

In the context of message-passing algorithms for probabilistic inference, the traditional focus has been on the graph structure and on establishing a valid schedule for message propagation through it. Less attention has been paid to the specific order of operations at the message level. For example, the JTA does not prescribe a particular order in which the product of factors and marginalization of variables within a message should be carried out. Although such an ordering does not affect the outcome of these operations, – thanks to the associative properties of multiplication and addition, – we will later demonstrate that it significantly impacts computational performance. The key point is that the total number of operations required to compute a message is heavily influenced by the order in which these operations are performed.

We now present a method to efficiently compute a message represented by (3) that translates into remarkable scalability for models of increasing complexity. In essence, this equation involves the multiplication of a series of factors, followed by the marginalization of a subset of the variables across all factors. Given this insight, we can distill this equation into a more concise form:

$$\phi_Y = \sum_{V \in \boldsymbol{V}} \prod_{X \in \chi} \phi_X, \tag{8}$$

where ϕ_Y is the output factor over the variables $\boldsymbol{Y}$. χ is a collection of sets of variables where each set $\boldsymbol{X}$ is associated with an input factor ϕ_X, and $\boldsymbol{V}$ is the set of variables to be marginalized. The total number of input factors is $|\chi|$.

3.1 Notation and terminology

To facilitate the explanation of our message computation method, new terminology and notation are introduced. Recall that a factor ϕ_V is essentially a multidimensional array of nonnegative numbers with labeled dimensions. These labels correspond to a set of random variables $\boldsymbol{V} = \{i, j, k, \ldots\}$ that serve as indices for the array, and hence also referred to as *indices*. The *shape* of a factor is a tuple that defines the size of each index. The number of indices $|\boldsymbol{V}|$ is the *rank* of the factor. A factor with shape (2,4,6) has $2 \times 4 \times 6$ *elements*, and a rank of 3. Many common multidimensional array operations can be formulated as *factor contractions*. A contraction is an operation that generates a new factor by summing the product of elements from two or more original factors. For instance, the matrix product of three 4×4 matrices is given by

$$D_{il} = \sum_{j,k=1}^{4} A_{ij} B_{jk} C_{kl}. \tag{9}$$

Here, the indices j and k are contracted, meaning that we sum over all possible values for each of these indices. Similarly, other types of operations between factors, such as outer products, diagonals, reductions, traces, and so on, can also be expressed in this manner. We can streamline (9) by employing the *einsum* notation:

$$\texttt{ij, jk, kl -> il}. \tag{10}$$

In this notation, we omit the summation symbols, as these are inherently accounted for across all possible values of indices that appear on the left side of the equation, but not on the right side. The factor identifiers are omitted as well. Furthermore, a comma-separated list of the input factors appears first, followed by an arrow (->), and finally the output factor.

3.2 Sum-product message optimization

We now present a two-step method for computing (8). First, we establish an order for the pairwise contraction of input factors that minimizes the overall memory consumption. Second, we formulate the pairwise factor contractions, derived from the previous step as batched *general matrix multiply* (GEMM) operations. Notably, these steps occur at compile-time, with the execution of the GEMM operations, deferred until runtime. Detailed explanations of these two steps follow.

3.2.1 Pairwise order optimization

The goals of this step are twofold: 1) Establish an order of pairwise contractions for the input factors, enabling their execution as batched GEMM operations, which we explore further in the next section, and 2) Reduce the overall number of operations required to compute a message by finding an order of pairwise contractions that minimizes memory consumption. We present a greedy method to achieve these goals.

Algorithm 1 Pairwise Factor Contraction Order Optimization

Input: Set of input factors Φ

Output: Order of pairwise contraction O for input set Φ

function ORDERCONTRACTIONS (Φ)
  Initialize an empty list O
  while $|\Phi| > 1$ do
    $(i, j) \leftarrow \arg\min_{(i', j') \in \Phi \times \Phi}$ MEMORYCOST$(\phi^{i'}, \phi^{j'})$
    Add (i, j) to O
    $\phi^k \leftarrow$ SIMULATECONTRACTION$(\phi^{i'}, \phi^{j'})$
    $\Phi \leftarrow (\Phi \setminus \{\phi^{i'}, \phi^{j'}\}) \cup \{\phi^k\}$
  return O

Given a set of input factors Φ, the algorithm iteratively selects the pair of factors $(\phi^i \phi^j)$ that, when contracted, minimizes the memory usage as evaluated by the MEMORYCOST function. This function calculates the *space complexity* of the contraction, which refers to the number of elements in the largest factor, resulting from the contraction's intermediate operations. Upon identifying the optimal factors at each step, the algorithm records them in the O list. Next, it simulates their contraction, which results in a new factor, ϕ^k. It then removes the original factors $(\phi^i \phi^j)$ from the set Φ, and adds to it the newly contracted factor, ϕ^k. Importantly, actual contractions are not performed; instead, the SIMULATECONTRACTION function is used to calculate the resulting factor's shape from each contraction. The algorithm ends once a single factor remains within the set Φ and returns the O list, which contains the optimal contraction sequence. As an example, invoking the ORDERCONTRACTIONS function with the message below as argument,

```
ij, kj, ki -> i,
```

yields the following result:

```
ij, ji -> i
├─ ij
└─ kj, ki -> ji
   ├─ kj
   └─ ki
```

The order is depicted in a tree-like structure, with the sequence of operations progressing from bottom to top.

This algorithm facilitates memory-efficient planning for factor contractions, which reduces data accesses and manipulation time, leading to improved execution at runtime. It also lays the groundwork for the following step.

3.2.2 Batched general matrix multiply

The goal of this step is to recast each pairwise factor contraction, obtained from the previous step, as a single batched GEMM operation. This allows us to leverage the computational advantages offered by this highly optimized operation, including vectorized operations, low-level hardware-specific optimizations, and the amortization of computational overhead through batching. A batched GEMM operation performs b independent multiplications of $n \times m$ and $m \times p$ matrices, yielding b $n \times p$ matrices. In einsum notation, this operation is described as

$$\texttt{bnm, bmp -> bnp}. \tag{11}$$

Because each factor in a pairwise contraction from the previous step could have an arbitrary number of indices, we must transform them into a suitable shape to execute a batched GEMM operation. To illustrate this process, consider the following example of a contraction of a pair of factors:

$$\texttt{ijrq, qrjks -> qikr}. \tag{12}$$

The necessary steps to prepare this pairwise contraction for execution as a batched GEMM operation are as follows:

1. *Contract the dangling indices*: The first step is to contract the indices that appear in only one factor, for example, index s in our running example. This step can be represented by

   $$\texttt{ijrq, (qrjks -> qrjk) -> qikr},$$

   where the parentheses denote the operations corresponding to the current step. Executing this step results in

   $$\texttt{ijrq, qrjk -> qikr}. \tag{13}$$

2. *Reshape the input factors*: The aim of this step is to recast the contraction so that it aligns with the form outlined in (11). This process involves reshaping the input factors into two 3-rank factors that, when contracted, yield another 3-rank factor. Reshaping each factor involves two steps: *grouping* and *permuting* the indices. The grouping of the indices is done based on their presence in each of the three factors as shown in Table 1. The reason why there are four groups is that each

index must appear in at least two of the three factors (given that the dangling indices have been removed in the previous step).

Table 1: Grouping of indices, based on their presence in each of the three factors of a pair-wise contraction. I_1 and I_2 correspond to the two input factors, O to the output factor, and ✓/ ✗ denote the presence/absence of an index in a factor.

Group	I_1	I_2	O
b	✓	✓	✓
n	✓	✗	✓
m	✓	✓	✗
p	✗	✓	✓

Finally, indices in each group are then flattened and permuted according to (11). Continuing with our running example, we can use this procedure to reshape each factor in (3.2.2) as follows:

$$((\texttt{ijrq->}\overbrace{\texttt{qr}}^{b}\overbrace{\texttt{i}}^{n}\overbrace{\texttt{j}}^{m}),\ \overbrace{\texttt{qr}}^{b}\overbrace{\texttt{j}}^{m}\overbrace{\texttt{k}}^{p}\ \texttt{->}\ \overbrace{\texttt{qr}}^{b}\overbrace{\texttt{i}}^{n}\overbrace{\texttt{k}}^{p})\texttt{->qikr}, \tag{14}$$

where we use braces to denote the grouping of the indices. The result is an operation that can be fed into a batched GEMM:

$$\overbrace{\texttt{qr}}^{b}\overbrace{\texttt{i}}^{n}\overbrace{\texttt{j}}^{m},\ \overbrace{\texttt{qr}}^{b}\overbrace{\texttt{j}}^{m}\overbrace{\texttt{k}}^{p}\ \texttt{->}\ \overbrace{\texttt{qr}}^{b}\overbrace{\texttt{i}}^{n}\overbrace{\texttt{k}}^{p}. \tag{15}$$

The computational cost of grouping indices is generally negligible; however, the permutation step can entail a memory copy, depending on the specific reordering involved. Although batched GEMM can handle transposes and strides in simple cases, some complex permutations are unsupported, potentially increasing the computational overhead.

3. *Dispatch the batched GEMM*: At this point, the batched GEMM operation, given by (15), is ready to be dispatched and executed. We reiterate that the steps being described occur at compile-time. Hence, at this stage, the operation is only being scheduled, with the actual execution deferred until runtime.
4. *Reshape the output factor*: The final step is to reshape the output factor to its original form. This involves ungrouping and permuting the result of the batched GEMM operation. In the context of our running example, this step corresponds to the operations of the output factor denoted in (15), namely,

$$\overbrace{\texttt{qr}}^{b}\overbrace{\texttt{i}}^{n}\overbrace{\texttt{k}}^{p}\ \texttt{->qikr} \tag{16}$$

4 A metaprogramming-based inference framework

A well-known strategy to improve the runtime inference performance of probabilistic inference algorithms involves partitioning the algorithm into two distinct phases: a *compilation* and a *runtime* phase, as detailed in [7] and [13]. The driving purpose is to minimize the runtime overhead by performing as many computations as feasible during the compilation phase. This strategy is particularly beneficial as it leverages the more flexible computing resources available during compile-time, which is especially advantageous for computations that need to be executed multiple times during the runtime phase. In contrast, the runtime phase is often constrained by time and hardware capacity, particularly when algorithms run on embedded devices.

The novelty of our approach lies in the adoption of *metaprogramming* [14] to enhance the aforementioned strategy. Metaprogramming is a programming technique where computer programs have the capability to read, generate, analyze, or transform other programs as data, as well as the ability to modify themselves during execution [15]. The key advantage of this technique is that it can perform intensive calculations ahead of time, during compilation, which leads to improved efficiency when the program is running. Furthermore, metaprogramming enables the seamless integration of optimization passes into existing programs, similar to how compilers operate. The power of metaprogramming to traverse and manipulate the abstract syntax tree of a program proves particularly beneficial in the domain of probabilistic inference algorithms, where it enables the implementation of optimization passes that aggressively exploit the domain's unique characteristics and assumptions. Figure 1 illustrates the phase in which each step of the JTA is executed.

Figure 7 depicts the architecture of our metaprogramming-based JTA framework. Initially, a base program is generated using the conventional JTA (as presented in Section 2), which processes the PGM alongside the query and evidence variables. The result, a metaprogram, is then passed through a series of optimization passes, each refining the input with the objective of enhancing the runtime performance. The result of the compilation phase is a specialized metaprogram that can then be evaluated online to compute the probability of each query variable, given the evidence. An example of an optimization pass utilizing this metaprogramming approach is the *partial evaluation* pass, described in [16], which reduces runtime overhead by precomputing expressions with static inputs during the compilation stage.

Another example of the benefits derived from separating an algorithm into two phases, within the context of probabilistic inference, is the employment of advanced tree decomposition solvers during compile-time to enhance runtime performance. This approach facilitates the use of such sophisticated solvers. Although this method may lengthen the compilation process, it typically results in tree decompositions that exhibit an induced tree width, close to the theoretical optimum. In the following section, we provide experimental results that compare two distinct tree decomposition

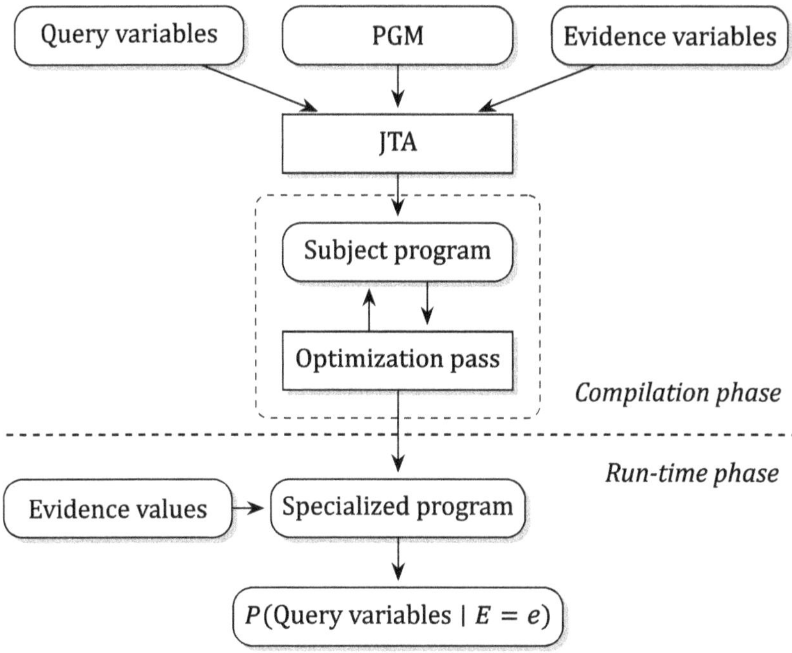

Figure 7: Overview of our metaprogramming-based framework for the generation of Bayesian inference algorithms, based on the junction tree algorithm.

methods, specifically focusing on their runtime performance. These experiments were conducted utilizing our metaprogramming-based framework.

5 Experimental evaluation

We now examine the performance impact of our message contraction method. First, we apply our method to a set of purposefully designed messages with varying complexity. This experiment is designed to evaluate the impact of our method on the efficiency of computing individual messages, thereby isolating the influence of other factors within the overall algorithm. Next, we compare the performance of our method against a conventional implementation for computations of messages within the context of the JTA. Subsequently, we evaluate the performance of our JTA implementation, relative to two other open-source libraries for probabilistic inference. The first two experiments were conducted on a high-end workstation, with an Intel Core i9-9900K CPU running at 3.60 GHz with 64 GB of RAM. The third experiment was performed on two different hardware platforms: the previously mentioned high-end workstation and a Raspberry Pi 4 Model B with an ARM Cortex-A72 CPU clocked at 1.5 GHz with 4 GB of RAM. Including the Raspberry Pi aims to validate that the results observed on the more powerful platform can be extrapolated to resource-constrained devices. This section concludes with a comparative study on the runtime performance of two prominent tree decomposition methods: the standard min-fill heuristic and Tamaki's algorithm. Our source code, licensed under the MIT open-source license, can be accessed on GitHub at `https://github.com/mroavi/JunctionTrees.jl`.

5.1 Messages of increasing complexity

In our initial experiment, we compare the execution time performance of two methods for computing sum-product messages, as defined by (8). These methods include our message-contraction optimization method, as detailed in Section 3, and a conventional method outlined in [5] and implemented in [16]. For this experiment, we generated messages of varying complexity by adjusting two parameters: 1) The total number of factors $|\chi|$ involved in the product of (8), and 2) The dimensionality of these factors, which was kept equal across all factors. These parameters, respectively correspond to the y and x axes of Figure 8. Each cell in this plot indicates the speedup factor achieved using our method, compared to the conventional approach.

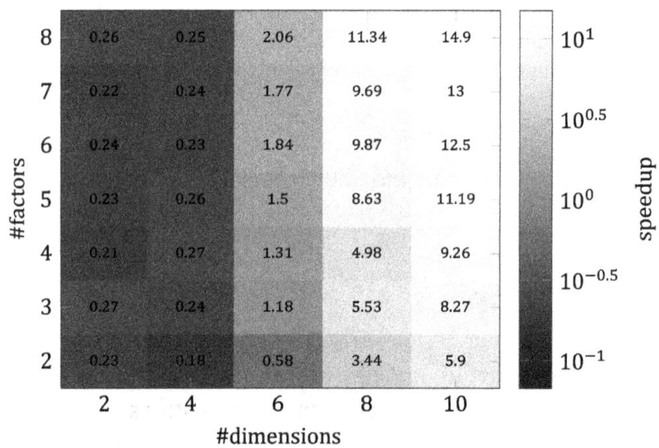

Figure 8: Speedup achieved by our message contraction method (Section 3), relative to a standard method utilized in [16], for purposefully generated messages. This experiment was carried out on a high-end workstation, equipped with an Intel Core i9–9900K CPU running at 3.60 GHz and 64 GB of RAM.

From this figure, we notice a substantial increase in the speedup factor as the number of dimensions in the factors increase, irrespective of the number of factors involved in the computation of the message. We also note that an increase in the number of low-dimensional factors results in a modest improvement in the speedup factor, whereas a higher number of high-dimensional factors produces a more pronounced improvement. Interestingly, this figure also discloses a complexity threshold for the message – determined by the combination of the number of factors and their dimensionality – which distinguishes between a speedup and slowdown effect of our method. We discuss these findings further in Section 6.

5.2 Probabilistic inference using the JTA

We now assess the performance impact of employing our message contraction method (Section 3) within the context of performing probabilistic inference using the JTA. We used the Julia language (version 1.8.5) to implement the JTA, as described in Section 2. In these experiments, we used the UAI 2014 inference competition's benchmark suite, which comprises 114 problems, in total, from various domains, including computer vision, signal processing, and medical diagnosis, which serve as a standardized testbed for algorithms dealing with uncertainty in AI. For each problem, we measured the execution time required to compute the posterior marginal probabilities of all variables in the network-given evidence. We verified the correctness of the results against other open-source libraries for probabilistic inference.

5.2.1 Comparison of message computation methods

For this experiment, the construction of the junction trees was carried out using a state-of-the-art solver, published by Hisao Tamaki in [17]. Figure 9 presents the speedup achieved by employing our message contraction method (Section 3), relative to the conventional method outlined in [5] and implemented in [16]. The benchmark problems are ordered along the x-axis, in increasing order of the largest cluster size of their corresponding junction tree. This factor predominantly dictates the complexity of performing probabilistic inference on these problems. Compared to the conventional message computation method, our message contraction method demonstrated an average speedup of 4.4 times across all problems, and this improvement increased to 14.7 times, when applied to the 10 most complex problems. From this figure, we observe that our method exhibits an improvement in performance for problems of greater complexity. Conversely, it shows a decrease in performance for less complex problems, with a notable exception being the *ObjectDetection* problem set. Perhaps most significantly, a correlation emerges between the problem's complexity and an increase in the speedup factor.

5.2.2 Library performance comparison

We now turn to the comparison results of our message computation method against two other open-source libraries for probabilistic inference in PGMs, namely, *Mer-*

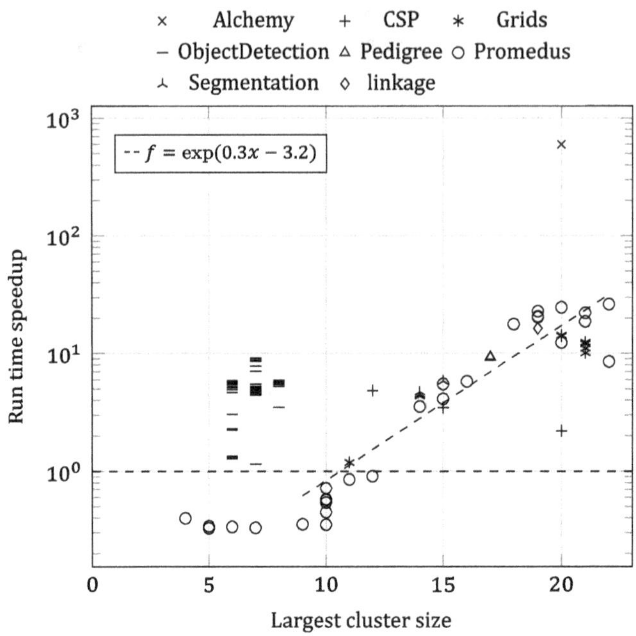

Figure 9: Speedup achieved by our message contraction method (Section 3), relative to a standard implementation presented in [16], for the UAI 2014 competition benchmark problems. This experiment was carried out on a high-end workstation, equipped with an Intel Core i9–9900K CPU running at 3.60 GHz and 64 GB of RAM.

lin [18] and *libDAI* [19]. To ensure a fair comparison of the probabilistic inference execution time, we employed the *min-fill heuristic* [20] for constructing the junction trees, as this technique is utilized by both referenced libraries.

Figure 10 illustrates a comparison of our message computation method's performance against the Merlin and libDAI libraries, conducted on the Intel-based high-end workstation described earlier. From this figure, we can observe that as the complexity of the problem increases, our method progressively outperforms both libraries. On average, our method achieves a speedup of 23 times across all problems, which increases to 64 times for the 10 most complex problems. We also compared the performance of the conventional method for computing messages implemented in [16] against the Merlin and libDAI libraries. This method yielded an average speedup of 1.9 times across all problems, and for the 10 most complex problems, it showed a speedup of 0.11 times, indicating a slowdown, in reality. In contrast with our method, the performance of this conventional method diminishes, relative to the two libraries, as the problem complexity increases. These results underscore the potential and significance of our proposed method for computing messages in the context of message-passing algorithms.

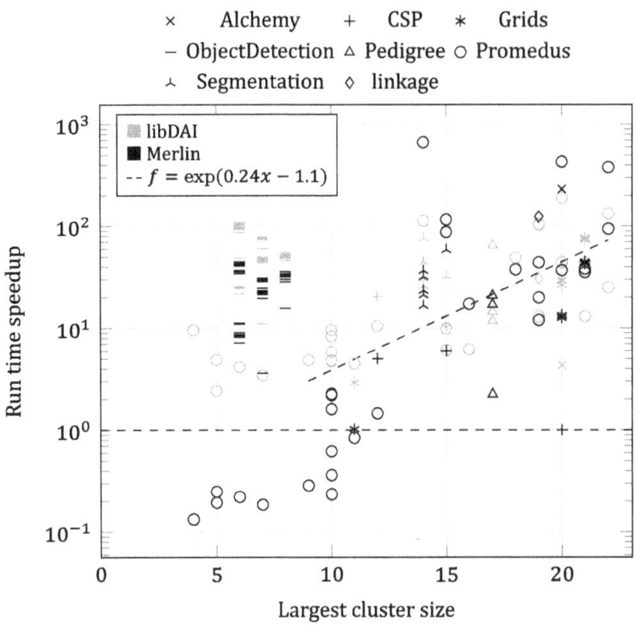

Figure 10: Speedup achieved by our message contraction method (Section 3), relative to *Merlin* [18] and *libDAI* [19], for the UAI 2014 inference competition benchmark problems. This experiment was carried out on a high-end workstation, equipped with an Intel Core i9-9900K CPU running at 3.60 GHz and 64 GB of RAM.

Figure 11 presents the results of the same experiment, but conducted on the Raspberry Pi 4 platform. The results corroborate the findings from the high-end workstation: our method continues to progressively outperform both Merlin and libDAI as the complexity of the problem increases. These findings validate the scalability of our proposed method on more resource-constrained hardware platforms. It is worth noting that of the 114 problems tested, our method successfully processed and yielded correct results for 99. The instances that could not be processed are primarily due to the memory constraints of the Raspberry Pi 4 platform. Nevertheless, our method still outperforms the reference libraries: Merlin was successful in only 60 cases and libDAI in 56, demonstrating that our method copes much better with limited resources. The figure excludes problems where Merlin or libDAI did not yield correct results.

5.2.3 Tree decomposition method comparison

To conclude this section, we present the experimental results from a comparative analysis of two tree decomposition methods in terms of runtime performance. Specifically, we examine the traditional min-fill heuristic, as referenced in [20], against Tamaki's algorithm, detailed in [17]. Both methods were integrated into the metaprogramming-

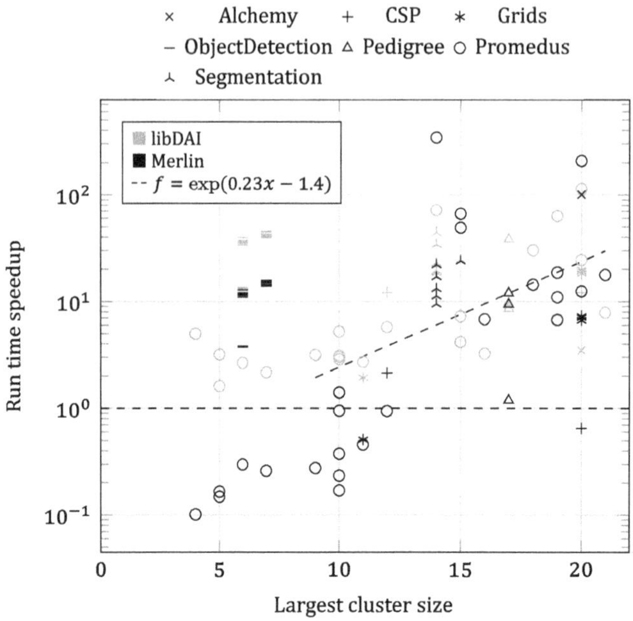

Figure 11: Speedup achieved by our message contraction method (Section 3), relative to *Merlin* [18] and *libDAI* [19], for the UAI 2014 inference competition benchmark problems. This experiment was performed on a Raspberry Pi 4 Model B, which is equipped with an ARM Cortex-A72 CPU running at 1.5 GHz and 4 GB of RAM.

based framework discussed in Section 4. The experimental methodology and the benchmarks employed remain consistent with those used in earlier experiments, presented in Section 5.2.

Figure 12 depicts the comparative runtime performance of Tamaki's tree decomposition method against the min-fill heuristic approach. Similar to previous results, in general, Tamaki's method demonstrates superior efficiency, particularly as the complexity of the benchmark problem escalates. This advantage in runtime performance, however, comes at the expense of an increased compilation time (a trade-off not visible in the figure). The metaprogramming-based framework significantly streamlined the experimentation process by enabling both tree decomposition methods to be performed during the compilation stage.

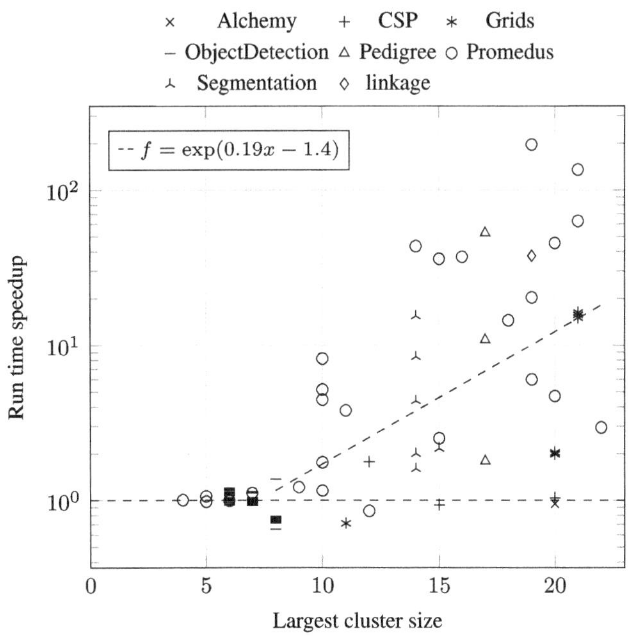

Figure 12: Speedup in runtime performance using Tamaki's tree decomposition algorithm over the min-fill heuristic method, benchmarked on UAI 2014 inference competition problems. The experiment was conducted on a high-end workstation with an Intel Core i9-9900K CPU at 3.60 GHz and 64 GB of RAM.

6 Discussion

A common trend emerges from Figures 8–11: our proposed message contraction method incurs a cost that can cause a slowdown in probabilistic inference when the problem's complexity is relatively low. However, as the problem complexity increases, this cost becomes negligible. In such cases, our method can often yield performance improvements that are several orders of magnitude higher.

The speedup improvements observed through the utilization of our message contraction method are attributed to two main factors. Firstly, we leverage the computational capabilities of batched GEMMs, which include vectorized operations, low-level hardware-specific optimizations, and the amortization of computational overhead, through batching. Secondly, we streamline the computation of a message by determining an optimized sequence for pairwise contractions between factors. The order chosen can significantly affect the memory usage necessary for a message's execution, subsequently leading to substantial differences in execution times. Currently, our setup does not allow us to quantify the individual contribution of each of these two factors to the overall speedup; however, this issue is on our agenda for future research.

With regard to the tree decomposition comparison presented in Figure 12, the results align with the exponential nature of the problem's complexity related to tree width. Since Tamaki's algorithm is adept at finding tree decompositions with an induced tree width near the theoretical optimum, it reinforces the trend where runtime performance improves exponentially with increasing problem complexity. However, it is important to note that the superior runtime performance of Tamaki's method comes at the cost of lengthier compilation times, an aspect not depicted in the figure but important to consider.

In this paper, our primary focus has been on the optimization of individual sum-product message computations. We believe that by employing a similar approach at a higher level – specifically, by using contraction optimization techniques to efficiently schedule the propagation of messages across the junction tree – we could enhance performance further. This presents an exciting direction for future exploration.

7 Conclusion

We introduced a method for executing sum-product operations that are crucial for the performance of inference within the framework of message-passing algorithms. Our approach first establishes an order for the pairwise contraction of the input factors, aiming to minimize overall memory consumption and consequently, execution time. Following this, we recast each pairwise contraction as a batched GEMM operation to leverage its computational advantages. Notably, our method demonstrated substantial speedups in inference performance compared to earlier methods, with the improvements becoming more pronounced as the model complexity increases. Through a comparative evaluation against two other open-source libraries for probabilistic inference, our method achieved an average speedup of 23 times using the UAI 2014 benchmark set, with the speedup reaching up to 64 times for the 10 most complex problems, demonstrating its scalability. Furthermore, we have validated the scalability of our method on more resource-constrained devices, such as the Raspberry Pi 4. In contrast with the reference libraries, which failed to process a subset of the problems, due to inherent memory limitations of the embedded platform, our approach consistently produced correct results across all problems. Additionally, we introduced a novel metaprogramming-based framework for probabilistic inference that leverages the generally larger compute power available during compile-time to alleviate runtime limitations, often encountered by embedded devices. These results underscore the potential of our method in broadening the tractability spectrum of exact inference for increasingly complex models.

While our discussion has primarily focused on the calculation of exact marginals for discrete variables, it is important to note that our proposed method is equally applicable to other types of algorithms within the realm of probabilistic inference. These include approximate variants of sum-product message-passing algorithms, as well as algorithms aimed at calculating the probability of evidence (also known as the *partition function*) or determining the most likely assignment to all variables in a model-given evidence (also known as a *maximum a posteriori* (MAP) estimate). Exploring these additional applications is a direction we plan to pursue in future research.

References

[1] G. F. Cooper, "The Computational Complexity of Probabilistic Inference Using Bayesian Belief Networks," *Artificial Intelligence*, vol. 42, no. 2, pp. 393–405, 1990. Available from: https://www.sciencedirect.com/science/article/pii/000437029090060D.

[2] S. E. Shimony, "Finding MAPs for Belief Networks is NP-hard," *Artificial Intelligence*, vol. 68, no. 2, pp. 399–410, 1994. Available from: https://www.sciencedirect.com/science/article/pii/0004370294900728.

[3] P. Dagum, and M. Luby, "Approximating Probabilistic Inference in Bayesian Belief Networks is NP-hard," *Artificial Intelligence*, vol. 60, no. 1, pp. 141–153, 1993. Available from: https://www.sciencedirect.com/science/article/pii/000437029390036B.

[4] A. M. Abdelbar, and S. M. Hedetniemi, "Approximating MAPs for Belief Networks is NP-hard and Other Theorems," *Artificial Intelligence*, vol. 102, no. 1, pp. 21–38, 1998. Available from: https://www.sciencedirect.com/science/article/pii/S0004370298000435.

[5] D. Koller, and N. Friedman, *Probabilistic Graphical Models: Principles and Techniques*. Cambridge, Massachusetts, USA: MIT press, 2009.

[6] S. L. Lauritzen, and D. J. Spiegelhalter, "Local Computations with Probabilities on Graphical Structures and their Application to Expert Systems," *Journal of the Royal Statistical Society: Series B (Methodological)*, vol. 50, no. 2, pp. 157–194, 1988.

[7] F. Jensen, S. Lauritzen, and K. Olesen, "Bayesian Updating in Causal Probabilistic Networks by Local Computations," *Computational Statistics Quarterly*, vol. 4, pp. 269–282, 1990.

[8] M. Roa-Villescas, J.G. Liu, P. W. A. Wijnings, S. Stuijk and H. Corporaal, "Scaling Probabilistic Inference Through Message Contraction Optimization," 2023 Congress in Computer Science, Computer Engineering, & Applied Computing (CSCE), Las Vegas, NV, USA, 2023, pp. 123–130, doi: 10.1109/CSCE60160.2023.00025.

[9] C. Huang, and A. Darwiche, "Inference in Belief Networks: A Procedural Guide," *International Journal of Approximate Reasoning*, vol. 15, no. 3, pp. 225–263, 1996. Available from: https://www.sciencedirect.com/science/article/pii/S0888613X96000692.

[10] P. P. Shenoy, and G. Shafer, "Axioms for Probability and Belief-function Propagation," R. D. Shachter, T. S. Levitt, L. N. Kanal, and J. F. Lemmer, Eds., *Uncertainty in Artificial Intelligence. Vol. 9 of Machine Intelligence and Pattern Recognition*. North-Holland, 1990, pp. 169–198. Available from: https://www.sciencedirect.com/science/article/pii/B9780444886507500196.

[11] S. Arnborg, D. G. Corneil, and A. Proskurowski, "Complexity of Finding Embeddings in a k-Tree," *SIAM Journal on Algebraic Discrete Methods*, vol. 8, no. 2, pp. 277–284, 1987.

[12] F. V. Jensen, and F. Jensen, "Optimal Junction Trees," *Proceedings of the Tenth International Conference on Uncertainty in Artificial Intelligence. UAI'94*. San Francisco, CA, USA: Morgan Kaufmann Publishers Inc., 1994, pp. 360–366.

[13] A. Darwiche, and G. M. Provan, "Query DAGs: A Practical Paradigm for Implementing Belief-network Inference," *ArXiv*, 1996. cs.AI/9705101.

[14] J. McCarthy, "Recursive Functions of Symbolic Expressions and their Computation by Machine, Part I," *Communications of the ACM*, vol. 3, no. 4, pp. 184–195, 1960 Apr. Available from: https://doi.org/10.1145/367177.367199.

[15] K. Czarnecki, and U. W. Eisenecker, *Generative Programming: Methods, Tools, and Applications*. USA: ACM Press/Addison-Wesley Publishing Co., 2000.

[16] M. Roa-Villescas, P. W. A. Wijnings, S. Stuijk, and H. Corporaal "Partial Evaluation in Junction Trees," *2022 25th Euromicro Conference on Digital System Design (DSD)*, 2022. pp. 429–437.

[17] H. Tamaki, "Positive-instance Driven Dynamic Programming for Treewidth," *arXiv*, 2017. Available from: https://arxiv.org/abs/1704.05286.

[18] R. M. Marinescu, Develop an Open, Easy-to-use, Extensible Framework, 2018. Accessed: 2022-02-25. Available at https://www.ibm.com/opensource/open/projects/merlin/.

[19] J. M. Mooij, "libDAI: A Free and Open Source C++ Library for Discrete Approximate Inference in Graphical Models," *Journal of Machine Learning Research*, vol. 11, pp. 2169–2173, 2010 Aug. Available from: http://www.jmlr.org/papers/volume11/mooij10a/mooij10a.pdf.

[20] U. Kjærulff "Triangulation of Graphs – Algorithms Giving Small Total State Space," Report (Aalborg universitetscenter. Afdeling for matematik og datalogi). University of Aalborg, Institute for Electronic Systems, Department of Mathematics and Computer Science, 1990. Available from: https://books.google.nl/books?id=CqfWHwAACAAJ.

Darshan Nayak, Abhijot Bedi, David Degbor, Shelley Zhang, and Eugene Chabot

Facilitating cooperative missions through information sharing in heterogeneous agent teams

Abstract: Sharing information and using the information in decision-making are important in multi-agent systems (MAS) teamwork. Effective sharing and using information among agents in a dynamic environment, such as a disaster response scenario, can lead to more successful cooperative missions. This paper discusses information sharing mechanisms among homogeneous and heterogeneous agents as well. We also present different decision-making mechanisms using shared information. These mechanisms to share and use information have been implemented with different platoon agents in a rescue simulation testbed from Robocop Rescue Simulation League (RRSL). Experiments have been conducted with various scenarios to examine the performance score difference with these information-sharing and using mechanisms compared with the baseline approach in rescue agents' decision-making under different configurations.

Keywords: multi-agent systems, cooperation, information-sharing, decision-making

1 Introduction

In Multi-Agent Systems (MAS), teamwork refers to agents' ability to collaborate toward a goal or action. It becomes crucial in several real-world applications, such as Urban Search and Rescue (USAR) and Supply Chain Management [6, 7]. In these applications, information-sharing becomes necessary, and agents in the works must communicate to share their local information to achieve common goals more efficiently in the dynamic environment [9].

A realistic testbed is particularly important to study multi-agent teamwork coordination. RoboCup Rescue Simulation League (RRSL) has been around in response to the Great Hanshin Earthquake of 1995 in Kobe, Japan, to promote research in disaster mitigation and response in USAR format [1]. We use the agent rescue testbed supplied as open source in the RoboCup Rescue Simulation (RCRS) [1] servers to develop a communication strategy among rescue platoon agents, improving overall simulation performance concerning civilians' rescues.

Darshan Nayak, Abhijot Bedi, David Degbor, Shelley Zhang, Computer and Information Science, University of Massachusetts Dartmouth, North Dartmouth, US
Eugene Chabot, NUWC Division Newport, Newport, RI

https://doi.org/10.1515/9783111344126-003

Three programmable platoon agents in the testbed (police force, ambulance team, and fire brigade) were used for this paper. Each of these agents has a high-level goal in the server. Police force agents handle removing blockages from the roads on the map, clearing the way for other platoon agents and civilians to move around freely. The ambulance team agents rescue injured civilians from the buildings and debris while the fire brigade agents work to extinguish the building fire to stop the spread. We will discuss more details about this testbed in Sections 2 and 6.

Communication and information-sharing are essentials in MAS, particularly USAR. Considering the effects of a disaster in real-time, there can be many instances where situations would demand coordinated actions from the teams participating in rescue missions. Environmental knowledge of one agent might help the other agent act effectively and efficiently, allowing them to coordinate and work towards the complex problems that aid disaster response [8]. This paper shares an information-sharing protocol within the testbed's simulation to facilitate these agents' abilities to share knowledge and collaborate on rescue tasks and an upgraded approach for multiple agents of the same type with improved performance in more advanced and complex environments.

This paper is organized as follows. First, we introduce the Agent rescue testbed with the sample agent in Section 2. Then, we present two different decision-making mechanisms in Section 3. Communication within heterogeneous agent and homogeneous agents are discussed in Section 4 and 5 respectively. Experiment setup is provided in Section 6 and results are presented in Section 7. Finally, Section 8 finishes with conclusions and future work.

2 RCRS agent rescue testbed and sample agents

The RCRS testbed comes with a set of sample agents whose behavior has been programmed. These agent programs may be used as examples for other developers, and these agents may also serve as a baseline to evaluate the performance of other teams.

The decision-making in the sample agents' programming is built on a collection of basic rules that enable them to respond rapidly to dynamically shifting simulation environmental conditions and obstacles, as discussed in [2]. For example, in this RCRS testbed, when an ambulance agent encounters an end-of-the-road dead zone or a blockage, it will try to navigate through it to continue moving toward its destination [3]. There are times when the road it turns onto is entirely blocked by debris, with no way of entering without clearing it; this is when it should call a police officer, sharing information on the road, for assistance.

The sample program for rescue agents in RCRS uses the simulation's line of sight (LOS) component to perceive the environment and situation around agents. LOS allows agents to detect and identify objects or entities in the range of visibility and not blocked by debris or other obstacles, for instance, buildings [1]. An illustration of LOS on the simulation test map can be seen in Figure 1. LOS is a crucial aspect of the simulation as it also helps detect and identify if any object obstructs the agent's viewing. In simple words, it can detect the blocked road if within the range of the LOS sensor.

The rescue testbed sample agents' actions are simple, providing a baseline for developing better-performing strategies. As discussed in Section 1, the programmable platoon agents have a sense of decision-making based on navigating through the obstacles to explore the uncharted territories on the map or carry their objectives, or high-level goals, to provide aid in response to the virtual disaster simulation.

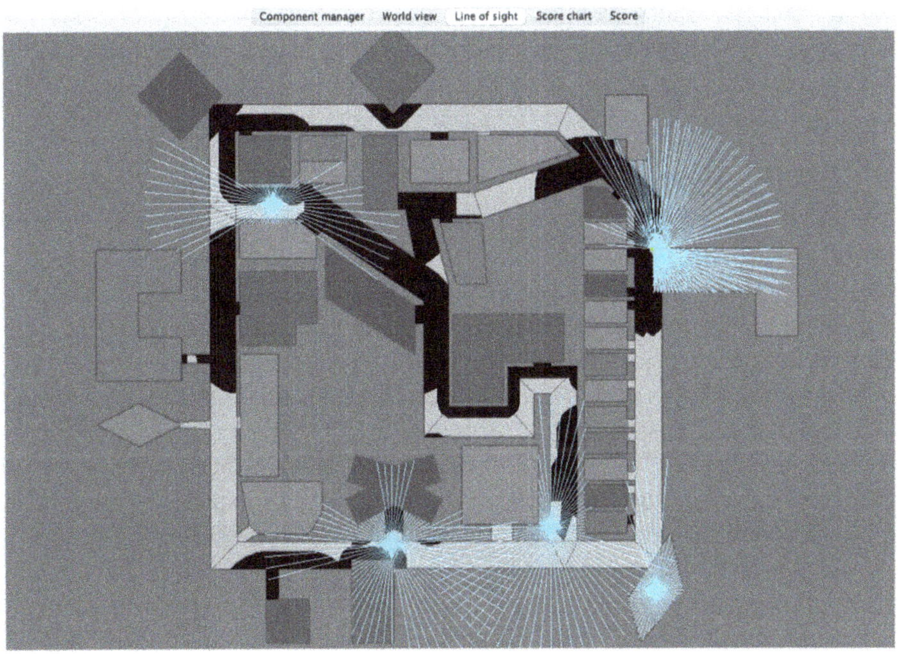

Figure 1: A line of sight (LOS) map for rescue agents and civilians.

The other human entity, the civilians, when trapped in the blockages, waits for police force agents to clear the path for them to make a run towards the refuge center. Whereas, when civilians are trapped in a building or under debris, the buriedness level of such civilians is greater than 0, which gradually affects their health points. In this case, even after clearing the blocked paths, this human entity cannot move independently and requires ambulance team agents to rescue it. In the upcoming sections of this paper, we share a communication and information-sharing-powered approach to deal with such situations [10].

3 Role of communication in decision-making

Incorporating communication into decision-making is one of the most challenging aspects of MAS. Within the RCRS framework, a solution to this problem would coincide flawlessly with such heterogeneous agents. The LOS for each individual agent is limited; thus, information available to this agent can be partial or inaccurate [2]. Information sharing between the agents could allow them to make better decisions quicker, which is of the utmost importance since the simulation terminates after 300 s by default. It is also essential that the communication module can withstand various constraints, such as the size and quantity of messages [4] and the distance between agents. We have worked on two algorithms depending upon the number of rescue agents of each type placed on the map: short distance approach, in which only single agent of each type is placed, and hybrid approach, in which a blend of short distance mechanism along with priority factor approach is used for multiple agents of each type in a far more complex environment.

3.1 Short distance approach

In this section, we analyze the two categories of information sharing, namely, the sender and receiver. The RCRS framework provides many methods to send data, for example, integers, strings, entities, properties, etc. In our case, having the sender class send an entity message gave us the most success because of the simplicity of receiving and processing the message.

Currently, for our configuration, every agent can send messages; however, we restrict communication from the police force to the ambulance team agent. This is done purposely, and the reason is apparent. The purpose of the ambulance team is to transport damaged civilians from broken buildings to refuge centers. When civilians are injured, they cannot move, requiring the ambulance team's assistance. It follows that any damage a building does to a civilian requires the fire brigade to rescue the civilian before the civilian can be transported. Since a civilian can be discovered in one of two states, damaged and buried or not damaged and not buried, it would be unnecessary for the police force to share this civilian entity with the ambulance team. The civilians would benefit more if the fire brigade rescued the civilians first, followed by the ambulance team to transport them. Therefore, communication is bidirectional between the fire brigade and police force, the fire brigade and ambulance team, and unidirectional from the ambulance team to the police force.

Depending on the two communicating agents, a message will either contain a blockade entity or a civilian entity. A blockade entity is sent to the police force when a fire brigade or ambulance team is blocked and cannot rescue or transport the civilian. Figure 4 gives us a sequential diagram of the police force receiving a message

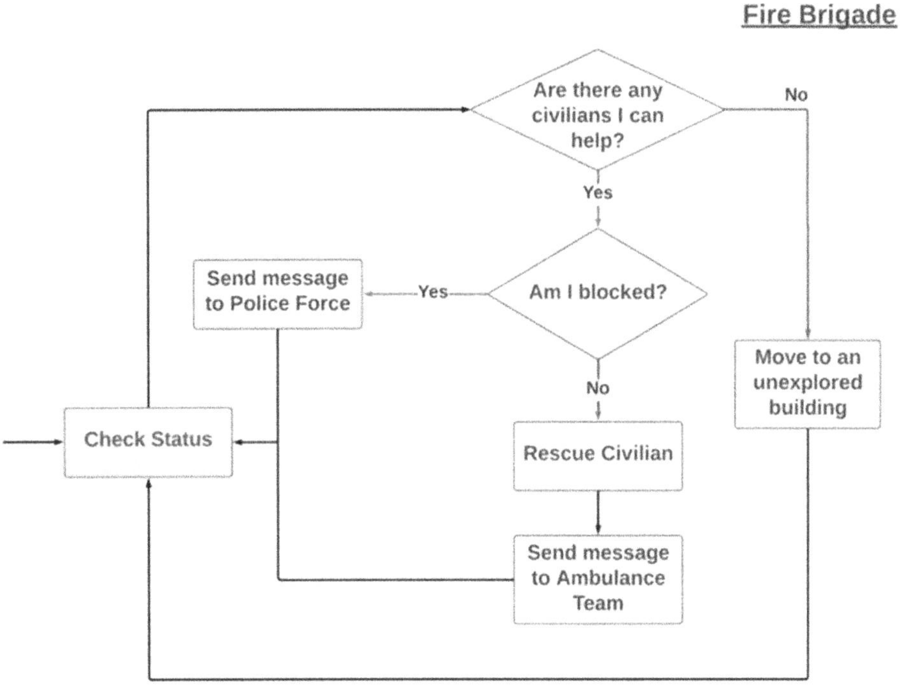

Figure 2: Flowchart for the sender side of the fire brigade communication module.

from the channel. Eventually, suppose the blockade sent from either the ambulance team or the fire brigade is in clearing range. In that case, the police force will clean the blockade and unblock the sending agent, and an unblocked agent will terminate sending repeated messages to this blockade entity.

A civilian entity is sent when a damaged civilian is discovered by one of the agents. Figures 2 and 3 outline this concept where a continuous cycle between updating other agents and moving their own location is established. Like prior reasoning, if an ambulance team agent perceives a civilian, it will notify the fire brigade. Once the entity message has been sent, processing this message should be able to smoothly integrate this new knowledge into the agent's "brain." Luckily, the fire brigade and ambulance team agents are equipped with a list-based structure for keeping track of civilian targets. This means that once a civilian message reaches one of those agents, the agent will add the incoming civilian to its tracking list.

Processing a blockade message from the police force side requires more involvement, primarily because they work with a clear-the-closest-blockade heuristic. We leverage this supplied heuristic by repetitively sending the blockade, causing the agent to be blocked. For example, if the fire brigade is blocked, they will continuously send the blockade to the police force until it is removed. Doing this forces the police force to the fire brigade's location to decide if it can remove the received blockade. If not,

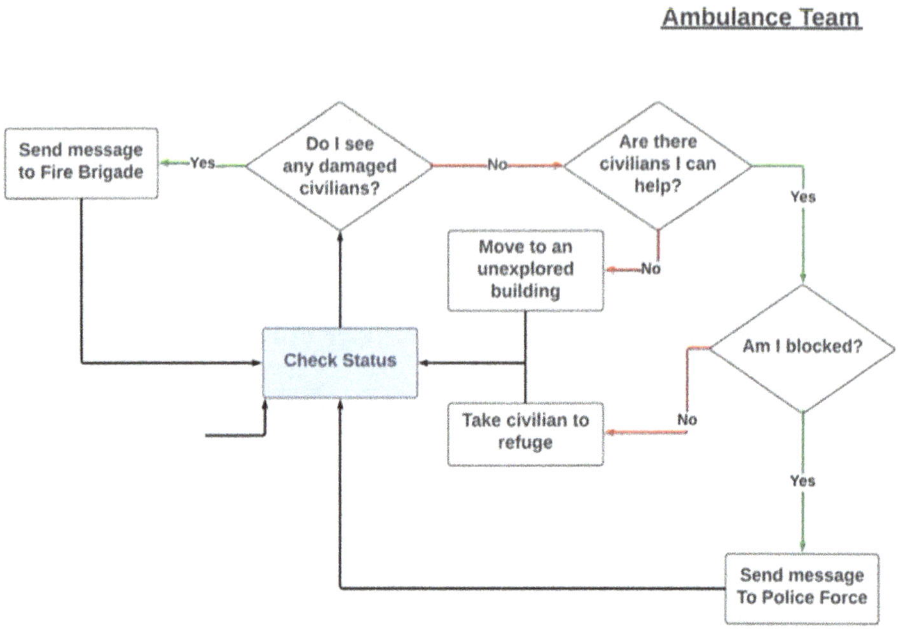

Figure 3: Flowchart for the sender side of the ambulance team communication module.

the police force will then iteratively switch off between moving towards the blockade and clearing any blockade on its path. The fire brigade then stops sending blocked messages once it can reach its desired location. Testing this algorithm has revealed that the police force sometimes stalls due to processing an influx of messages; therefore, revising our strategy is discussed further in our future work.

3.2 Hybrid approach with a priority factor

As the name suggests, this algorithmic approach works on a priority basis – a little different than the one discussed earlier. For instance, when examining the police's handling of road blockades, we uncover a complex process, mainly because they prioritize clearing the nearest blockade first [5]. To ensure effective blockade clearance, our strategy revolves around continuous notifications to the police when a blockade hinders another group, like a fire brigade. For instance, if a blockade impedes the fire brigade's progress, it keeps informing the police until the blockade is removed. This approach prompts the police to head to the fire brigade's location, assess the situation, and attempt to clear the blockade. If immediate clearance is not possible, the police alternate between reaching the blockade and addressing other barriers along the way. The fire brigade ceases notifications once it can proceed unhindered to its destination.

Figure 4: Flowchart for the receiver side of the police force communication module.

However, in our tests, we sometimes saw that the police seemed to get overwhelmed by a lot of notifications, which caused delays. This issue provides an opportunity for further improvement and will be the focus of our upcoming research.

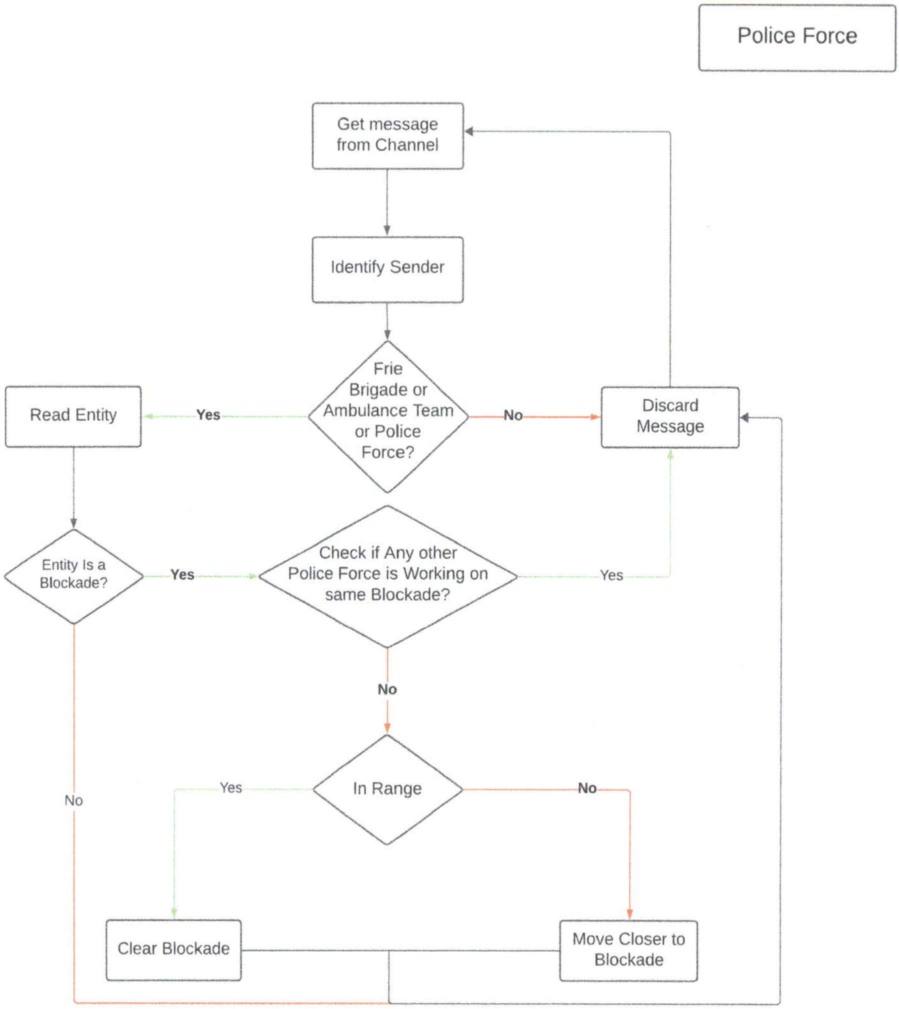

Figure 5: Flowchart for the receiver side of police force communication module in hybrid communication.

The flowchart shows how the ambulance team makes decisions during emergencies. First, they look around to see if there are injured people nearby [6]. If there are no visible casualties, they start searching in buildings they have not checked yet to expand their rescue efforts. If they find someone who needs help, they quickly check if there are others nearby who also need assistance. They also check if another ambulance team is already taking care of the person. If the person is not being helped, the team moves them to a safe place for medical attention. If they are already being assisted, the team removes them from their list to make the rescue operations more efficient.

While moving through the area, the ambulance team may encounter obstacles. When this happens, they send a message to the police force for immediate help. They also maintain communication with other groups. If they see a fire, they let the fire brigade know, and they stay in touch with other ambulance teams as needed. This process ensures a quick, organized, and coordinated response to emergencies, increasing the chances of saving lives.

The fire brigade follows a methodical approach to handling emergencies during rescue missions to ensure that every action is carried out with precision and efficiency [7]. When they receive the call to respond, the first thing they do is a “status check.” This step is critical as it helps them confirm that they are ready, that they have all the necessary equipment, and that they review any important information before they start the mission.

Once they arrive at the scene, their main goal is to find people in distress. This involves a thorough search to assess the seriousness of the situation. If they locate people who need help, they carefully determine where exactly they are. This is particularly important when there are multiple people trapped in one place because it changes how they prioritize their rescue efforts. The level of urgency for each person in a group depends on the total number of individuals there. For example, if there are five people in one location, it becomes a higher priority for rescue compared to a spot with just one person. This system helps ensure that they save as many lives as possible in the shortest time, especially when there are many people involved, which can make the situation more dangerous.

However, rescue missions are often not straightforward. The fire brigade may come across obstacles or challenges like debris or barricades that block their path. In such cases, they quickly contact the police, who have the training and equipment to remove or work through these obstacles. This ensures that the fire brigade can continue their mission without unnecessary delays.

Lastly, once they successfully rescue a person, it is crucial that the individual receives immediate medical attention. So, the fire brigade works closely with the ambulance team and keeps them updated in real-time. This ensures that as soon as a person is safely removed from a dangerous situation, they are transferred for medical care without delay, increasing their chances of recovery and survival.

4 Communication among heterogeneous agents

In this section, we will explore the distinctions between the functioning of “short distance” and the “priority factor” approaches in decision-making during emergencies. We will delve into how these two methods influence how agents, such as first responders, prioritize their actions when dealing with crises. Furthermore, in the subsequent section, we will investigate the indirect effects of communication between

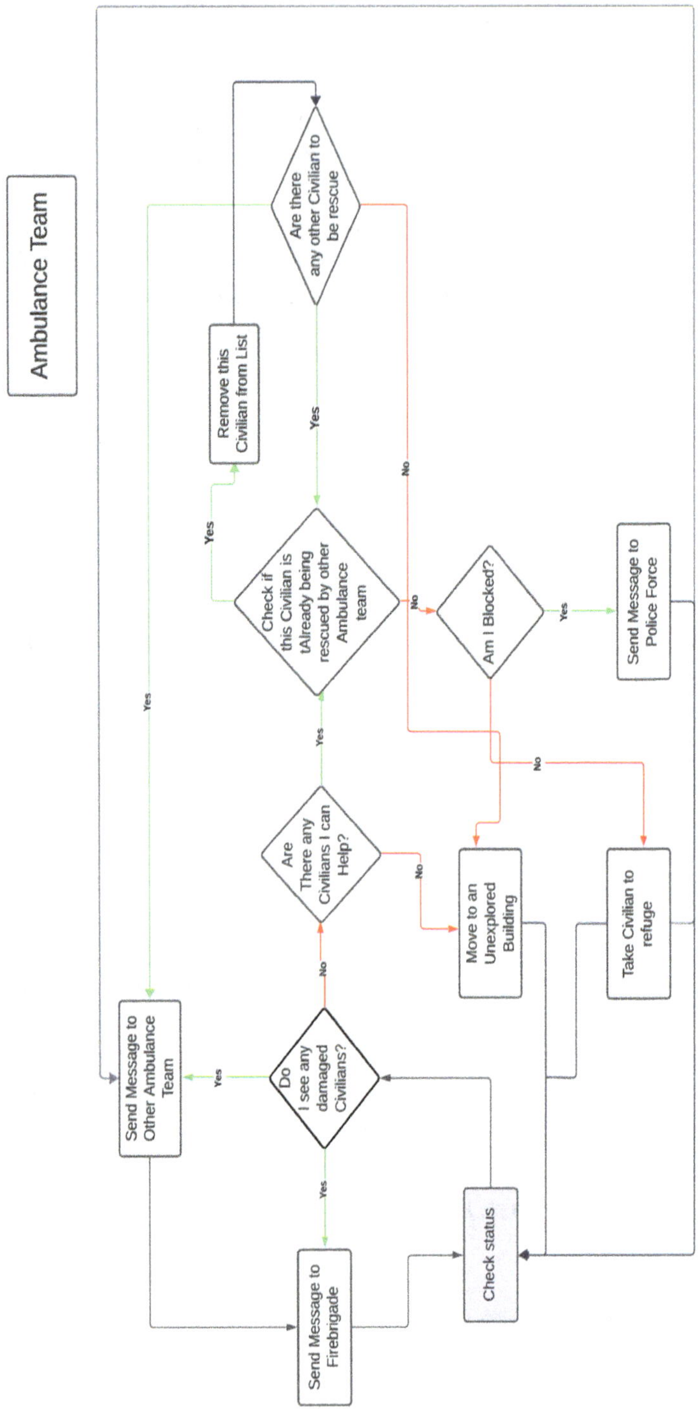

Figure 6: Flowchart for the sender side of the ambulance team communication module in hybrid communication.

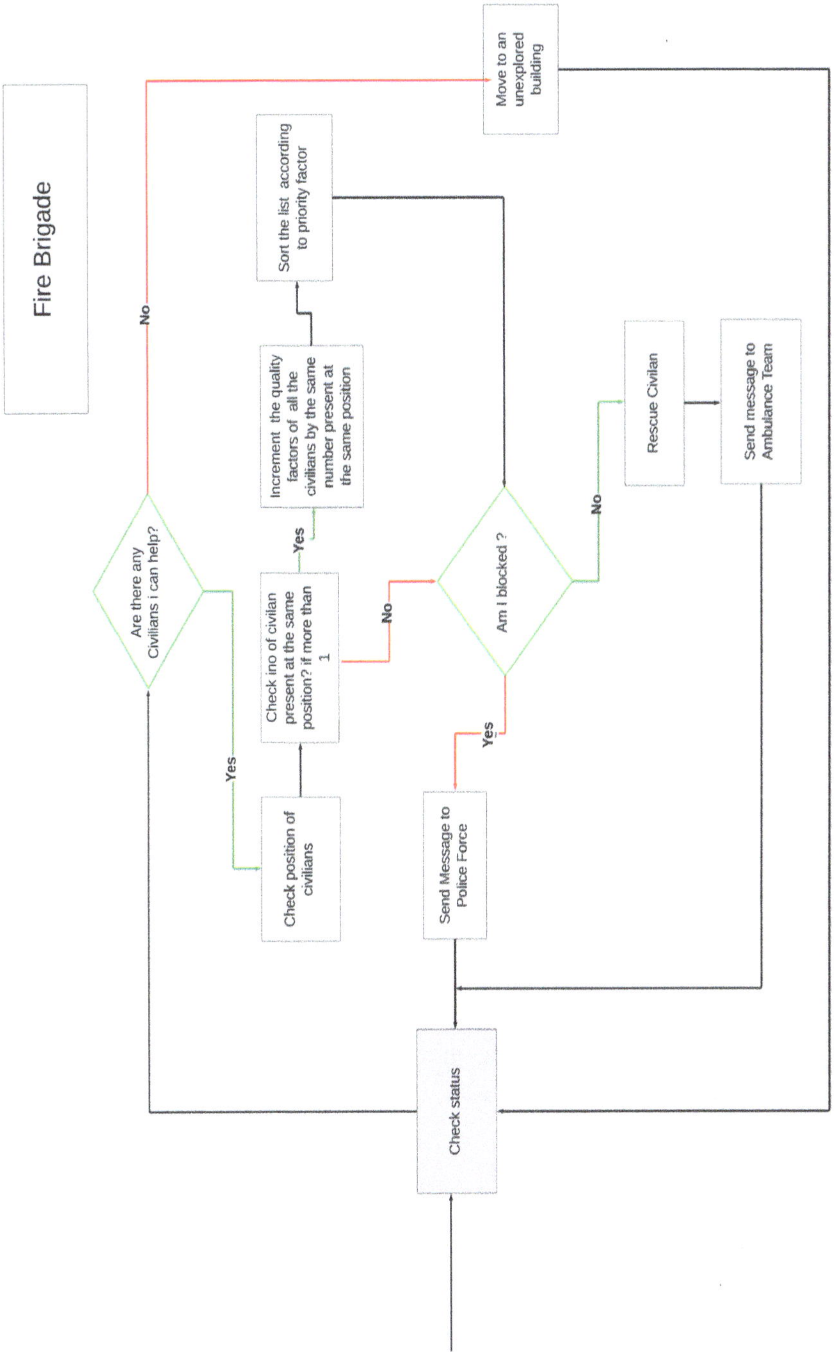

Figure 7: Flowchart for the sender side of the fire brigade communication module in hybrid communication.

agents of the same group, focusing on how it impacts the decision-making processes of entities like the police force. This will provide insights into how intragroup communication can shape response strategies during emergency operations.

4.1 Short distance vs. priority factor

In decision-making systems, especially during emergencies, there is a choice between focusing on short distances and using a priority factor. The short distance approach means dealing with the closest targets first, aiming to minimize the physical distance between the agent and the target [12]. However, this only sometimes guarantees the most efficient or effective solution.

On the other hand, the priority factor approach takes a more layered approach. It considers numerous factors like how severe the incident is, how many civilians are involved, their health status, and other dynamic variables that affect how urgent the situation is. For example, even if a civilian is closer to an agent, there might be another civilian farther away who is in a more critical condition or surrounded by more dangers. Addressing the latter situation could save more lives or prevent a larger disaster, making the priority factor a more comprehensive approach.

4.2 Indirect effects on temporary targets

Even if we do not change how the police make decisions, the way other teams like the fire brigade and ambulance make their choices can indirectly be affected. This happens because of the "priority factor" approach. This approach looks at how serious the situation is and how many people are involved when deciding what to do first. So, when the fire brigade and ambulance teams follow this approach, they might save people in a different order. This also means that the order of dealing with obstacles, like blockades, can change. The obstacles blocking the most important people (according to the "priority factor") get removed first. This can significantly impact how well the emergency response works.

5 Communication among homogeneous agents

Communication among similar groups, like the ambulance team and the fire brigade, is crucial and complex. These flowcharts show a detailed system where each step in their decision-making process is based on shared information and collective knowledge**[11]**. In the ambulance team, for example, messages are constantly exchanged with other ambulance units to prevent duplication of efforts and make the best use of resources. Similarly, the fire brigade does not work in isolation; they rely on information about where civilians are potential obstacles and prioritizing rescues based on how many people are in danger.

Additionally, when an agent faces a challenge they cannot handle on their own, like encountering obstacles, there is a procedure to communicate with other units, such as the police. This emphasizes the importance of agents from different groups talking to each other to streamline their efforts. Even though each agent has its main role, they are continuously informed and updated by their peers, making the response during rescue missions more coordinated and efficient.

6 Experimentation setup

The simulation has a functionality called scenario editor [1], which lets users place rescue agents and other entities in desired locations on the map. Utilizing this feature, we came up with five different scenarios in total to test the designed communication-based programs. The first three scenarios were used for single agents of each type, and two other scenarios were used for multiple agents of each type for a much more complicated environment.

Table 1: Visual representation of human entities in the simulation.

Icons	**Visual representation of:**
	Civilian
	Fire brigade
	Police force
	Ambulance team

Important entities observed on the map are civilians, fire brigade, police force, and ambulance team, as illustrated in Table 1. Black polygons on the map are blockages between buildings on roads, as observed in Figures 8–12.

6.1 Scenarios with single agent of each type

The experiment scenarios we tested for the short distance approach with single rescue agents of each type consisted of 6 civilians, 1 fire brigade, 1 police force, and 1 ambulance team. We wanted to challenge the agents and test their abilities to communicate in scenarios they have not been in before. Thus, by initializing different starting locations for civilians and agents, we were able to see exciting results. The different configurations we tested the simulation on yielded an average score increase of 2.165 points out of 6 maximum points. This can be considered significant progress since a perfectly healthy civilian is equivalent to 1 point. Using these test cases can be used to conclude that, on average, our communication-based implementation in the agents warrants a higher score in the simulation. Scenarios can be observed in Figures 8–10.

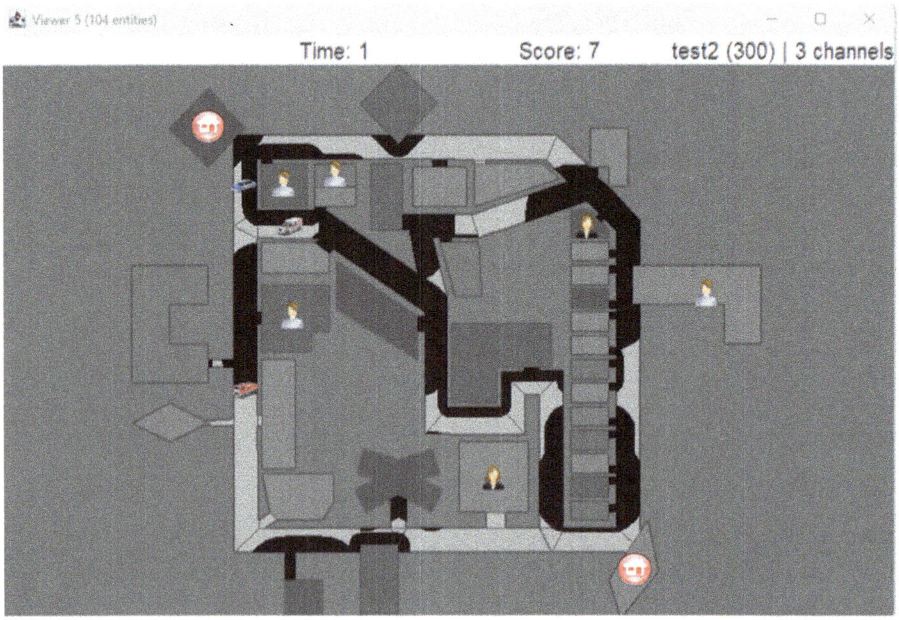

Figure 8: Image of configuration 1.

6.2 Scenarios with multiple agents of the same type

In the simulation scenario, we assessed the interaction dynamics between 16 civilians and an assembled team of agents, including 3 fire brigades, 3 police forces, and 3 ambulance teams, as illustrated in Figure 11. The goal was to gauge the proficiency of agents in leveraging communication and teamwork in a setting peppered with unforeseen challenges. Civilians and agents were placed in diverse initial locations, aiming

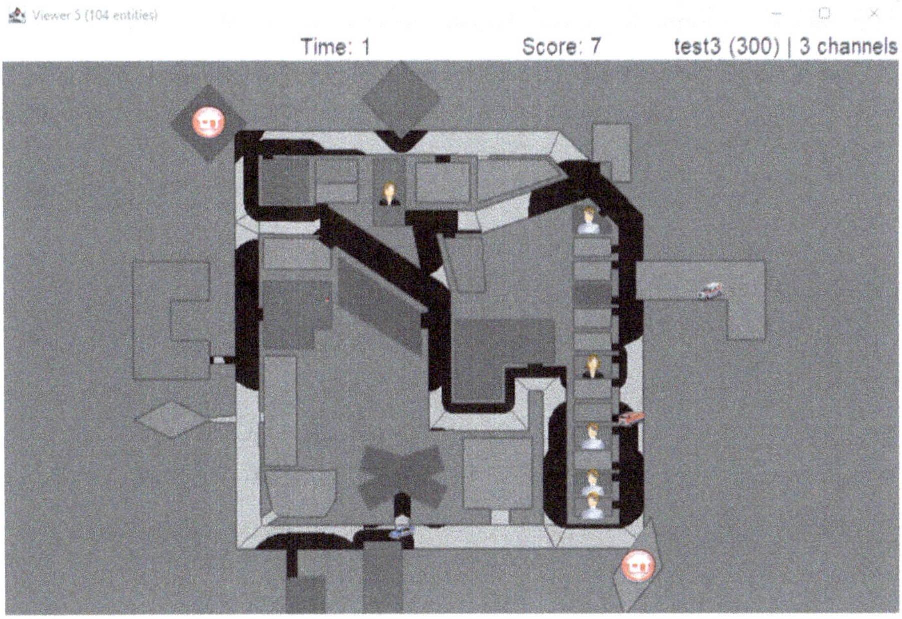

Figure 9: Image of configuration 2.

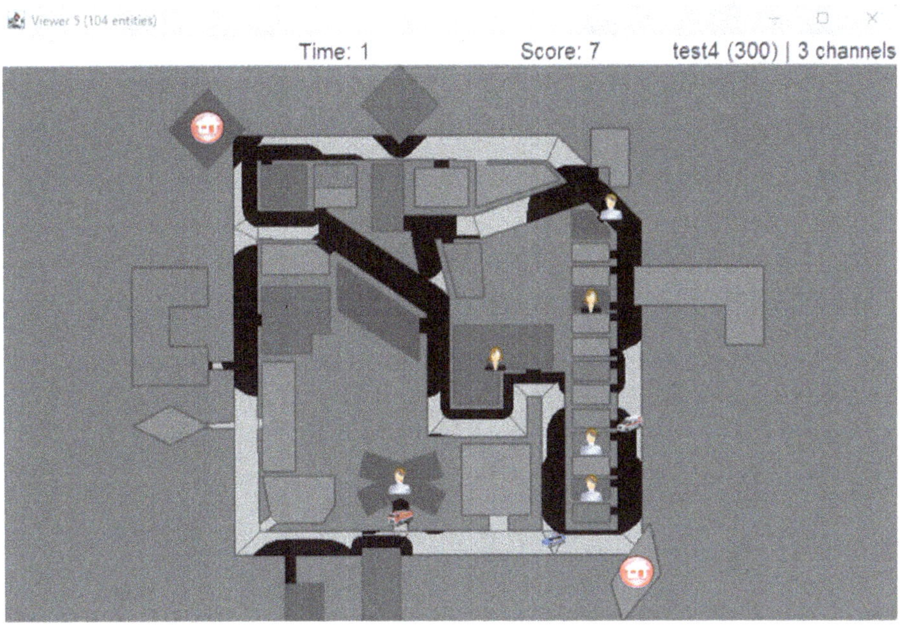

Figure 10: Image of configuration 3.

to replicate real-world unpredictable emergencies. These varying test conditions noted an encouraging average score elevation of 2 points out of a possible 13 keeping in mind that each point signifies a civilian in prime health; this outcome vividly emphasizes the pivotal role played by effective inter-agent communication in achieving commendable rescue results in the simulation.

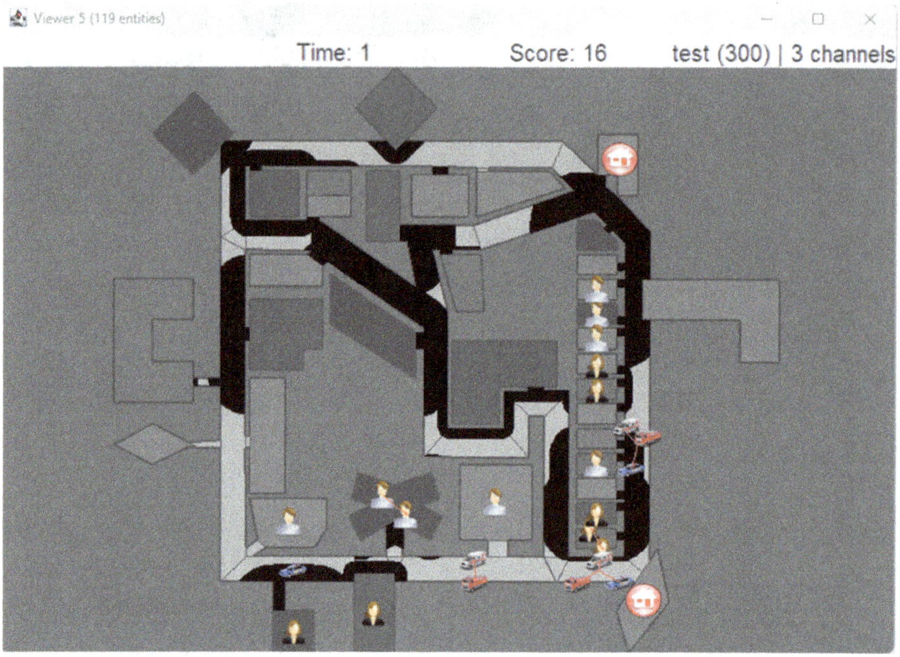

Figure 11: Image of the configuration with agents in close proximity.

Figure 12 depicts a distinct simulation setup used in the experiment featuring 10 civilians; the agent team included 3 fire brigades, 3 police forces, and 3 ambulance teams. The objective of strategically dispersing civilians and agents across varying initial positions was to simulate unpredictable emergency scenarios. Despite the challenges, the communication-enabled agents registered an average score boost of 2 points out of the potential 10, reinforcing the importance of collaborative efforts in such emergency contexts.

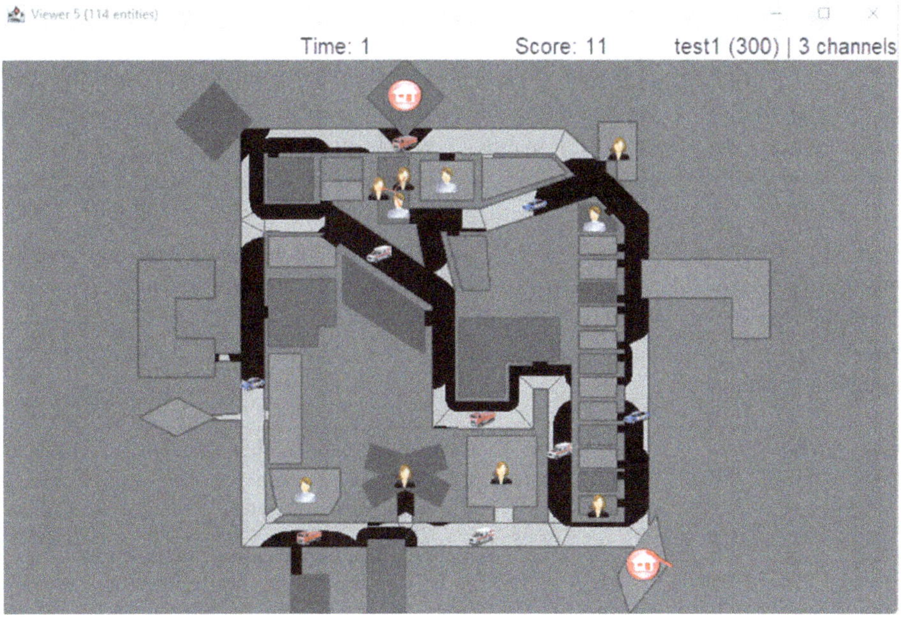

Figure 12: Image of the configuration with agents far apart in the initial run.

7 Results

Results were calculated to compare the sample and communication-focused codes for both types of experimentation setup in two charts. The mathematical formula we used to achieve this is expression 1, where $S(N)$ denotes the score for the non-communicating agents, and $S(C)$ represents the score for the communicating agents:

$$100 * \left(\frac{S(C) - S(N)}{S(N)} \right) \tag{1}$$

Table 2: Chart comparing communicating agents and noncommunicating agents.

Configuration	No communication	Communication-based (short distance)	Improvement (%)
1	2.333	5.567	138.6
2	3.5	4.557	30.2
3	0	2.204	Infinite

Table 2 presents the scores for each configuration. A configuration's score is calculated by adding the total number of civilians to the proportion of the civilian's health; hence, in the three scenarios depicted in Figures 5–7, we have an upper bound score of 7 points. In Table 2, the communicating agents could save 5 civilians, whereas the noncommunicating agents only saved 2 civilians, meaning our agents performed 138.6% better than the original agents. Additionally, to evaluate the second approach, where we have multiple communicating agents of each type, we plot another table, Table 3, to compare the results on two scenarios with no communication code, communication-based (short distance) and communication-based (hybrid).

Table 3: Chart comparing communicating agents (both short distance and hybrid) and noncommunicating agents.

Configuration	No communication	Communication-based (short distance)	Communication-based (hybrid)	Improvement hybrid over short distance approach (%)
Agents in close proximity at Initial run	11.7	12.6	13.7	8.7
Agents far apart at Initial run	5.4	5.4	7.5	38.2

In the case of agents close by, it is evident that the "hybrid communication" configuration outperforms both the "no communication" and "communication-based (short distance)" setups, showing a substantial improvement of 8.7% over short distance communication-based setup. This suggests the hybrid communication code is significantly more effective in optimizing agent interactions when close to them. Conversely, when agents are far apart, the same configuration demonstrates a remarkable 38.2% improvement over the "short distance" approach. This substantial improvement underscores the enhanced efficiency of the communication system in scenarios with a larger distance between agents. The results highlight the considerable advantages of the "hybrid communication" configuration over both "no communication" and the "communication-based (short distance)," particularly in scenarios involving agents who are either close by or far apart.

8 Conclusion and future work

For the future of our research in MAS, we are dedicated to improving the solutions we have presented thus far, improving our focus, and resolving shortcomings within the agent's path planning, response time, and collaboration. Additionally, in a previous section, we explained the difficulty with the police force agent processing incoming messages. Due to the repetitive sending of blockade entities from other agents, the police force will likely experience an overflow of messages. It has been shown that the police force will drop some of these messages or ignore them entirely [5]. As a result, the police force will receive a thorough rework and a completely new structure for how it operates, in addition to finding an efficient way to send messages. Another concern lies within the center's involvement and how we can effectively incorporate it into the simulation. We believe revisiting these topics will allow us to continue utilizing our communication-based implementation, develop more ideas, and augment our knowledge in MAS.

References

[1] RoboCup Agent Rescue Simulation Competition Homepage, http://rescuesim.robocup.org/competitions/agent-simulation-competition/, last accessed 2023/04/10.

[2] H. Kitano, and S. Tadokoro, "RoboCup Rescue: A Grand Challenge for Multiagent and Intelligent Systems," *AI Magazine*, pp. 22–39, 2001.

[3] H. L. Akin, N. Ito, A. Jacoff, A. Kleiner, J. Pellenz, and A. Visser, "Robocup Rescue Robot and Simulation Leagues," *AI Magazine*, vol. 34, no. 1, pp. 78, 2012.

[4] M. Nanjanath *et al.*, "Decision and Coordination Strategies for Robocup Rescue Agents," in *Simulation, Modeling, and Programming for Autonomous Robots: Second International Conference, SIMPAR 2010, Darmstadt, Germany, November 15–18, 2010.* Springer Berlin Heidelberg, Proceedings 2, pp. 473–484.

[5] S. B. M. Post, and M. L. Fassaert, *A Communication and Coordination Model for 'Robocup rescue' Agents.* Department of Computer Science, University of Amsterdam, 2004.

[6] S. H. Kim, J. Y. Lee, J. Lee, and B. Choi, "A Multi-agent System for Disaster Management Using Social Network Analysis," *IEEE Transactions on Systems, Man, and Cybernetics: Systems*, vol. 44, no. 6, pp. 693–704, June 2014.

[7] Z. Xu, X. Zhao, and Z. Chen, "Multi-agent System for Supply Chain Management: A Review," *IEEE Transactions on Industrial Informatics*, vol. 10, no. 3, pp. 2103–2115, Aug 2014.

[8] D. Nardi, A. Bondavalli, A. Ceccarelli, and L. Falai, "Communication Middleware for Multi-Agent Systems: A Review," *IEEE Communications Surveys and Tutorials*, vol. 20, no. 1, pp. 672–707, 2018.

[9] A. Bedi, S. Zhang, and E. Chabot, "A Collaborative Approach to Robocup Rescue Challenge," in *The Proceedings of the 24th International Conference on Artificial Intelligence (ICAI'22).* USA, July 25–28 2022.

[10] A. Visser, N. Ito, and A. Kleiner, "Robocup Rescue Simulation Innovation Strategy," in *RoboCup 2014: Robot World Cup XVIII 18*. Springer International Publishing, 2015, pp. 661–672.

[11] W. Niu, and J. Wu 2015. Rescue Simulation League SEU_Jolly Team Description. [5] P. Ardestan, M. Taherian, P. Mohammad Ali Zadeh, and E. JazebNikoo 2017. Rescue Simulation League MRL Team Description.

[12] D. Arthur, and S. Vassilvitskii, "K-Means++: The Advantages of Careful Seeding," *Proceedings of the Annual ACM-SIAM Symposium on Discrete Algorithms*, vol. 8, pp. 1027–1035, 2007.

Sait Alp, Taymaz Akan, and Mohammad Alfrad Nobel Bhuiyan

Transferring knowledge: CNNs in Martian surface image classification

Abstract: The exploration of Mars, a planet with potential for scientific discovery, has seen increasing attention in recent years, leading to numerous missions with robotic rovers. However, landing on Mars is a challenging endeavor with a high risk of failure, especially during the landing phase. One of the ongoing projects is the classification of the Mars surface from orbital images. We propose a transfer learning approach that classifies the Martian orbital imagery, reducing the need for manual feature engineering and improving generalization across various surface conditions. To classify Martian images, our proposed method involves fine-tuning pre-trained CNN models, including ResNet152V2, EfficientNetV2M, and VGG19. By modifying these models to accommodate custom layers, we can adapt them to the specific task of Martian imagery classification. Transfer learning allows retaining knowledge learned from the original dataset, improving accuracy, and reducing the need for extensive Martian surface image data. We utilized the HiRISE dataset provided by NASA's Planetary Data System, which contains images of Martian landmarks. While the HiRISE dataset is imbalanced, with one category dominating the others, we employ alternative performance measures like precision, recall, and F1-Score to evaluate our models' performance. Our study demonstrates the potential of transfer learning with CNNs to enhance the classification of Martian surface features, contributing to the success of future Mars exploration missions. The results indicate that the EfficientNetV2M outperforms the other models in terms of accuracy, making it a top candidate for applications where high precision, recall, and F1-Score are essential.

Keywords: convolutional neural networks, image classification, Mars, planetary science

1 Introduction

Enhancing spacecraft and robot autonomy is essential to expand the reach of solar system exploration [1]. Mars, the planet in the solar system most similar to Earth, is regarded as an optimal candidate for planetary exploration [2, 3]. Recent years have

Sait Alp, Department of Computer Engineering, Erzurum Technical University, Erzurum, Turkey
Taymaz Akan, Department of Software Engineering, Istanbul Topkapi University, Istanbul, Turkey; Department of Medicine, Louisiana State University Health Sciences Center at Shreveport, Shreveport, LA, USA
Mohammad Alfrad Nobel Bhuiyan, Department of Medicine, Louisiana State University Health Sciences Center at Shreveport, Shreveport, LA, USA

https://doi.org/10.1515/9783111344126-004

witnessed an all-time high in the number of expeditions to Mars and the amount of research focused on these missions [4]. Humans have conducted near-distance investigations of Mars exploration missions since the 1960s. Multiple attempts have been made to explore Mars since the first one, each informed by massive technological development. As aerospace science and technology have advanced, landing and roving explorations have replaced flyby and orbiting methods of Mars exploration. Over the past 50 years, 46 Mars exploration spacecraft have been launched, but the overall success rate is only 41.3%. Out of 20 landing attempts, only seven robotic rovers have been successful, with a 35% success rate for landing missions, with most failures occurring during the landing phase [1].

Due to the limited human involvement in Mars expeditions, successful missions primarily rely on automated devices such as rovers to gather crucial ground-level data. Mars landing exploration is also vital and is expected to soon be one of the most popular tasks of human deep space exploration due to its high potential scientific returns and exploration capabilities. Since its launch in January 2004, the Mars Exploration Rover (MER) mission has achieved remarkable results. However, many operational tasks for the current MER mission and the 1997 Mars Pathfinder mission are decided upon on Earth, where the process is manual and time-consuming. Accurate position and attitude information received by a rover and the support of simultaneous localization and mapping are necessary for the successful automation of rover navigation and localization during extraterrestrial exploration [5].

Mars' thin and dense atmosphere covers about 125 km [6]. The landing mission involves the final approach, entry, descent, and landing (AEDL) phases. The final approach phase occurs 12 h before the spacecraft reaches the upper layer of the atmosphere, requiring navigation to accurately estimate entry conditions and adjust attitude and trajectory [7]. The entry phase begins when the entry vehicle reaches the atmosphere and ends when the parachute is deployed. The entry phase is the most dangerous and unpredictable, with the vehicle potentially getting peak deceleration and dynamic pressure. The descent phase can be divided into parachute descent and powered descent phases. The parachute descent phase decelerates the vehicle's velocity, while the powered descent phase eliminates horizontal and vertical velocity for the final landing. This phase requires obstacle detection and guidance to ensure stronger maneuverability.

The landing phase is the final stage of the AEDL phase, requiring the rover to land safely on the ground for the following scientific exploration mission. The landing procedure appears to be the most challenging and risky part of the Mars landing mission. Landing at specific sites on Mars may be necessary for future missions to ensure their success and collect more data for scientists. Previous robotic lander missions relied on conventional devices like inertial measurement units (IMUs) and Doppler radar due to the absence of infrastructure such as the global positioning system (GPS). The issue lies in the inherent imprecision of these devices, with potential errors spanning several kilometers. This imprecision arises from the cumulative impact of noise, biases, and initialization errors [1, 6]. Because of these factors, most major space agencies agree that systems that land themselves using vision are needed. In vision-guided navigation systems, craters and other known features on the surface can be used to get a rough idea of where the lander is. In many space missions, cameras onboard spacecraft capture images of the surfaces of celestial bodies, such as Mars. These images are then analyzed to detect and identify surface landmarks critical for navigation, scientific exploration, and various mission objectives.

Spacecraft and rovers on Mars use “descent landmarks,” which are features or locations on the Martian surface that serve as points of reference during the descent phase of the mission. These points of reference are essential for landing safely and navigating the area. The surface of Mars is made up of only natural landforms. There are no buildings or living things on it. Monumental things that can be seen on Mars' surface include big features like mountains, dunes, and craters, as well as smaller features like rocks [5]. Landing site features from previous missions are one type of artificial feature that can serve as a descent landmark, along with natural features like craters, valleys, and distinctive rock formations.

NASA developed a content-based search capability for Mars rover surface images and Mars orbital images so that users could more easily locate images that interest them. The Planetary Data System (PDS) at NASA stores the acquired data and makes it accessible to the public. Much of the data consists of photographs, which PDS stores. Automated image analysis to detect surface features such as craters has been common for several decades. For several decades, automated image analysis has been used to detect surface features in orbital images automatically [8–11]. These surface features include craters [12] and dune fields [13].

This paper employs transfer learning, which offers several advantages over traditional image processing methods for classifying Martian surface imagery. Unlike conventional methods requiring manual feature extraction and engineering, transfer learning leverages pre-trained deep learning models, allowing automatic feature learning from diverse and extensive datasets. This reduces the need for domain-specific knowledge and enhances the system's ability to adapt to various surface conditions and lighting on Mars. Since pre-trained models bring much prior knowledge, transfer learning reduces data requirements, which is especially useful for limited HiRISE v3 image datasets. This efficiency and improved generalization improve classification accuracy and reduce model development time and computational resources.

2 Proposed approach

Transfer learning involves modifying an existing deep learning model to suit a new task or dataset; this is accomplished through fine-tuning a pre-trained Convolutional Neural Network (CNN) for image classification [14]. The initial step involves the choice of an existing CNN model, such as ImageNet, which has undergone training on a substantial and varied dataset. Pre-trained models that have acquired the ability to identify a diverse array of image features and patterns consist of widespread alternatives such as VGG [15], ResNet [16], Inception [17], and MobileNet [18].

The fine-tuning procedure commences by removing the uppermost layers of the pre-trained model. These top layers are responsible for classifying images into specific categories, which may not be relevant to the new task. Consequently, the operational model excludes these classification layers, including the fully connected layers. Following this, specialized image classification tasks are incorporated into the model by adding custom layers. A certain number of neurons equal to the number of classes in the target dataset are added to these newly created layers, which are usually fully connected. Dropouts and supplementary layers for batch normalization were integrated to augment the model's generalizability. To retain the knowledge learned from the original dataset, the lower layers of the pre-trained model are typically frozen. Subsequently, only the weights of the custom layers were adjusted throughout the fine-tuning procedure, as the weights of the lower layers remained unchanged. As the lower layers maintained their capability to identify general features, this methodology averted the potential for the model to be overfitted to the new, reduced dataset.

In this study, we used the power of three well-known CNN architectures – ResNet152V2, EfficientNetV2M [19], and VGG19 – to solve the complex problem of classifying images of the surface of Mars. These architectures are renowned for their exceptional performance and robustness for various computer vision tasks. Each of these CNN models offers distinct advantages and characteristics that cater to different requirements in the domain of image classification.

ResNet152V2 is a CNN architecture introduced in 2017 that is a more profound and complex version of the original ResNet152 architecture. It is based on the residual network architecture, which uses residual blocks to learn more profound and more complex representations of input data. Residual blocks consist of two or more convolutional layers, allowing the network to skip over layers and learn directly from the input data. This allows ResNet152V2 to achieve better accuracy than previous CNN architectures without sacrificing efficiency.

EfficientNetV2M, introduced in 2022, is a more efficient version of the original EfficientNet architecture. It is based on the EfficientNet architecture and uses techniques such as compound scaling, mobile inverted bottleneck convolution, and squeeze-and-excitation to improve input data representation. Both architectures have achieved state-of-the-art results on image classification and object detection tasks while using fewer parameters and computations. This model mixes training-aware neural architecture search (NAS) and scaling to make training go faster and better use parameters. NAS has been used to improve the architecture of networks for image classification by automating the process of designing networks.

VGG19 [20] is a deep convolutional neural network architecture with 19 layers, including 16 convolutional layers and three fully connected layers. Developed by the Visual Geometry Group, VGG19's simplicity and uniform structure, using 3 × 3 convolutional filters and 2 × 2 max-pooling layers, made it a prominent model in computer vision. While it achieved strong performance in the ImageNet competition 2014, it has been surpassed by more efficient architectures in recent years. VGG19 is often used for transfer learning and as a reference in deep learning.

Compared to EfficientNet and ResNet152V2, VGG19's advantage lies in its straightforward and understandable architecture, making it a suitable choice for educational and research purposes and transfer learning applications. However, it falls short in performance, efficiency, and scalability, as both EfficientNet and ResNet152V2 offer superior accuracy, parameter efficiency, and the ability to handle more complex tasks in deep learning and computer vision.

We started the fine-tuning procedure by removing the top layers of the aforementioned pre-trained model. Then, we added a global average pooling layer to summarize feature maps. Subsequently, we added a batch normalization layer to address the internal covariate shift problem. We then added a dropout layer with a dropout rate 0.2 to prevent overfitting. Finally, we incorporated two fully connected layers as the classifier. Moreover, a dropout layer with a dropout rate 0.5 was placed between these fully connected layers.

2.1 Dataset

The NASA Planetary Data System (PDS) maintains archives of data collected by NASA missions, including millions of images from Mars [21]. To aid users in finding images of interest, NASA has developed content-based classification and search capabilities for Mars orbital and surface images. The PDS Image Atlas is publicly accessible. NASA has created two new labeled data sets to train and evaluate the latest versions of its Mars image classifiers. The HiRISE [22] images were collected by the High-Resolution Imaging Experiment (HiRISE) instrument onboard the Mars Reconnaissance Orbiter (MRO).

In contrast, the MSL images were collected by the Mast Camera (Mastcam) and Mars Hand Lens Imager (MAHLI) instruments mounted on the Mars Science Laboratory (MSL) Curiosity rover. To ensure high quality, the labels for both data sets were acquired using crowdsourcing with local volunteers who received specific training for each data set. The HiRISE dataset contains 73,031 landmarks, extracted from 180 HiRISE browse images and augmented from 10,433 original landmarks. Each original landmark was cropped into a square bounding box, resized to 227 × 227 pixels, and augmented to generate six additional landmarks using various methods such as 90 degrees clockwise rotation, 180 degrees clockwise rotation, 270 degrees clockwise rotation, horizontal flip, vertical flip, and random brightness adjustment. The information about the collected data is listed in Table 1.

Table 1: Information of the collected dataset.

	Class	No. of images			
		Train ins: 46,739	Validation ins: 11,685	Test ins: 14,607	Total ins: 73,031
C1	Bright_dune	1,120	280	350	1,750
C2	Crater	3,136	784	980	4,900
C3	Dark_dune	730	183	228	1,141
C4	Impact_ejecta	148	37	46	231
C5	Other	39,074	9,768	12,212	61,054
C6	Slope_streak	1492	373	466	2,331
C7	Spider	305	76	95	476
C8	Swiss_cheese	734	184	230	1,148

3 Experimental results and discussion

All experiments were conducted utilizing Python on a Windows 10 system with 16 GB of RAM, an 8GB NVIDIA GeForce RTX 2080, and an Intel (R) Core (TM) i5-6400 processor operating at 2.70 GHz.

CNNs have demonstrated significant utility in computer vision-related tasks such as object detection, classification, and segmentation. The CNN concept involves training large multilayer networks to classify large input images efficiently. CNN includes various models such as AlexNet, DenseNet, ResNet, VGGNet, ResNet, etc. In this paper, we applied three well-known EfficientNet, VGG19, and ResNet152V2 pre-trained CNN models. The details of the models and their learning parameters are listed in Tables 2 and 3. Adam optimizer was used to train the models for 100 epochs; the batch size was 128, and the learning rate was 1e-4. As a loss function, we employed sparse categorical cross-entropy (see Table 4).

Table 2: Baseline model's properties.

Model	Architecture	Depth	Parameters
EfficientNetV2M (EF)	EfficientNet	–	20M +
VGG19 (VG)	VGG	19	143M +
ResNet152V2 (RE)	ResNet	307	60M +

When dealing with unbalanced data, accuracy can be a deceptive metric [23]. Accuracy alone might not be a reliable indicator of the model's performance when working with imbalanced datasets in which one class substantially outnumbers the others. This is because a highly accurate classifier may not adequately address the minority class, which is frequently of greater interest if it predicts the majority class in every

Table 3: Transfer learning model parameters.

Base-Model	Total parameters	Trainable parameters	Non-trainable parameters
EfficientNetV2M	53,165,756	12,808	53,152,948
VGG19	20,030,536	5,128	20,025,408
ResNet152V2	58,356,232	20,488	58,335,744

instance. As it is clear from Table 1, the HiRISE dataset is imbalanced, where the number of instances in the category with the label "other" is much higher than the rest of the labels. The algorithm's accuracy has been computed using the confusion matrix results. Due to the uneven distribution of samples, alternative performance measures such as precision, recall, and F1Score were employed for evaluation (see Table 5). The table has four primary columns: Accuracy, Precision, Recall, and F1-Score. Each of these primary columns is further divided into three sub-columns, labeled "EF (EfficientNetV2M)," "VG (VGG19)," and "RE (ResNet152V2)."

Table 4: Model compilation parameters.

Parameter	Values
Image size	224 × 224
Batch size	128
Optimization	Adam (learning rate = 0.001)
Loss function	Categorical Cross entropy
Epochs	100
Metrics	Categorical Accuracy
Call back	Save best Only

Table 5: Comparison of the classifiers using a variety of performance metrics.

Class	Accuracy			Precision			Recall			F1-score		
	EF	VG	RE	EF	VG	RE	EF	VG	RE	EF	VG	RE
C1	–	–	–	0.99	0.99	0.78	0.99	0.98	0.27	0.99	0.99	0.40
C2	–	–	–	0.83	0.82	0.77	0.75	0.61	0.14	0.79	0.70	0.24
C3	–	–	–	0.95	0.93	0.46	0.92	0.82	0.28	0.94	0.87	0.35
C4	–	–	–	0.95	0.76	0.50	0.80	0.41	0.13	0.87	0.54	0.21
C5	–	–	–	0.97	0.95	0.86	0.98	0.98	0.98	0.97	0.96	0.92
C6	–	–	–	0.83	0.83	0.32	0.74	0.57	0.06	0.78	0.68	0.10
C7	–	–	–	0.99	0.96	0.47	0.98	0.72	0.18	0.98	0.82	0.26
C8	–	–	–	0.99	0.97	0.37	0.98	0.92	0.06	0.99	0.94	0.10
All Class	**0.96**	0.94	0.85	0.94	**0.90**	0.57	**0.89**	0.75	0.32	**0.91**	0.81	0.38

The data is organized into four primary columns: Accuracy, Precision, Recall, and F1-Score. Accuracy measures the performance of a classification model. Recall measures the number of correctly predicted positives. F1-Score represents the harmonic mean of precision and recall, providing a balanced measure of a model's performance. Data rows contain the actual numerical values for each sub-column, representing a different data set and corresponding to a specific model (EF, VG, or RE).

According to Precision, EfficientNetV2M consistently demonstrates high precision, ranging from 0.83 to 0.99. This indicates that when EfficientNetV2M makes predictions, it is correct 83% to 99% of the time. VGG19 also maintains high precision, with values varying between 0.76 and 0.99. This means that VGG19 achieves precision levels of 76–99%. ResNet152V2 exhibits varying precision values from 0.32 to 0.86, with the lowest precision of 32%.

In terms of Recall, EfficientNetV2M achieves a recall between 0.74 and 0.99. This indicates that EfficientNetV2M effectively captures between 74% and 99% of relevant instances specific to the model. VGG19 demonstrates recall values in the range of 0.41–0.98, suggesting that VGG19 excels at identifying instances specific to VGG19, with a minimum recall of 41%. ResNet152V2 exhibits recall values ranging from 0.06 to 0.98. ResNet152V2 has the lowest recall values among the models, with a minimum recall of 6%, indicating its performance in identifying instances specific to ResNet152V2.

The F1-Score values for EfficientNetV2M range from 0.78 to 0.99. This model consistently achieves high F1-scores, indicating a strong balance between precision and recall in classifying data. The highest F1-score of 0.99 demonstrates that EfficientNetV2M excels in both accurately predicting positive instances and capturing relevant instances, making it a robust model for the given classification task. VGG19 displays a range of F1-scores from 0.54 to 0.99. F1-scores for ResNet152V2 range from 0.10 to 0.40. These F1-scores are relatively lower than EfficientNetV2M and VGG19, indicating a different trade-off between precision and recall. The lowest F1-score of 0.10 suggests that ResNet152V2 may face challenges in accurate predictions and capturing relevant instances for the given classification task.

The macro average values of each model across different experiments or datasets provide a global perspective on their performance. EfficientNetV2M shows high precision (0.94) and good recall (0.89), indicating a strong balance between precision and recall. VGG19 has slightly lower precision (0.90) but maintains good recall and F1-score (0.81), indicating overall solid classification performance. ResNet152V2 has significantly lower precision (0.57) and recall (0.32), indicating challenges in achieving high precision and recall across different experiments or datasets. Overall, EfficientNetV2M and VGG19 consistently show strong overall classification performance.

The EfficientNetV2M model has the highest accuracy of 0.96. The VGG19 model follows closely with an accuracy of 0.94, while the ResNet152V2 model achieves an accuracy of 0.85, indicating reasonable performance.

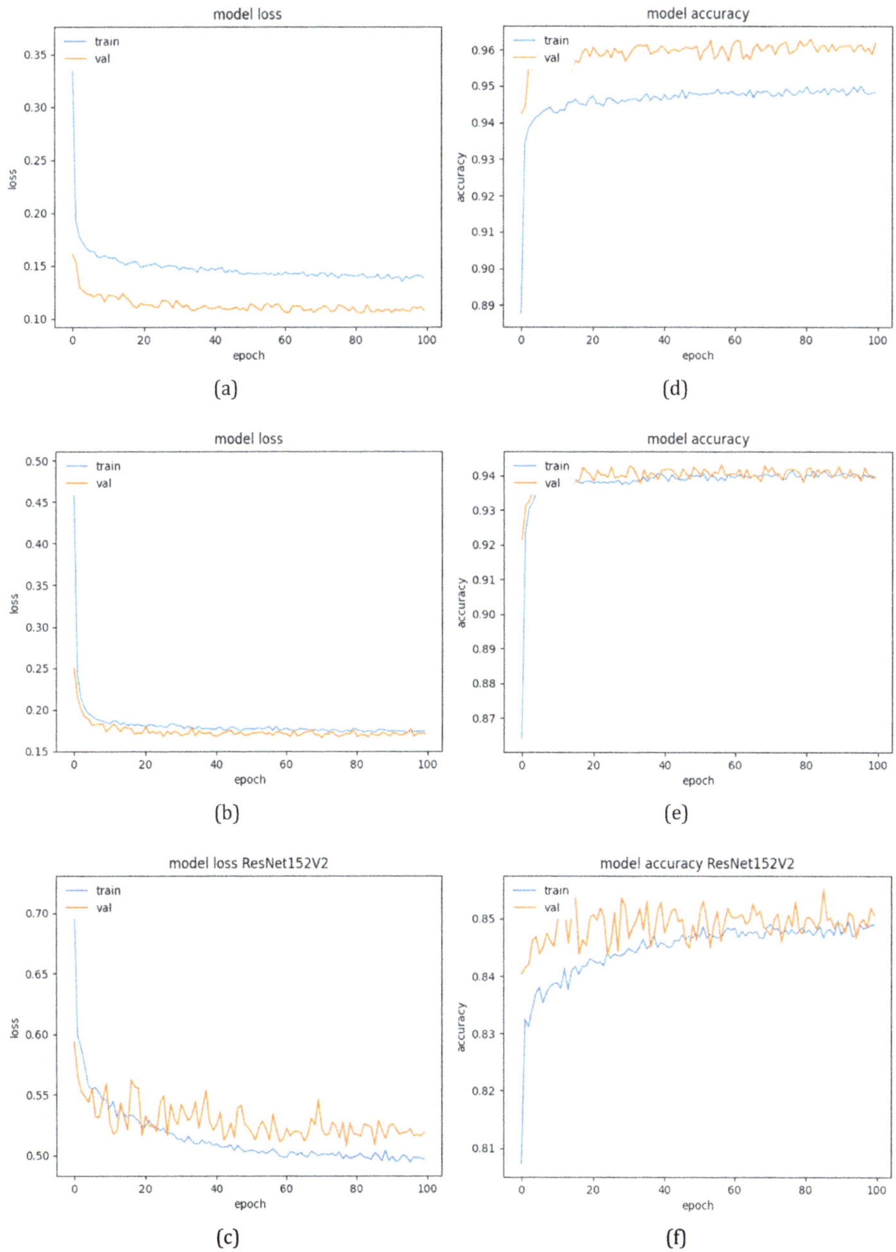

Figure 1: a, b, and c show the training and validation loss, and d, e, and f show the training and validation accuracy of fine-tuned EfficientNetV2M, VGG19, and ResNet101 models, respectively.

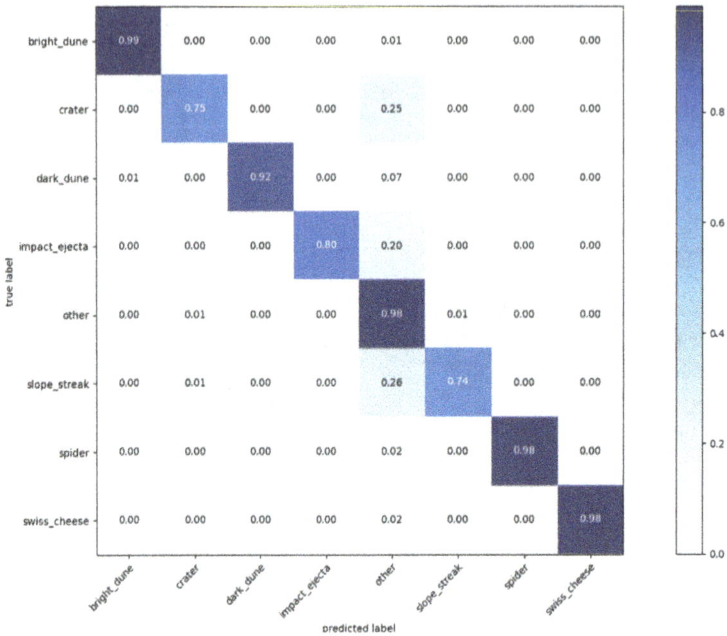

(a)

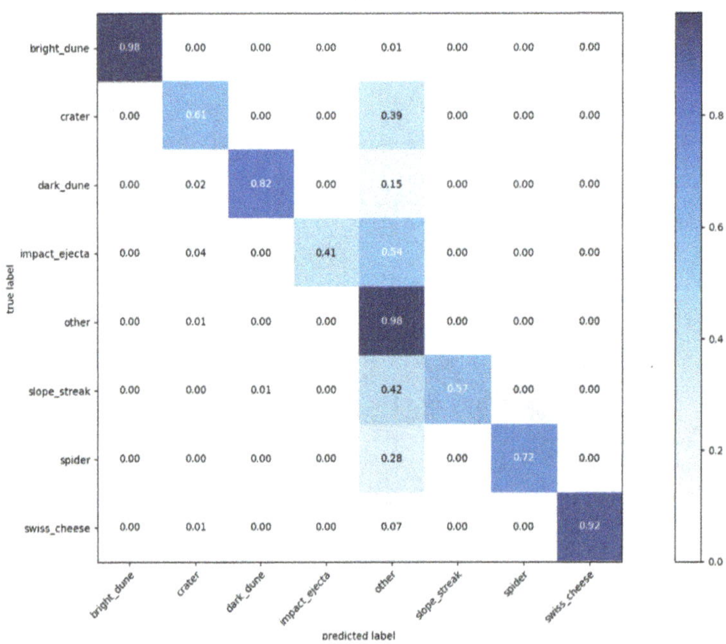

(b)

Figure 2: a, b, and c show the confusion matrixes of fine-tuned EfficientNetV2M, VGG19, and ResNet152V2 models, respectively.

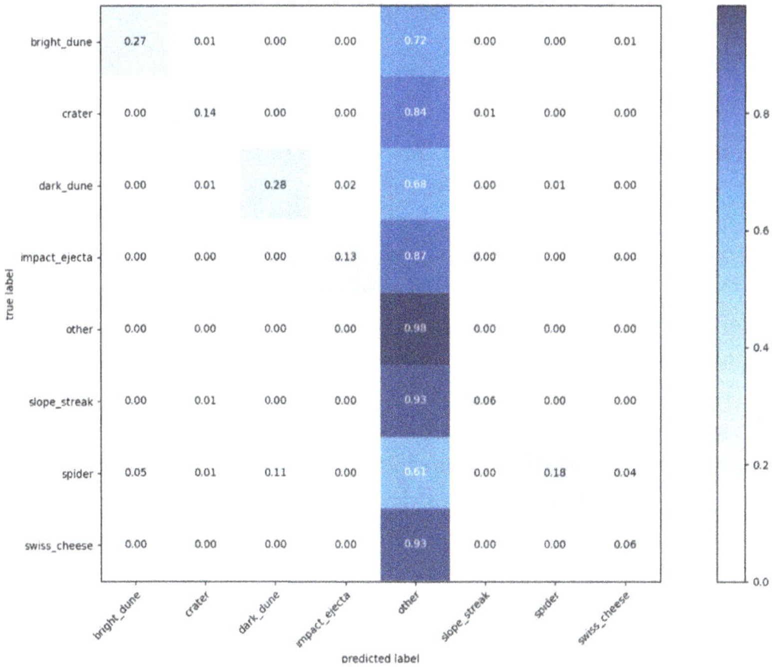

(c)

Figure 2 (continued)

Figure 1 illustrates the training and validation loss and training and validation accuracy of fine-tuned EfficientNetV2M, VGG19, and ResNet101 models. It is clear that EfficientNetV2M quickly reaches a satisfactory level of performance during the training process. Moreover, Figure 2 shows the confusion matrixes of fine-tuned models.

We provided a comprehensive and well-structured summary of the key performance metrics for the models EfficientNetV2M, VGG19, and ResNet152V2. This summary effectively conveys the models' performance in terms of precision, recall, F1-score, macro average values, and accuracy. Based on the provided performance metrics, "EfficientNetV2M" is the preferred choice for achieving the highest accuracy and balanced precision-recall performance.

4 Conclusion

The study suggests using transfer learning to classify Martian surface imagery, reducing the need for manual feature engineering and improving generalization across surface conditions. The method entails fine-tuning pre-trained CNN models such as ResNet152V2, EfficientNetV2M, and VGG19 to accommodate custom layers, improving accuracy and reducing the need for extensive Martian surface image data. The HiRISE dataset from NASA's Planetary Data System, which contains images of Martian landmarks, is used in the study. The study shows how transfer learning with CNNs can classify the Martian surface features, which will help future Mars exploration missions. EfficientNetV2M and VGG19 models consistently demonstrate high precision and recall in classifying data. EfficientNetV2M achieves 83–99% accuracy in predictions, while VGG19 maintains high precision levels of 76–99%. ResNet152V2 has the lowest precision of 32%. Both models achieve high F1 scores, indicating a balance between precision and recall. EfficientNetV2M excels at predicting positive instances and capturing relevant instances, making it a robust model. VGG19 maintains good recall and F1-score, while ResNet152V2 faces high precision and recall challenges across different experiments or datasets. Overall, EfficientNetV2M and VGG19 show strong overall classification performance. The EfficientNetV2M model has the highest accuracy of 0.96, followed closely by VGG19 and ResNet152V2. According to the results, the EfficientNetV2M model outperforms other models in terms of accuracy, making it an excellent candidate for applications requiring high precision, recall, and F1-score.

Acknowledgment: Dr. Steven A. Conrad, Mulsow Endowed Professor, for his support.

Competing Interests: The authors declare no potential competing interests.

References

[1] U. Galassi, "Landmark Detection for Autonomous Spacecraft Landing on Mars," *Lecture Notes in Computer Science (Including Subseries Lecture Notes in Artificial Intelligence and Lecture Notes in Bioinformatics)*, vol. 6804 LNAI, pp. 653–662, 2011, doi: 10.1007/978-3-642-21916-0_69/COVER.

[2] P. Y. Cui, Z. S. Yu, and S. Y. Zhu, "Research Progress and Prospect of Autonomous Navigation Techniques for Mars Entry Phase," *Journal of Astronautics*, vol. 34, no. 4, pp. 447–456, 2013.

[3] Z. Yu, P. Cui, and J. L. Crassidis, "Design and Optimization of Navigation and Guidance Techniques for Mars Pinpoint Landing: Review and Prospect," 2017, doi: 10.1016/j.paerosci.2017.08.002.

[4] A. Nandi, A. Mallick, A. De, A. I. Middya, and S. Roy, "Mars-TRP: Classification of Mars Imagery Using Dynamic Polling between Transferred Features," *Engineering Applications of Artificial Intelligence*, vol. 114, p. 105014, Sep. 2022, doi: 10.1016/J.ENGAPPAI.2022.105014.

[5] J. Wang, R. Li, T. Schenk, Y. Alper, and S. Advisor, "Modeling and Matching of Landmarks for Automation of Mars Rover Localization," 2008.

[6] R. D. Braun, and R. M. Manning, "Mars Exploration Entry, Descent and Landing Challenges," *IEEE Aerospace Conference Proceedings*, vol. 2006, 2006, doi: 10.1109/AERO.2006.1655790.

[7] E. Glenn Lightsey, S. A. Todd Ely, W. T. Fowler, D. G. Hull, and C. Ocampo, "Real-Time Navigation for Mars Final Approach using the Mars Network".

[8] B. Rothrock, J. Papon, R. Kennedy, M. Ono, M. Heverly, and C. Cunningham, "SPOC: Deep Learning-based Terrain Classification for Mars Rover Missions," *AIAA Space and Astronautics Forum and Exposition, SPACE 2016*, 2016, doi: 10.2514/6.2016-5539.

[9] V. T. Bickel, S. J. Conway, P. A. Tesson, A. Manconi, S. Loew, and U. Mall, “Deep Learning-Driven Detection and Mapping of Rockfalls on Mars,” *IEEE Journal of Selected Topics in Applied Earth Observations and Remote Sensing*, vol. 13, pp. 2831–2841, 2020, doi: 10.1109/JSTARS.2020.2991588.
[10] J. Li *et al.*, “Autonomous Martian Rock Image Classification Based on Transfer Deep Learning Methods,” *Earth Science Informatics*, vol. 13, no. 3, pp. 951–963, Sep. 2020, doi: 10.1007/S12145-019-00433-9/FIGURES/13.
[11] A. S. Chakravarthy, R. Roy, and P. Ravirathinam, “MRSCAtt: A Spatio-Channel Attention-Guided Network for Mars Rover Image Classification.” pp. 1961–1970, 2021.
[12] E. R. Urbach, and T. F. Stepinski, “Automatic Detection of sub-km Craters in High Resolution Planetary Images,” *Planetary and Space Science*, vol. 57, no. 7, pp. 880–887, Jun. 2009, doi: 10.1016/J.PSS.2009.03.009.
[13] L. Bandeira, J. S. Marques, J. Saraiva, and P. Pina, “Automated Detection of Martian Dune Fields,” *IEEE Geoscience and Remote Sensing Letters*, vol. 8, no. 4, pp. 626–630, Jul. 2011, doi: 10.1109/LGRS.2010.2098390.
[14] S. J. Pan, and Q. Yang, “A Survey on Transfer Learning,” *IEEE Transactions on Knowledge and Data Engineering*, vol. 22, no. 10, pp. 1345–1359, 2010, doi: 10.1109/TKDE.2009.191.
[15] K. Simonyan, and A. Zisserman, “Very Deep Convolutional Networks for Large-Scale Image Recognition,” *3rd International Conference on Learning Representations, ICLR 2015 – Conference Track Proceedings*, Sep. 2014, Accessed: Nov. 06, 2023. [Online]. Available: https://arxiv.org/abs/1409.1556v6.
[16] K. He, X. Zhang, S. Ren, and J. Sun, “Deep Residual Learning for Image Recognition.” pp. 770–778, 2016. Accessed: Nov. 06, 2023. [Online]. Available: http://image-net.org/challenges/LSVRC/2015/.
[17] C. Szegedy, V. Vanhoucke, S. Ioffe, J. Shlens, and Z. Wojna, “Rethinking the Inception Architecture for Computer Vision.” pp. 2818–2826, 2016.
[18] A. G. Howard *et al.*, “MobileNets: Efficient Convolutional Neural Networks for Mobile Vision Applications,”Apr. 2017. Accessed: Nov. 06, 2023. [Online]. Available: https://arxiv.org/abs/1704.04861v1.
[19] M. Tan, and Q. V. Le, “EfficientNetV2: Smaller Models and Faster Training.” PMLR, pp. 10096–10106, Jul. 01, 2021. Accessed: Nov. 06, 2023. [Online]. Available: https://proceedings.mlr.press/v139/tan21a.html.
[20] K. Simonyan, and A. Zisserman, “Very Deep Convolutional Networks for Large-Scale Image Recognition,” *3rd International Conference on Learning Representations, ICLR 2015 – Conference Track Proceedings*, Sep. 2014, Accessed: Nov. 11, 2023. [Online]. Available: https://arxiv.org/abs/1409.1556v6.
[21] K. Wagstaff *et al.*, “Mars Image Content Classification: Three Years of NASA Deployment and Recent Advances,” *Proceedings of the AAAI Conference on Artificial Intelligence*, vol. 35, no. 17, pp. 15204–15213, May 2021, doi: 10.1609/AAAI.V35I17.17784.
[22] G. Doran, S. Lu, L. Mandrake, and K. Wagstaff, “Mars orbital image (HiRISE) labeled data set version 3,” 2019, doi: 10.5281/ZENODO.2538136.
[23] A. Géron, *Hands-on Machine Learning with Scikit-Learn, Keras, and TensorFlow*. O’Reilly Media, Inc., 2022.

Alireza Bagheri Rajeoni, Breanna Pederson, Ali Firooz, Hamed Abdollahi, Andrew K. Smith, Daniel G. Clair, Susan M. Lessner, and Homayoun Valafar

Vascular system segmentation using deep learning

Abstract: Peripheral Arterial Disease (PAD) is a progressive disorder of the blood vessels supplying blood to lower limbs and affects millions of people each year. Computed tomographic angiograms (CTA) are often used to identify the severity of PAD and the location of occlusions and stenoses. However, the manual analysis of diagnostic images of the vascular system is time-consuming and tedious making it currently impractical for clinical use. To address this challenge, we propose a deep learning model designed to segment the vascular system in CTA images of patients undergoing surgery for PAD. Our study specifically aims to achieve accurate segmentation of the vascular system, both (1) from the descending thoracic aorta to the iliac bifurcation and (2) from the descending thoracic aorta to the patella in CTA images, using deep learning techniques. Our approach demonstrates impressive performance, achieving average Dice accuracies of 93.5% and 80.64% in the test dataset for (1) and (2), respectively. These results underscore the high accuracy and potential clinical utility of our proposed method. The use of deep learning techniques in this context emerges as an efficient and precise tool for medical professionals to analyze the health of the vascular system. For further details and access to the codebase, please visit our GitHub page: https://github.com/pip-alireza/TransOnet.

Acknowledgments: This work was funded by NIH grant number P20 RR-016461 to Dr. Valafar and HL145064-01 to Dr. Lessner. This work was also partially supported by the National Science Foundation EPSCoR Program under NSF Award # OIA-2242812.

Alireza Bagheri Rajeoni, Computer Science and Engineering, University of South Carolina, Columbia, SC, USA, e-mail: alirezab@email.sc.edu
Breanna Pederson, Biomedical Engineering, University of South Carolina School of Medicine, Columbia, SC, USA, e-mail: pedersob@email.sc.edu
Ali Firooz, Computer Science and Engineering, University of South Carolina, Columbia, SC, USA, e-mail: ali.firooz@sc.edu
Hamed Abdollahi, Computer Science and Engineering, University of South Carolina, Columbia, SC, USA, e-mail: ha25@mailbox.sc.edu
Andrew K. Smith, Computer Science and Engineering, University of South Carolina, Columbia, SC, USA, e-mail: aks3@email.sc.edu
Daniel G. Clair, Surgery, Vanderbilt University Nashville, TN, USA, e-mail: dan.clair@vumc.org
Susan M. Lessner, Cell Biology and Anatomy, University of South Carolina School of Medicine, Columbia, SC, USA, e-mail: susan.lessner@uscmed.sc.edu
Homayoun Valafar, Computer Science and Engineering, University of South Carolina, Columbia, SC, USA, e-mail: homayoun@cse.sc.edu

https://doi.org/10.1515/9783111344126-005

Keywords: medical image segmentation, computer vision, aorta, artificial intelligence, peripheral arterial disease, computed tomography angiogram

1 Introduction

The use of machine learning (ML) algorithms in the medical field has shown remarkable potential to aid healthcare professionals. For example, ML techniques have been shown to be advantageous in evaluating patient response to cardiac resynchronization therapy [1], prediction of patient response to medication administration [2], and diagnostic of brain images [3, 4]. The segmentation of the vascular system is another promising potential for advancement in medicine delivery, as demonstrated by [5], where machine learning was used in a medical device to automatically locate peripheral vessels in ultrasound images. In this research paper, we propose a tailored deep learning algorithm designed to identify and extract the aorta and lower body arteries in CT scans as the first step in automating diagnostic and prognostic approaches by ML techniques.

Despite several attempts in segmenting the vascular system [6–8], a comprehensive analysis of the vascular system from the aorta to the patella remains unexplored. The vascular system of the lower extremities carries crucial information about a person's health, as a complete blockage in these arteries can result in the need for amputation. Furthermore, analyzing the vascular calcification in the lower extremities may serve as a supplementary tool for assessing the risk of cardiovascular morbidity and mortality [2].

In this study, our objective is to trace the vascular system as it descends from the thoracic aorta, bifurcates into the iliac arteries, and further extends into the femoral arteries until it reaches the patella, as illustrated in Figure 1. The aorta, originating at the left ventricle, is the largest artery in the human body, running along the abdominal wall before branching into the common iliac arteries, which supply blood to the legs. Tracing the vascular system becomes increasingly challenging beyond the point of bifurcation as the size of the arteries diminishes, and they branch off multiple times, with the location of these smaller arteries varying from one patient to another. Additionally, blockages can occur in these arteries. It is worth noting that the area of interest also significantly decreases, often reducing to only a few pixels, making it challenging for ML models to accurately detect and segment the arteries, particularly if there is an occlusion.

Considering the complexities involved, our study aims to address these challenges and contribute to a better understanding of the vascular system in the lower extremities. By employing our application specific deep learning model, we strive to improve the accuracy of vascular tracing and segmentation beyond the point of bifurcation, which can have significant implications for diagnosis, treatment, and patient care.

The structure of this paper is as follows: the Background and methods section offers a comprehensive exploration of prior and related works in AI. It underscores the importance of the vascular system and the complexities associated with Peripheral Arterial Diseases (PADs) within this system. Additionally, it highlights the potential of AI for the automated analysis of vascular systems. The section places a significant emphasis on the important role of data standardization, thoroughly explores the characteristics of the dataset employed in this study, and concludes by introducing the ML models used for vascular system segmentation. The subsequent section details the obtained results, followed by comprehensive discussions and insights into future research directions.

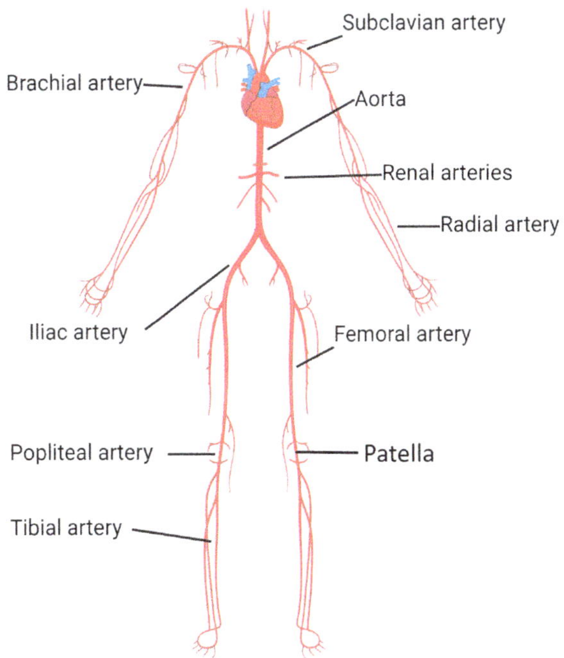

Figure 1: Human vasculature map.[1]

1 Created with BioRender.com.

2 Background and methods

2.1 Previous and related work

In recent years, the landscape of machine learning (ML) has undergone a remarkable transformation, marked by significant improvements and cross-domain inspirations. Innovations and techniques originating in one domain often transcend their original boundaries, inspiring diverse applications across various fields. A notable example is the trajectory from the AlexNet model, which laid the foundation for the Fully Convolutional Network, subsequently inspiring the development of popular U-net model for medical image segmentation [9–11].

The evolution of language models, notably with the advent of the transformer architecture, stands as a cornerstone in this journey [12]. This architectural innovation, fundamental to models like ChatGPT that revolutionized generative AI for conversations, has transcended its initial applications [13]. Its influence extends across a broad spectrum of domains, including image processing, where it facilitates the extraction of global context in image data. This transformative journey has given rise to a plethora of state-of-the-art models in computer vision, particularly in the specialized domain of medical image segmentation.

The dynamic evolution of ML algorithms is pivotal in broadening the application of ML to an extensive array of tasks [14–17]. This inherent versatility and adaptability underscore the transformative potential of ML, reshaping and automating tasks across diverse domains [18–21]. As we bear witness to these ongoing advancements, the boundaries that once defined manual processes are continually being redefined, unlocking new possibilities for the application of machine learning. This perpetual redefinition of possibilities highlights the ever-expanding role of ML in shaping the future of problem-solving across various fields.

A critical factor in training ML models is the availability of substantial datasets. The evolution of ML is not merely confined to algorithmic advancements but dependent on the richness and diversity of the datasets on which these models are trained. The utilization of pre-trained models, exemplified by ResNet [22], has become a pivotal practice in the ML landscape mitigating the need for extensive dataset. This approach leverages the knowledge gained from one dataset to bootstrap the learning process on a new dataset, illustrating a transfer learning paradigm that facilitates the training of the ML model.

However, the journey from problem formulation to ML solution is a process heavily influenced by the inclusiveness of data. For optimal performance from AI models, it is important to have a dataset that encapsulates all possible variations, mitigating biases towards specific outputs while preventing overfitting and underfitting. As new problems emerge, the acquisition of datasets tailored to specific tasks becomes imperative. This resource-intensive process, punctuated by the meticulous annotation of data, is fundamental to achieving the desired accuracy at the conclusion of the training cycle.

In the healthcare domain, the application of ML techniques has appeared in many forms. ML techniques combined with sensory information from smart devices has led to real-time observation of human behavior including eating, medication taking, and smoking to name a few [23–25]. More directly related to this work, the application of ML for medical image segmentation has been widely prevalent. Automated and semiautomated models have been proposed to segment medical images to investigate organs such as lungs, aorta, liver, pancreas, brain, and others [26–29]. Classically, ML techniques and threshold-based approaches were applied to hand-engineered features [26, 30]. Many of these approaches require expert human intervention. Human intervention including hand-engineering features is prone to introduce unrepresentative bias into the segmentation process. Further, modern deep learning systems have proven to be highly effective in learning complex nonlinear relationships, with much focus on image analysis [9, 22, 31, 32]. For these reasons, it is desirable to develop new end-to-end deep learning approaches for CT scan segmentation.

A few end-to-end deep learning approaches have been proposed to segment the aorta from CT scans. One such approach uses a dilated three-dimensional convolutional neural network (CNN) [33]. Others commonly use the U-net structure to segment the aorta from CT scans [34, 35]. Of the end-to-end deep learning approaches that segment CT scans to investigate bodily organs, none have specifically employed a block of Multi-Head Attention [12] between a pre-trained ResNet encoder and decoder to segment aorta and lower body vessels. Multi-Head Attention is one of the best learning models in deep learning and serves as the fundamental building block in transformer architectures.

Community engagement through public dataset challenges, like ImageNet [31], has proven to be a powerful strategy for advancing the state-of-the-art in various ML domains. A notable example of this collaborative approach is the Beyond the Cranial Vault (BTCV) challenge [36], which aims to identify the best models for organ segmentation in medical images. Participating models in this challenge, such as Swin UNET [37], FCT [38], nnFormer [39], and TransUnet [40], employ a combination of transformers and CNNs to tackle the complexities of medical image segmentation. These models showcase the power of integrating different architectural elements for enhanced performance and accuracy. However, there are only a few approaches focusing solely on vascular segmentation. Guidi et al. [6] uses an expert system with CNNs to segment the vascular system from aorta to iliac arteries.

2.2 Anatomical background and clinical application

The human vascular system plays a crucial role in facilitating nutrient and metabolite transfer throughout the body. A 70-kg man has approximately a total vascular surface area of 30–70 m^2 depending on his current activity level [41]. Blood flow, driven by a pressure gradient, exhibits pulsatile behavior in arteries, with pressures ranging from 80 to 120 mmHg. Arterioles, vessels with diameters of 20–80 μm, regulate blood flow and vessel resistance. As atherosclerosis progresses, fatty plaques develop, leading to vessel occlusion and changes in blood vessel mechanical properties. This progression impacts cardiovascular morbidity and mortality risk, emphasizing the need for tools quantifying blood vessel properties [42].

PAD is a progressive disorder resulting from atherosclerosis, affecting arteries that supply blood to the lower extremities, in which the vessels become stenotic or occluded leading to spectrum of symptoms, which reduce physical capacity and functional status [43, 44]. Current estimates suggest that over 200 million people worldwide may have PAD, with approximately 8–10 million in the United States, with prevalence increasing with age [43, 45]. PAD is underrecognized due to the range of symptoms as only approximately 3 million people in the United States had typical claudication symptoms [45], The Peripheral Arterial Disease Awareness Risk and Treatment: New Resources for Survival study showed that 29% of high-risk individuals have PAD, where high risk includes people over 70 without additional risk factors and people between 50 years and 69 years of age with additional cardiovascular risk factors including history of cigarette smoking and diabetes [43]. Overall, approximately 2–7% of men and 1–2% of women over the age of 50 have intermittent claudication [45]. PAD can present with a variety of symptoms and can have asymptomatic, acute, or chronic presentations [46]. Chronic limb-threatening ischemia (CLTI) is the severe state, carrying the risk of limb loss [47]. Classification systems like Fontaine and Rutherford categorize PAD symptoms, emphasizing the need for accurate diagnosis.

The ankle-brachial index (ABI) is commonly used for diagnosis of PAD [48]. ABI is the ratio of systolic blood pressure in the tibial artery to the higher systolic blood pressure of the brachial arteries [43, 48]. Resting ABI is classified into four categories: below 0.9 is abnormal, 0.91–0.99 is borderline, 1–1.4 is normal, and greater than 1.4 is noncompressible [43]. Typically, an ABI value of 0.9 has been used as the cutoff for PAD, as it has been shown that patients with an ABI below 0.9 have an increased risk of cardiovascular disease (CVD) morbidity and mortality events. However, this definition may in part explain why PAD is underrecognized, as an ABI greater than 1.4, in the "noncompressible" category, is associated with increased levels of CVD risk factors and coronary artery calcium [48]. In a cardiovascular health study, patients with an ABI greater than 1.4 had a 60% greater risk for all-cause mortality [48]. However, ABI is a representative single value, which cannot account for detailed risk stratification. As PAD prevalence increases and the significant associated risk of cardiovascular morbidity and mortality is recognized there is an increasing need for more detailed risk assessment for individual patient-focused strategies [49]. Therefore, ABI is typically a starting point for further tests and imaging. Patients in whom revascularization is being considered for treatment frequently undergo imaging such as computed tomography and angiography (CTA) to identify the anatomic location of occlusions and severity of stenosis [49], wherein ML techniques to analyze CTA images can offer valuable assistance.

PAD significantly predicts overall mortality, which is true across many potentially relevant categories such as men and women, community vs. medical cohorts, the elderly, and others. Asymptomatic PAD is also important to recognize as it may indicate risk for future ambulation difficulties, lower extremity ulcers, revascularization, or amputation. PAD, both asymptomatic and symptomatic, is a powerful independent predictor of coronary artery disease (CAD), cerebrovascular disease events, and mortality. Across categories, both symptomatic and asymptomatic, PAD is associated with increased risk for nonfatal myocardial infarction. Using noninvasive criteria, PAD is also significantly associated with fatal myocardial infarctions and CAD death. In addition, it correlates with stroke and transient ischemic attack, as well as with worse outcomes in stroke patients. There is a 10-year mortality gradient as PAD severity increases, with normal subjects having 14% mortality, asymptomatic PAD patients 45%, and severe symptomatic PAD patients 75%. Additionally, mortality increases with lower ABI, averaging a 3.1% increased risk per 0.50 decrease in ABI. Beyond mortality and cardiovascular events, PAD correlates with non-cardiovascular events and perioperative mortality, as evidenced in renal transplant patients with low toe blood pressure [50]. These insights underscore the importance of continuous monitoring of PAD stages, a task well-suited for the capabilities of AI.

Recent studies utilizing CT scans to quantify total lower extremity calcification volume have revealed the significance of Lower Limb Arterial Calcification (LLAC) scoring in predicting adverse clinical outcomes in PAD patients, thereby identifying high-risk individuals [49]. Elevated LLAC levels are associated with limited success in revascularization procedures, particularly in the "extreme" tibial arterial calcification category, which is linked to an increased risk of unplanned amputation and Major Adverse Cardiovascular Events (MACE) [51]. Lower extremity arterial calcification also has potential implications for the success of drug-eluting balloons (DEB). It has been shown that the presence of calcium lowers the drug activity of DEBs [52, 53]. Previous studies had focused on length of calcium deposits rather than circumferential extent, which is now being shown to have greater impact on DEB efficacy than length [52]. As calcification of the coronary arteries is a recognized indicator of cardiovascular morbidity and mortality and PAD is a strong predictor of CAD and overall mortality, evaluating calcification extent in PAD could potentially provide predictive value for both all-cause and cardiovascular morbidity and mortality. The integration of AI for quick and economical analysis of the vascular system, extending from the aorta to the lower limbs, facilitates comprehensive analysis of the arterial system in the lower extremities and rapid assessment of calcification within the arteries [54].

Coronary artery calcification (CAC) is a recognized indicator of cardiovascular morbidity and mortality. The amount of atherosclerosis and the rate of future cardiac events are both strongly correlated with the CAC quantity. CAC is evaluated noninvasively using CT imaging; therefore, tracing the aorta through the lower vascular tree and quantifying calcium in PAD patients in a manner analogous to CAC scoring could become a useful tool for clinicians. There have been a few different methods used in single-center studies; however, making such a tool useful to clinicians and patients on a large scale requires development and validation of a single, reliable, standardized method that can be used with all imaging protocols and CT scanners [55, 56].

CTA with intravenous contrast is a useful high-resolution imaging technique for visualizing pathological changes in the arterial tree. Its usefulness lies in the three-dimensionality of the multidetector imaging, which allows for volumetric analysis, as the scans can be viewed in multiple two-dimensional planes and as 3D reconstructions [57, 58].

In PAD treatment, CTAs are frequently used in presurgical planning to document the patient's vascular anatomy and to visualize the distribution of diseased, stenotic vessels. However, clinicians must take time during their office hours to do this annotation and it is not necessarily consistent. ML algorithms and artificial intelligence (AI) systems are promising prospects for multiple fields of health care but are still in their infancy in clinical application [59]. For PAD, such techniques could be used to detect the disease, refine risk stratification, aid prognosis, and help with treatment choices [59, 60]. Some previous studies have applied AI/ML analyses to CTA scans of patients, demonstrating the potential to use CT imaging to visualize lower extremity inflow and runoff, integrate data in order to make predictions about the presence/absence of PAD, and predict risk of future cardiovascular events [58, 60]. There is emerging evidence that this type of model may predict risk better than conventional risk prediction scores based on linear modeling.

This study focuses primarily on the arteries that supply blood to the lower extremities, primarily the external iliac artery as it becomes the femoral artery and branches into the deep femoral artery in the leg. We begin tracking at the descending thoracic aorta as that is the largest artery in the body and supplies blood from the abdomen to the rest of the body. The study aims to evaluate the automated segmentation of the arterial system starting from the descending thoracic aorta all the way to patella. To evaluate the performance of the segmentation system, the data is annotated under the supervision of medical professionals, and the dataset is split into training, validation, and testing sets to assess the model's performance.

While there have been several approaches to the segmentation of the aorta by ML/AI techniques, there have been relatively fewer attempts at the segmentation of the arterial system past the iliac branching. In a recent study [54], CNNs were employed to extract the vascular system and measure calcification, yielding promising results. In contrast [61] utilized an object tracking technique in identifying the arterial system in the lower extremities. While successful in identifying the majority of the arterial system, it exhibited a shortcoming: if the object of interest was lost in one slice, or when intensity goes beyond the defined threshold, it would lose the object in all other subsequent slices of the CTA scan. This limitation was the primary cause of reduced performance in some instances.

2.3 Data standardization

Inherent variability in medical images necessitates standardization. Such variability may include differences in imaging equipment, settings, and protocols across hospitals and imaging centers. Resultant medical images may vary with respect to contrast or brightness due to operator settings or patient-specific factors. Further, differences in patient anatomy and wide-ranging pathologies may affect the appearance of tissues and organs. The purpose of developing standards for medical images is to remove variability that would not influence the downstream task.

Such techniques have been well-studied across diverse domains. Here we provide some standardization techniques for correcting some prototypical medical image variability. Biased or scaled pixel intensity can be corrected with intensity normalization techniques such as z-score normalization or min-max scaling. Removing variability with respect to bias and scale assumes that these parameters do not contain information to support the downstream task. Variability across size, orientation, or resolution of images can be corrected with resampling, cropping, or affine transformations. Neural network architectures are commonly optimized for images of the same resolution. Not only is it sometimes critical to standardize the image itself, but commonly the interpersonal anatomy (i.e., torso length, breadth, etc.) should be standardized for constraining network optimization. Modern networks can learn variation across anatomy; however, constraining learning with standardized datasets may more easily generalize within that standardized frame. Moreover, learning on medical image datasets with standardized anatomical structures can succeed on far smaller datasets.

Apart from pixel bias and scaling, image resolution and size, and geometric anatomical variation, medical images may include elements not characteristic of typical distributions (i.e., outliers). Outliers may result from motion artifacts during medical imaging, for example. These artifacts may be removed/replaced via interpolation techniques. Finally, though not necessarily a standardization technique, data augmentation can increase diversity and robustness of medical imaging data by applying transformations like rotation, flipping, and scaling. Certain data augmentation techniques may not be used if the anatomical frame is being geometrically standardized (i.e., scaling).

In conclusion, the standardization of data represents a pivotal preprocessing step in the realm of medical image segmentation utilizing deep learning methodologies. This process is instrumental in mitigating the inherent variability and inconsistencies prevalent in medical imaging datasets, which stem from diverse imaging protocols, equipment, and patient-specific factors. Through the implementation of techniques such as intensity normalization, geometric standardization, artifact reduction, and strategic data augmentation, researchers and practitioners can substantially enhance the homogeneity and quality of the datasets, thereby paving the way for more robust and accurate deep learning models.

2.4 Data description

The dataset used in this study consists of CTA images, obtained with informed consent from 11 patients who underwent femoral endarterectomy for PAD at Prisma Health Midlands (IRB protocol 1852888). There are over 500 slices of images with height and width of 512 for each patient, extending from the descending thoracic aorta to the patella. Annotation is performed using ITK-SNAP [62] software under the supervision of a medical professional. ITK-SNAP provides a semiautomatic annotation tool that enables users to differentiate the area of interest from other regions by adjusting the intensity threshold. Users can create bubbles within the region of interest, and ITK-SNAP will automatically expand the selected region by following similar intensity patterns until the intensity drops at the edges. While ITK-SNAP offers a quick segmentation tool, it faces challenges when encountering complex geometries and vessel blockages, requiring more user intervention in these areas. Users must also carefully set the threshold for each patient and ensure there is no overlap between the area of interest and other regions having similar intensity.

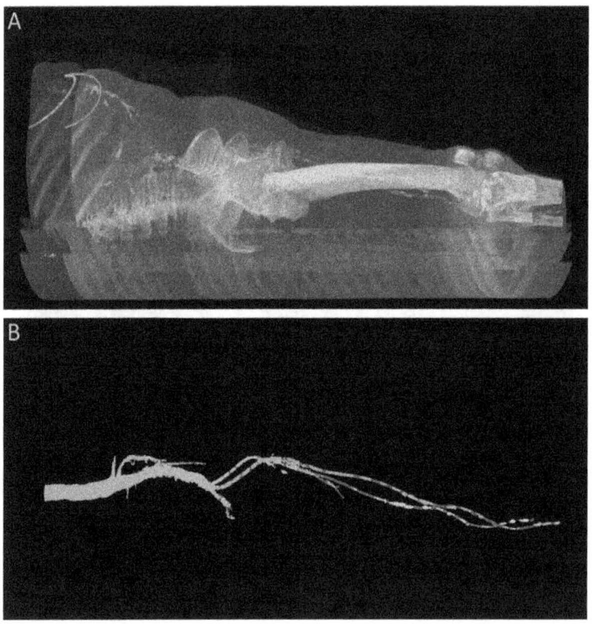

Figure 2: A) is the 3D side view of human CT scan image and B) is the manually annotated vascular system.

An example of a dataset used for training is shown in Figures 2 and 3. Figure 2A shows a 3D reconstruction of a patient data sample, created using ImageJ software, with the vessels annotated using ITK-SNAP shown separately in Figure 2B. Figure 3A shows a typical transverse slice, with the annotated aorta in Figure 3B.

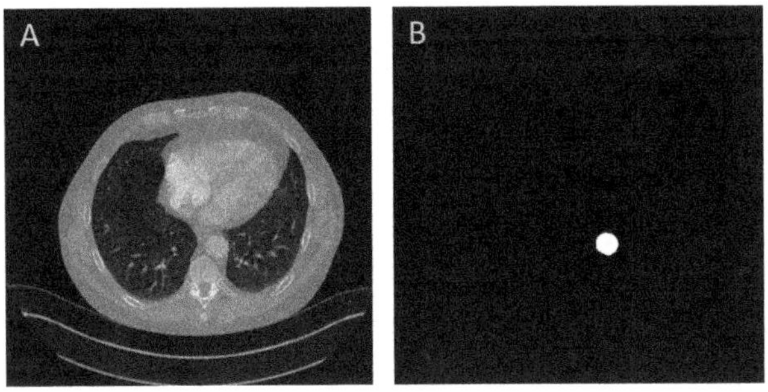

Figure 3: A) Human CTA transverse view. B) Annotation of the aorta.

We increased the training dataset to 12,000 images using an augmentation approach. The augmented images were produced by adding noise, flipping images horizontally, skewing the shape, and downscaling the images. The Scikit-learn library [63] in Python was used to perform these manipulations. As a result, the augmented images exhibited a small similarity with the original images and were used to pretrain our network. Figure 4A shows an example of an augmented transverse slice, with the isolated aorta shown in Figure 4B.

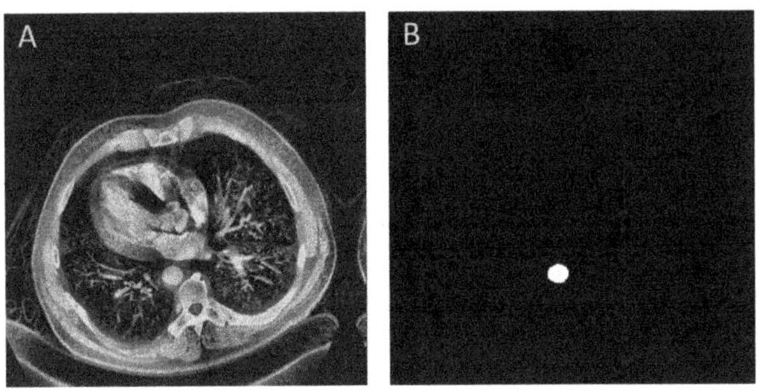

Figure 4: A) Data set after augmentation. In this augmentation, noise, downscaling, grid distortion, and enlargement are applied. B) Annotation of the aorta in the augmented image.

2.5 Image segmentation method

Our model architecture follows an encoder-decoder structure similar to U-net [11] as presented in [64], incorporating skip connections from the encoder to decoder. Additionally, a transformer [65] is utilized in the bridge between these two main blocks as illustrated in Figure 5.

The encoder uses ResNet34 [22] with residual connections, from which four skip connections are extracted and fed to the decoder section. Given input x, the model accepts $x \in \mathbb{R}^{W \times H \times C}$ where H, W, and C represent height, width, and the number of channels, respectively. To make the pretrained ResNet-34 model compatible with our dataset, the model's input is set to three channels. After feature extraction, the features undergo linear projection and reshaping. They are then passed through a bridge consisting layers of transformers. Subsequently, the features are reshaped back to $(H/3 \times W/3 \times 512)$ and processed by a layer normalization and 2D convolutional layer. The purpose of this architecture is to utilize both high-resolution spatial information from CNN features as well as the global context encoded by the transformer. Transformers can serve as strong encoders for medical image segmentation tasks, with the combination of U-net to enhance finer details by recovering localized spatial information [40]. This custom-built structure utilizes pretrained ResNet-34 to improve training of deeper networks and solve vanishing gradients, especially when dealing with limited datasets. It allows leveraging the valuable initial feature representations learned from a larger dataset, which in turn contributes to improving the overall performance of the model.

The decoder, which has four blocks, receives the output of the encoder, and uses an expanding path to construct a segmentation map from the encoded features. Each decoding block duplicates the spatial resolution while halving the number of feature channels, performs upsampling, and concatenates the output with the symmetric block's output in the encoder. After that, the block applies a convolutional layer with the same number of filters. The output of the last decoder block undergoes a 1×1 convolution that calculates the final map. The structure of the TransONet model is shown in Figure 5.

3 Results

To evaluate TransONet performance, we employed fourfold cross-validation, as depicted in Figure 6. The training process involved using Binary Cross Entropy loss function and Adam [66] optimizer. The inputs to the model were images with H and W of 512 and three channels, and the output was a mask with the same size but one channel.

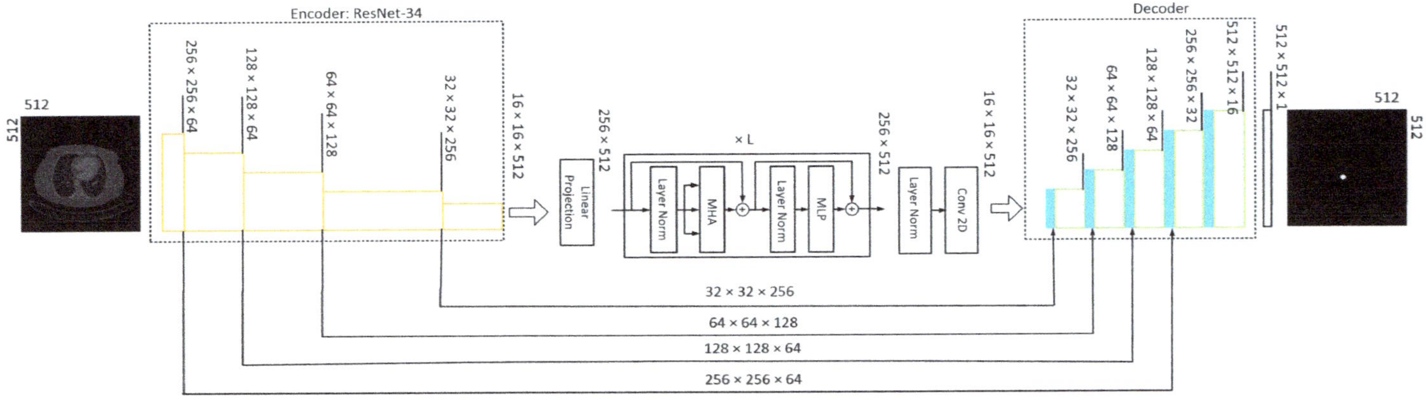

Figure 5: TransONet structure. The decoder section utilizes ResNet-34. The output from the encoder undergoes linear projection and reshaping before passing through transformer. The transformed features are then reshaped and processed by layer normalization and 2D convolution. Subsequently, they are fed into the decoder to construct the segmentation mask. Skip connections from various stages of the encoder are sampled and incorporated into the decoder to contribute to the mask construction.

Figure 6: Cross-validation experiment. We have access to 11 patients illustrated as P1–P11. In every training fold, three patients are excluded, except in fold 4 where only two patients were excluded.

We trained the model using a batch size of 40, for 400 epochs. During training and validation, we utilized Binary Cross Entropy plus Jaccard (BCEJ) for loss and Intersection over Union (IOU) as the evaluation metric, while Dice score was used for testing. The results are shown in Figure 7.

We evaluated our model's performance in comparison to state-of-the-art models and found promising results, particularly in aorta segmentation as shown in Figure 8. It is important to acknowledge the variations in dataset quality and size; therefore, we conducted training on our dataset with a selection of models to ensure a comprehensive evaluation of our proposed model.

Figure 8 illustrates the comparative analysis between our model and other architectures on our in-house dataset. For training, we utilized BCEJ loss and IOU metric, conducting training for 200 epochs using a batch size of 10 and image dimensions of 256 in height and width. The graph in Figure 9 displays the IOU score during both training and validation stages. Notably, TransONet highlights superior performance on our dataset. Our model also achieved an average 80.6% Dice score in segmenting the vascular system from the descending thoracic aorta to the patella. Lower accuracy in this area is due to decreasing size of the arteries in the lower extremities. Additionally, contrast intensity drops due to stenoses in the lower body. In other words, the

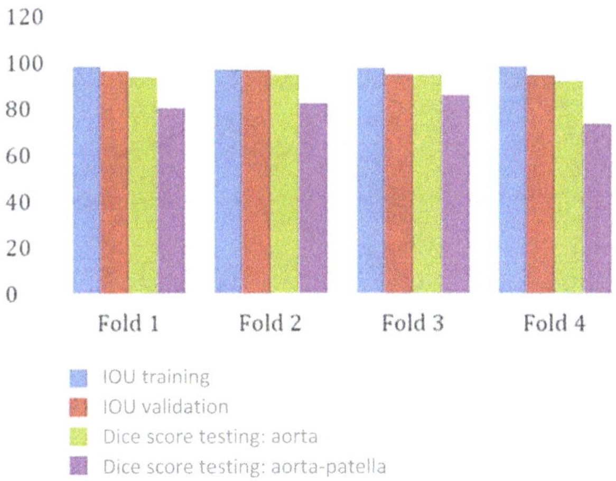

Figure 7: Cross-validation accuracy results.

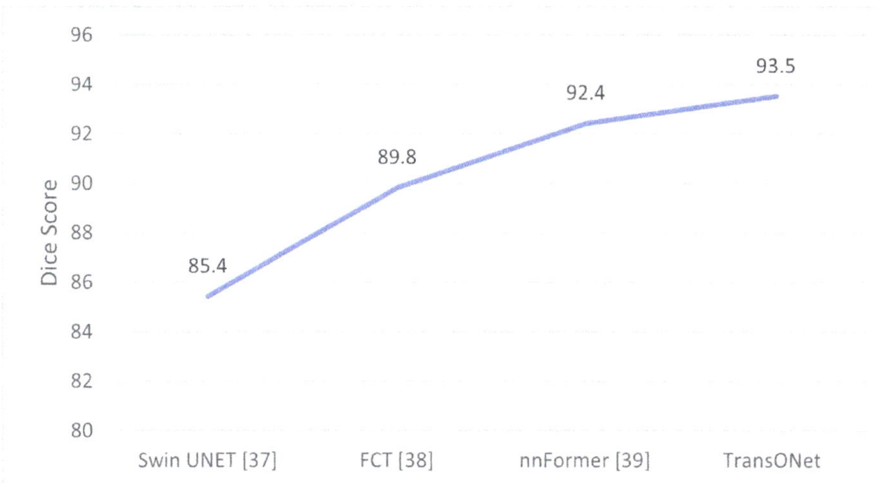

Figure 8: Segmentation results on aorta segmentation. Dice coefficient is reported. It is important to note that TransONet was trained and tested on a different dataset.

aortic cross-section in the upper body typically occupies more than 100 pixels, while the cross-sectional area of an artery in the leg may be less than 20 pixels.

In comparison to the object tracking model [61] using the same dataset, our approach demonstrates a significant improvement in vasculature segmentation. Figure 10 illustrates the results obtained by testing the model trained in fold 1 (Figure 6) on the patient P10 data. Compared to object tracking, TransONet demonstrated exceptional performance with a Dice score of 91.2% in segmenting the vascular system from the thoracic aorta up to the patella.

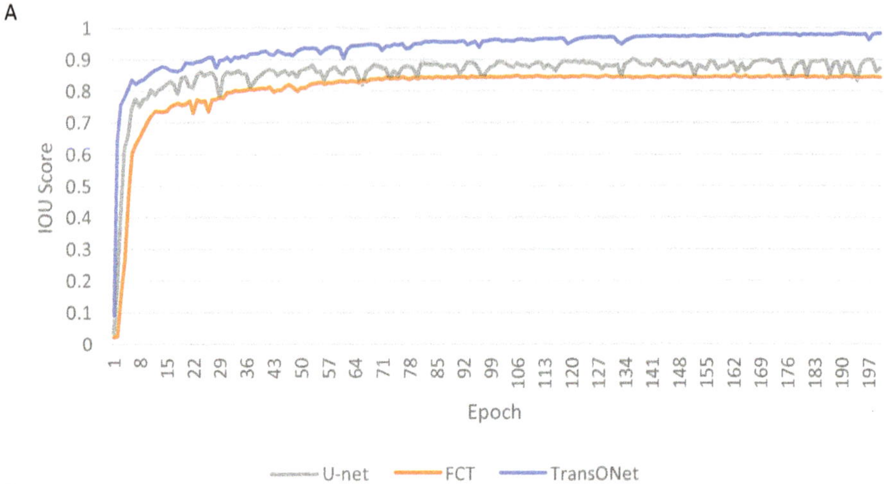

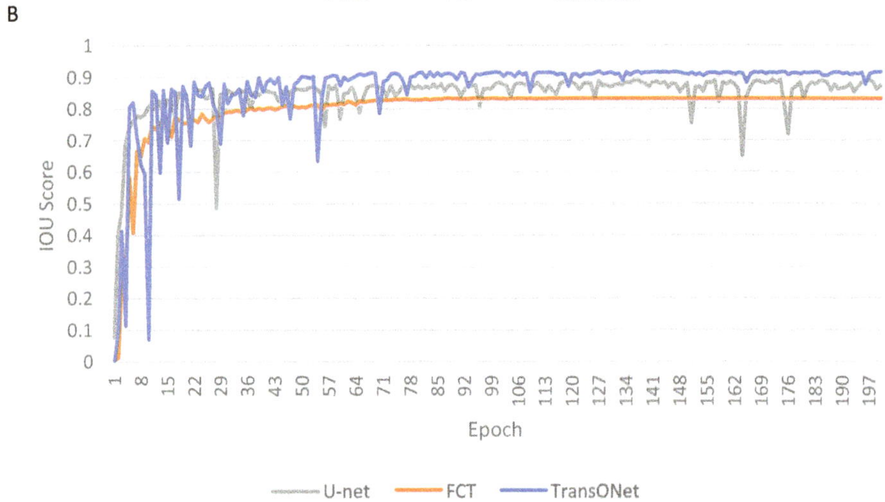

Figure 9: Comparison of TransONet with FCT and U-net for aorta segmentation on our in-house dataset. Figure 7A illustrates the IOU score in training, and Figure 7B showcases the IOU score in validation across epochs. It is important to note that the validation set was entirely distinct from the training data.

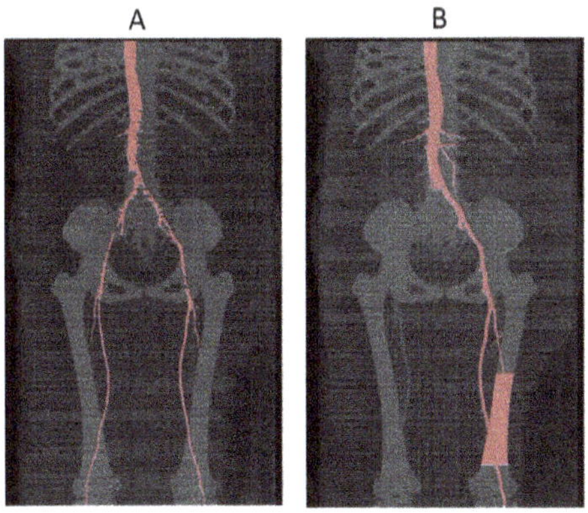

Figure 10: Performance of TransONet 8A compared to object tracking 8B [61]. While object tracking methods may lose track of the vascular system or fail to avoid the skeleton, TransONet achieves successful tracking. However, TransONet may miss certain parts of the vascular system during the segmentation of each slice.

In contrast, object tracking clearly failed to accurately track the vascular system of the right leg. Object tracking follows the intensity and when blockages happen in lower extremities causing intensity to drop, object tracking loses track of the vasculature. Also, in some cases the femoral arteries come in contact with bone structure and since they share similar intensity spectrum, object tracking identifies the bone structures as the vasculature, which leads to the failure of the object tracking.

4 Discussion and future work

Our study underscores the successful application of deep learning techniques in effectively segmenting the vascular system from the thoracic aorta to the patella. The obtained results signify the potential of the proposed model in automating vascular system analysis. To enhance its capabilities, future improvements can be made by augmenting the complexity of the deep learning structure and leveraging larger, more diverse datasets. The developed model holds promise for expeditious and reliable calcium scoring within the vascular system, enabling risk assessment, personalized interventions, and effective management of PAD.

While state-of-the-art methods, particularly those showcased in the Synapse beyond the Cranial Vault dataset, have primarily focused on segmenting multiple organs, our study reveals a distinctive emphasis on vascular segmentation [37, 38, 40]. Notably, the lower body poses new challenges due to a significant reduction in the area of vascular branches. Despite this, our approach extends beyond previous efforts by including the femoral arteries in our analysis, achieving promising results. It is crucial to acknowledge the comparative advantage of our study despite the smaller dataset, which may have influenced performance, particularly when contrasted with the study by Lareyre et al. [7].

Beyond vascular segmentation, our model exhibits future potential for evaluating stenosis and occlusion in PAD patients. CTAs have proven reliable for assessing lower extremity arterial occlusive disease, offering valuable insights for PAD management. While our study achieved high accuracy with a relatively small dataset, future developments shall incorporate larger datasets encompassing anatomical variations in humans. A limitation involves the challenge of tracking smaller vessels, such as the tibial artery, for comprehensive lower extremity calcium content quantification. Another potential solution may involve identifying and subtracting bones, enabling intensity thresholding to reveal all calcium present in smaller vessels without the need for individual tracking.

5 Conclusion

By accurately extracting and analyzing the vascular system, we can detect pathological conditions such as aneurysms and vascular calcification, among others. In the future, our focus will be on enhancing segmentation accuracy to precisely identify and measure calcification within the vascular system. This advancement will contribute to more accurate diagnosis, proactive treatment, and improved patient outcomes in the field of vascular health.

References

[1] B. E. Odigwe, A. B. Rajeoni, C. I. Odigwe, F. G. Spinale, and H. Valafar, "Application of Machine Learning for Patient Response Prediction to Cardiac Resynchronization Therapy," *Proceedings of the 13th ACM International Conference on Bioinformatics, Computational Biology and Health Informatics*, Northbrook Illinois: ACM, Aug. 2022, pp. 1–4, doi: 10.1145/3535508.3545513.

[2] M. M. Chowdhury *et al.*, "Lower Limb Arterial Calcification (LLAC) Scores in Patients with Symptomatic Peripheral Arterial Disease are Associated with Increased Cardiac Mortality and Morbidity," *PLOS ONE*, vol. 12, no. 9, p. e0182952, Sep. 2017, doi: 10.1371/journal.pone.0182952.

[3] M. Saeidifar, M. Yazdi, and A. Zolghadrasli, "Performance Improvement in Brain Tumor Detection in MRI Images Using a Combination of Evolutionary Algorithms and Active Contour Method," *Journal of Digital Imaging*, vol. 34, no. 5, pp. 1209–1224, Oct. 2021, doi: 10.1007/s10278-021-00514-6.

[4] T. Akan, S. Alp, and M. A. N. Bhuiyanb, *Vision Transformers and Bi-LSTM for Alzheimer's Disease Diagnosis from 3D MRI.* 2024, doi: 10.48550/ARXIV.2401.03132.

[5] A. B. Rajeoni, "Portable Autonomous Venipuncture Device," US20220160273A1, May 26, 2022 Accessed: May 31, 2023. [Online]. Available: https://patents.google.com/patent/US20220160273A1/en.

[6] L. Guidi *et al.*, "Automatic Measurement of Vascular Calcifications in Patients with Aorto-Iliac Occlusive Disease to Predict the Risk of Re-intervention After Endovascular Repair," *Annals of Vascular Surgery*, vol. 83, pp. 10–19, Jul. 2022, doi: 10.1016/j.avsg.2022.02.013.

[7] F. Lareyre *et al.*, "Automatic Detection of Visceral Arterial Aneurysms on Computed Tomography Angiography Using Artificial Intelligence Based Segmentation of the Vascular System," *EJVES Vascular Forum*, vol. 59, pp. 15–19, 2023, doi: 10.1016/j.ejvsvf.2023.05.001.

[8] O. Bernard *et al.*, "Deep Learning Techniques for Automatic MRI Cardiac Multi-Structures Segmentation and Diagnosis: Is the Problem Solved?," *IEEE Transactions on Medical Imaging*, vol. 37, no. 11, pp. 2514–2525, Nov. 2018, doi: 10.1109/tmi.2018.2837502.

[9] A. Krizhevsky, I. Sutskever, and G. E. Hinton, "ImageNet Classification with Deep Convolutional Neural Networks," *Communications of the ACM*, vol. 60, no. 6, pp. 84–90, May 2017, doi: 10.1145/3065386.

[10] J. Long, E. Shelhamer, and T. Darrell, *Fully Convolutional Networks for Semantic Segmentation*. 2014, doi: 10.48550/ARXIV.1411.4038.

[11] O. Ronneberger, P. Fischer, and T. Brox, "U-Net: Convolutional Networks for Biomedical Image Segmentation," in N. Navab, J. Hornegger, W. M. Wells, and A. F. Frangi, Eds., *Medical Image Computing and Computer-Assisted Intervention – MICCAI 2015*, Lecture Notes in Computer Science. Cham: Springer International Publishing, 2015, pp. 234–241, doi: 10.1007/978-3-319-24574-4_28.

[12] A. Vaswani *et al.*, "Attention is All You Need," in I. Guyon, U. V. Luxburg, S. Bengio, H. Wallach, R. Fergus, S. Vishwanathan, and R. Garnett, Eds., *Advances in Neural Information Processing Systems*. Curran Associates, Inc., 2017. [Online]. Available: https://proceedings.neurips.cc/paper_files/paper/2017/file/3f5ee243547dee91fbd053c1c4a845aa-Paper.pdf.

[13] "OpenAI. ChatGPT (Mar 14 version) [Large language model], 2023. https://chat.openai.com/chat."

[14] T. Akan, S. Alp, and M. A. N. Bhuiyan, *ECGformer: Leveraging Transformer for ECG Heartbeat Arrhythmia Classification*. 2024, doi: 10.48550/ARXIV.2401.05434.

[15] J. Akhavan, J. Lyu, and S. Manoochehri, "A Deep Learning Solution for Real-time Quality Assessment and Control in Additive Manufacturing Using Point Cloud Data," *Journal of Intelligent Manufacturing*, Apr. 2023, doi: 10.1007/s10845-023-02121-4.

[16] P. Hosseini, S. Taheri, J. Akhavan, and A. Razban, "Privacy-preserving Federated Learning: Application to Behind-the-meter Solar Photovoltaic Generation Forecasting," *Energy Conversion and Management*, vol. 283, p. 116900, May 2023, doi: 10.1016/j.enconman.2023.116900.

[17] H. B. Tabrizi, and C. Crick, *Brain-Inspired Visual Odometry: Balancing Speed and Interpretability through a System of Systems Approach*. 2023, doi: 10.48550/ARXIV.2312.13162.

[18] R. E. Haamer *et al.*, "Changes in Facial Expression as Biometric: A Database and Benchmarks of Identification," *2018 13th IEEE International Conference on Automatic Face & Gesture Recognition (FG 2018)*, Xi'an: IEEE, May 2018, pp. 621–628, doi: 10.1109/FG.2018.00098.

[19] N. Imanpour, A. R. Naghsh-Nilchi, A. Monadjemi, H. Karshenas, K. Nasrollahi, and T. B. Moeslund, "Memory- and Time-efficient Dense Network for Single-image Super-resolution," *IET Signal Process*, vol. 15, no. 2, pp. 141–152, Apr. 2021, doi: 10.1049/sil2.12020.

[20] K. S. Panagiotidis, A. Tagka, I. A. Vezakis, I. Kakkos, A. Kyritsi, and G. K. Matsopoulos, "Allergic Contact Dermatitis Detection with Machine Learning," *Preprints*, preprint, Jan. 2024, doi: 10.22541/au.170536831.19871463/v1.

[21] A. Bagheri Rajeoni, "Analog Circuit Sizing Using Machine Learning Based Transistor circuit Model," M.S., The University of Akron, United States, Ohio, 2021. Accessed: May 08, 2023. [Online]. Available: https://www.proquest.com/docview/2543477165/abstract/CDC9DAB8F50D49B8PQ/1.

[22] K. He, X. Zhang, S. Ren, and J. Sun, "Deep Residual Learning for Image Recognition," *2016 IEEE Conference on Computer Vision and Pattern Recognition (CVPR)*. IEEE, Jun. 2016, doi: 10.1109/cvpr.2016.90.

[23] C. A. Cole, B. Janos, D. Anshari, J. F. Thrasher, S. Strayer, and H. Valafar, *Recognition of Smoking Gesture Using Smart Watch Technology*. 2020, doi: 10.48550/ARXIV.2003.02735.

[24] C. O. Odhiambo, L. Ablonczy, P. J. Wright, C. F. Corbett, S. Reichardt, and H. Valafar, "Detecting Medication-Taking Gestures Using Machine Learning and Accelerometer Data Collected via Smartwatch Technology: Instrument Validation Study," *JMIR Human Factors*, vol. 10, p. e42714, May 2023, doi: 10.2196/42714.

[25] C. O. Odhiambo, S. Saha, C. K. Martin, and H. Valafar, *Human Activity Recognition on Time Series Accelerometer Sensor Data Using LSTM Recurrent Neural Networks*. 2022, doi: 10.48550/ARXIV.2206.07654.

[26] A. A. Duquette, P.-M. Jodoin, O. Bouchot, and A. Lalande, "3D Segmentation of Abdominal Aorta from CT-scan and MR Images," *Computerized Medical Imaging and Graphics*, vol. 36, no. 4, pp. 294–303, Jun. 2012, doi: 10.1016/j.compmedimag.2011.12.001.

[27] S. Almotairi, G. Kareem, M. Aouf, B. Almutairi, and M. A.-M. Salem, "Liver Tumor Segmentation in CT Scans Using Modified SegNet," *Sensors*, vol. 20, no. 5, p. 1516, Mar. 2020, doi: 10.3390/s20051516.

[28] A. D. Weston *et al.*, "Automated Abdominal Segmentation of CT Scans for Body Composition Analysis Using Deep Learning," *Radiology*, vol. 290, no. 3, pp. 669–679, Mar. 2019, doi: 10.1148/radiol.2018181432.

[29] A. G. Oskouei, M. A. Balafar, and T. Akan, "A Brain MRI Segmentation Method Using Feature Weighting and a Combination of Efficient Visual Features," in *Applied Computer Vision and Soft Computing with Interpretable AI*. New York, NY, USA: Chapman and Hall/CRC. 2024, [Online]. Available: https://www.taylorfrancis.com/chapters/edit/10.1201/9781003359456-2/brain-mri-segmentation-method-using-feature-weighting-combination-efficient-visual-features-amin-golzari-oskouei-mohammad-ali-balafar-taymaz-akan?context=ubx&refId=b075f77f-e1ca-4c98-87dc-1f5e2dbf8621 pp. 15–34.

[30] E. M. Van Rikxoort, and B. Van Ginneken, "Automated Segmentation of Pulmonary Structures in Thoracic Computed Tomography Scans: A Review," *Physics in Medicine and Biology*, vol. 58, no. 17, Art. no. 17, Sep. 2013, doi: 10.1088/0031-9155/58/17/R187.

[31] J. Deng, W. Dong, R. Socher, L.-J. Li, K. Li, and L. Fei-Fei, "ImageNet: A Large-scale Hierarchical Image Database," *2009 IEEE Conference on Computer Vision and Pattern Recognition*. IEEE, Jun. 2009, doi: 10.1109/cvpr.2009.5206848.

[32] Y. LeCun, Y. Bengio, and G. Hinton, "Deep Learning," *Nature*, vol. 521, no. 7553, pp. 436–444, May 2015, doi: 10.1038/nature14539.

[33] J. M. H. Noothout, B. D. de Vos, J. M. Wolterink, and I. Išgum, "Automatic Segmentation of Thoracic Aorta Segments in Low-dose Chest CT," in *Medical Imaging 2018: Image Processing*. SPIE, Mar. 2018, pp. 446–451, doi: 10.1117/12.2293114.

[34] S. Bonechi *et al.*, "Segmentation of Aorta 3D CT Images Based on 2D Convolutional Neural Networks," *Electronics*, vol. 10, no. 20, p. 2559, Oct. 2021, doi: 10.3390/electronics10202559.

[35] L. Cao *et al.*, "Fully Automatic Segmentation of Type B Aortic Dissection from CTA Images Enabled by Deep Learning," *European Journal of Radiology*, vol. 121, p. 108713, Dec. 2019, doi: 10.1016/j.ejrad.2019.108713.

[36] (Author Name Not Available), *Segmentation Outside the Cranial Vault Challenge*. 2015, doi: 10.7303/SYN3193805.

[37] H. Cao *et al.*, "Swin-Unet: Unet-like Pure Transformer for Medical Image Segmentation," *arXiv*, May 12, 2021. Accessed: Dec. 07, 2023. [Online]. Available: http://arxiv.org/abs/2105.05537.

[38] A. Tragakis, C. Kaul, R. Murray-Smith, and D. Husmeier, "The Fully Convolutional Transformer for Medical Image Segmentation," *Presented at the Proceedings of the IEEE/CVF Winter Conference on Applications of Computer Vision*, 2023, pp. 3660–3669. Accessed: May 07, 2023. [Online]. Available: https://openaccess.thecvf.com/content/WACV2023/html/Tragakis_The_Fully_Convolutional_Transformer_for_Medical_Image_Segmentation_WACV_2023_paper.html.

[39] H.-Y. Zhou, J. Guo, Y. Zhang, L. Yu, L. Wang, and Y. Yu, "nnFormer: Interleaved Transformer for Volumetric Segmentation," *arXiv*, 04 Feb. 2022, doi: 10.48550/arXiv.2109.03201.

[40] J. Chen *et al.*, "TransUNet: Transformers Make Strong Encoders for Medical Image Segmentation," *arXiv*, 08 Feb. 2021. doi: 10.48550/arXiv.2102.04306.

[41] B. E. Sumpio, J. Timothy Riley, and A. Dardik, "Cells in Focus: Endothelial Cell," *The International Journal of Biochemistry & Cell Biology*, vol. 34, no. 12, pp. 1508–1512, Dec. 2002, doi: 10.1016/S1357-2725(02)00075-4.

[42] C. M. Jones *et al.*, "Measurement Science in the Circulatory System," *Cellular and Molecular Bioengineering*, vol. 7, no. 1, pp. 1–14, Mar. 2014, doi: 10.1007/s12195-013-0317-4.

[43] J. Shu, and G. Santulli, "Update on Peripheral Artery Disease: Epidemiology and Evidence-based Facts," *Atherosclerosis*, vol. 275, pp. 379–381, Aug. 2018, doi: 10.1016/j.atherosclerosis.2018.05.033.

[44] J. G. Regensteiner *et al.*, "The Impact of Peripheral Arterial Disease on Health-related Quality of Life in the Peripheral Arterial Disease Awareness, Risk, and Treatment: New Resources for Survival (PARTNERS) Program," *Vascular Medicine*, vol. 13, no. 1, pp. 15–24, Feb. 2008, doi: 10.1177/1358863X07084911.

[45] A. T. Hirsch, W. R. Hiatt, and PARTNERS Steering Committee, "PAD Awareness, Risk, and Treatment: New Resources for Survival – The USA PARTNERS Program," *Vascular Medicine*, vol. 6, no. 1_suppl, pp. 9–12, Feb. 2001, doi: 10.1177/1358836X0100600i103.

[46] R. Hardman, O. Jazaeri, J. Yi, M. Smith, and R. Gupta, "Overview of Classification Systems in Peripheral Artery Disease," *Seminars in Interventional Radiology*, vol. 31, no. 04, pp. 378–388, Nov. 2014, doi: 10.1055/s-0034-1393976.

[47] H. J. Jung, S. C. Lee, K. Y. Kim, and S. S. Lee, "Simultaneous Hybrid Operation Common Femoral Endarterectomy and Endovascular Treatment in Multilevel Peripheral Arterial Disease with Critical Limb Ischemia," *Indian Journal of Surgery*, vol. 80, no. 2, pp. 140–145, Apr. 2018, doi: 10.1007/s12262-016-1570-2.

[48] M. A. Allison, W. R. Hiatt, A. T. Hirsch, J. R. Coll, and M. H. Criqui, "A High Ankle-Brachial Index Is Associated with Increased Cardiovascular Disease Morbidity and Lower Quality of Life," *Journal of the American College of Cardiology*, vol. 51, no. 13, pp. 1292–1298, Apr. 2008, doi: 10.1016/j.jacc.2007.11.064.

[49] M. M. Chowdhury *et al.*, "Lower Limb Arterial Calcification (LLAC) Scores in Patients with Symptomatic Peripheral Arterial Disease are Associated with Increased Cardiac Mortality and Morbidity," *PLOS ONE*, vol. 12, no. 9, p. e0182952, Sep. 2017, doi: 10.1371/journal.pone.0182952.

[50] B. A. Golomb, T. T. Dang, and M. H. Criqui, "Peripheral Arterial Disease: Morbidity and Mortality Implications," *Circulation*, vol. 114, no. 7, pp. 688–699, Aug 2006, doi: 10.1161/CIRCULATIONAHA.105.593442.

[51] I. S. Kang *et al.*, "Semiquantitative Assessment of Tibial Artery Calcification by Computed Tomography Angiography and its Ability to Predict Infrapopliteal Angioplasty Outcomes," *Journal of Vascular Surgery*, vol. 64, no. 5, pp. 1335–1343, Nov. 2016, doi: 10.1016/j.jvs.2016.04.047.

[52] F. Fanelli *et al.*, "Calcium Burden Assessment and Impact on Drug-Eluting Balloons in Peripheral Arterial Disease," *CardioVascular and Interventional Radiology*, vol. 37, no. 4, pp. 898–907, Aug. 2014, doi: 10.1007/s00270-014-0904-3.

[53] S. Mori *et al.*, "Impact of Calcification on Clinical Outcomes after Drug-coated Balloon Angioplasty for Superficial Femoral Artery Disease: Assessment Using the Peripheral Artery Calcification Scoring System," *Catheterization and Cardiovascular Interventions*, vol. 101, no. 5, pp. 892–899, Apr. 2023, doi: 10.1002/ccd.30622.

[54] A. Bagheri Rajeoni, B. Pederson, D. G. Clair, S. M. Lessner, and H. Valafar, "Automated Measurement of Vascular Calcification in Femoral Endarterectomy Patients Using Deep Learning," *Diagnostics*, vol. 13, no. 21, p. 3363, Nov. 2023, doi 10.3390/diagnostics13213363.

[55] Y. Dong *et al.*, "Lower Limb Arterial Calcification and Its Clinical Relevance with Peripheral Arterial Disease," *Frontiers in Cardiovascular Medicine*, vol. 10, p. 1271100, Nov. 2023, doi: 10.3389/fcvm.2023.1271100.

[56] H. Yan, Z. Chang, and Z. Liu, "The Risk Factors for Calcification Vary among the Different Sections of the Lower Extremity Artery in Patients with Symptomatic Peripheral Arterial Disease," *BMC Cardiovascular Disorders*, vol. 20, no. 1, p. 333, Dec. 2020, doi 10.1186/s12872-020-01615-w.

[57] D. T. Boll, J. S. Lewin, T. R. Fleiter, J. L. Duerk, and E. M. Merkle, "Multidetector CT Angiography of Arterial Inflow and Runoff in the Lower Extremities: A Challenge in Data Acquisition and Evaluation," *Journal of Endovascular Therapy*, vol. 11, no. 2, pp. 144–151, Apr. 2004, doi: 10.1583/03-1098.1.

[58] M. S. Conte *et al.*, "Global Vascular Guidelines on the Management of Chronic Limb-threatening Ischemia," *Journal of Vascular Surgery*, vol. 69, no. 6, pp. 3S–125S.e40, Jun. 2019, doi: 10.1016/j.jvs.2019.02.016.

[59] A. M. Flores, F. Demsas, N. J. Leeper, and E. G. Ross, "Leveraging Machine Learning and Artificial Intelligence to Improve Peripheral Artery Disease Detection, Treatment, and Outcomes," *Circulation Research*, vol. 128, no. 12, pp. 1833–1850, Jun. 2021, doi: 10.1161/circresaha.121.318224.

[60] E. G. Ross, N. H. Shah, R. L. Dalman, K. T. Nead, J. P. Cooke, and N. J. Leeper, "The Use of Machine Learning for the Identification of Peripheral Artery Disease and Future Mortality Risk," *Journal of Vascular Surgery*, vol. 64, no. 5, pp. 1515–1522.e3, Nov. 2016, doi: 10.1016/j.jvs.2016.04.026.

[61] L. Zhao, B. Odigwe, S. Lessner, D. Clair, F. Mussa, and H. Valafar, "Automated Analysis of Femoral Artery Calcification Using Machine Learning Techniques," *2019 International Conference on Computational Science and Computational Intelligence (CSCI)*, Las Vegas, NV, USA: IEEE, Dec. 2019, pp. 584–589, doi: 10.1109/CSCI49370.2019.00110.

[62] P. A. Yushkevich *et al.*, "User-guided 3D Active Contour Segmentation of Anatomical Structures: Significantly Improved Efficiency and Reliability," *NeuroImage*, vol. 31, no. 3, pp. 1116–1128, Jul. 2006, doi: 10.1016/j.neuroimage.2006.01.015.

[63] F. Pedregosa *et al.*, *Scikit-learn: Machine Learning in Python*. 2012, doi: 10.48550/ARXIV.1201.0490.

[64] A. B. Rajeoni *et al.*, *TransONet: Automatic Segmentation of Vasculature in Computed Tomographic Angiograms Using Deep Learning*. 2023, doi: 10.48550/ARXIV.2311.10328.

[65] A. Dosovitskiy *et al.*, "An Image is Worth 16x16 Words: Transformers for Image Recognition at Scale," *no. arXiv:2010.11929. arXiv*, 03 Jun. 2021, doi: 10.48550/arXiv.2010.11929.

[66] D. P. Kingma, and J. Ba, "Adam: A Method for Stochastic Optimization," *no. arXiv:1412.6980. arXiv*, 29 Jan. 2017, doi: 10.48550/arXiv.1412.6980.

Afsaneh Shams, Kyle Becker, Drew Becker, Soheyla Amirian, and Khaled Rasheed

Evolutionary CNN-based architectures with attention mechanisms for enhanced image classification

Abstract: This extended study builds upon prior research, serving as an extension of our previous study presented in "Evolving Efficient CNN-Based Model for Image Classification" [19]. Here, we delve deeper into Convolutional Neural Network (CNN) architectures and their performance on the CIFAR-10 dataset, expanding upon the insights gained from our earlier work. Beginning with an ECNNB [19] model that excelled on simpler datasets, we progress to examine advanced iterations featuring attention mechanisms like CBAM and MobileViTv2. Our empirical analysis demonstrates consistent performance improvement with each enhancement, culminating in the AECNNB with MobileViTv2 as the most efficient model. Notably, CBAM integration alone led to a substantial 7.27% improvement in average accuracy, and the final AECNNB model achieved an 86.89% average accuracy, marking a significant 9.77% improvement over the ECNNB [19]. This underscores the significance of architectural sophistication and the potential of advanced attention mechanisms, particularly MobileViTv2, for optimizing CNNs in complex image classification tasks. Our findings provide valuable insights for future neural network development and applications, building upon the foundation established in our earlier study.

Keywords: neural networks, convolutional neural network, CBAM, MobileVitv2, neural evolutionary, CIFAR-10

1 Introduction

Neural networks, functioning similarly to the human brain, map inputs to outputs based on complex interneuronal connections and logic, utilizing training algorithms like gradient descent and backpropagation. This paper applies backpropagation, fo-

Afsaneh Shams, School of Computing, University of Georgia, Athens, USA, e-mail: Afsaneh.Shams@uga.edu
Kyle Becker, Institute of Artificial Intelligence, University of Georgia, Athens, USA, e-mail: Kyle.Becker@uga.edu
Drew Becker, Institute of Artificial Intelligence, University of Georgia, Athens, USA, e-mail: Andrew.Becker@uga.edu
Soheyla Amirian, School of Computing, University of Georgia, Athens, USA, e-mail: amirian@uga.edu
Khaled Rasheed, School of Computing and Institute of Artificial Intelligence, University of Georgia, Athens, USA, e-mail: khaled@uga.edu

https://doi.org/10.1515/9783111344126-006

cusing on neural network layers – hidden units and connections where each has a weight and activation function [1].

In the realm of computational intelligence, evolutionary computation stands out, addressing tasks ranging from optimization to training neural networks [16]. It employs models like genetic algorithms and evolutionary programming [2], distinguished by their multi-solution approach compared to traditional single-solution methods. Shapiro highlights this distinction, emphasizing the need for data-driven solution creation or existing solution optimization, such as weight adjustment in neural networks using gradient descent [3].

An attention mechanism involves processing a query alongside a series of key-value pairs, all in vector form, to produce an output. This output is derived from a weighted summation of the values, where each value's weight is ascertained by a function evaluating the compatibility of the query with its associated key [25]. In this study two attention mechanisms are implemented: Convolutional Block Attention Module (CBAM) and MobileVitv2.

This study further delves into advanced attention mechanisms within Convolutional Neural Networks (CNNs), particularly focusing on the CBAM [24] and MobileViTv2 [26]. CBAM enhances CNNs by applying sequential channel and spatial attention filters, improving image classification and detection performance. MobileViTv2, optimized for mobile applications, incorporates a separable self-attention mechanism, boosting efficiency and performance on resource-constrained devices. Building on our prior work in "Evolving Efficient CNN Based Model for Image Classification," we extend our exploration to the CIFAR-10 dataset, employing these sophisticated attention mechanisms within the AECNNB model. This comprehensive study aims to bridge the performance gap in evolutionary neural networks across varying dataset complexities, offering insights into CNN optimization for improved image classification. This contributes to a deeper understanding of CNN evolution and architecture, providing a valuable reference for future neural network research and applications in diverse image recognition domains.

In this extended study, we expand upon our prior work detailed in "Evolving Efficient CNN-Based Model for Image Classification" [19], exploring CNNs for image classification further. Our initial research focused on the ECNNB model, demonstrating its effectiveness on grayscale datasets like MNIST, Fashion MNIST, and EMNIST Digits. Now, we delve into the challenging CIFAR-10 dataset, assessing both the original ECNNB model and advanced iterations, such as AECNNB with attention mechanisms like CBAM and MobileViTv2. These enhancements are expected to boost the model's performance on CIFAR-10. Our paper's main contribution lies in examining evolutionary neural network adaptability to complex color image datasets. By integrating advanced attention mechanisms, this study bridges the gap between simpler and more complex datasets, offering insights into optimizing CNN architectures for improved image classification. Our findings contribute to the broader understanding of CNN evolution and architectural sophistication, benefiting future neural network development and diverse image recognition tasks.

2 Related work

In this section, we summarize papers showcasing the efficiency of integrating CBAM [24] and MobileViTv2 [26] into diverse neural network architectures. These studies highlight substantial improvements in target detection, image classification, specialized tasks, and defect identification accuracy. We also emphasize the potential for further enhancements through additional iterations and experimentation with these integrated models.

The study by Fu, Song, and Wang on integrating CBAM with YOLOv4 [4] demonstrated significant enhancements in target detection accuracy. The integration focused on better identifying crucial target areas and minimizing irrelevant information, leading to improved detection precision. Experimental results showed that the modified YOLOv4 model outperformed the original, achieving a 2.02% increase in mAP50 and a 1.85% improvement in mAP75, while maintaining rapid detection speeds suitable for real-time applications.

As Woo et al. mention in their research [26], the CBAM significantly improved performance across various neural network architectures and tasks. In image classification on ImageNet-1 K, CBAM integrated into ResNet50 reduced Top-1 and Top-5 errors by 1.90% and 1.19%, respectively. In object detection tasks, CBAM applied to Faster-RCNN with ResNet50 and ResNet101 as baselines showed an increase in mean Average Precision (mAP) by up to 1.7%. Additionally, in the VOC 2007 object detection task, CBAM enhanced the mAP by 0.4% when used with VGG16 and StairNet. These results demonstrate CBAM's effectiveness in enhancing the representational power of CNNs with a minimal computational overhead.

In the study by Chen et al. [5], the combination of the CBAM with an enhanced RetinaNet model is investigated. The enhanced model achieved an mAP of 90.38%, surpassing the standard RetinaNet model's mAP by 2.61%. This demonstrates the efficacy of CBAM in refining feature extraction and significantly boosting recognition accuracy in specialized tasks like fly species identification.

Luo and Wang's paper [6] presents an enhanced ResNet algorithm integrated with the CBAM for flower recognition. This approach significantly improved the model's accuracy to nearly 98%, particularly excelling in scenarios with limited training data. The CBAM addition effectively enhanced the model's focus on relevant features for fine-grained image classification, marking a notable advancement in deep learning applications for specific image recognition tasks.

In their study, Zhang et al. [7] enhanced the DeepLabv3 + model [30] with the CBAM [24], achieving significant improvements in winter wheat detection. The addition of CBAM increased accuracy rates (OA, mAP, mIoU) by 1.52%, 1.51%, and 2.99%, respectively, over the original model. CBAM played a key role in refining the focus on important features, leading to more efficient and precise agricultural monitoring.

Ma et al. [8] improved MobileNetV2 with an advanced CBAM named I_CBAM, for maize seed variety identification. This led to a substantial accuracy increase of 4.88%

over the original MobileNetV2, achieving 98.21% accuracy. The I_CBAM_MobileNetV2 model surpassed other networks like Xception, showing the efficacy of attention mechanisms in enhancing lightweight neural models for agricultural use.

In the study "YOLOv5s-CBAM-DMLHead," [9] the authors focused on identifying weedy rice efficiently. Incorporating the CBAM [24] into the YOLOv5s model [31] significantly improved accuracy. Specifically, the integration of CBAM resulted in an increase in mAP from 98.2 to 98.9%, indicating a substantial improvement of approximately 0.7% in accuracy.

Lin et al. [10] conduct research to enhance the accuracy of power fingerprint identification by utilizing color-encoded V–I trajectory images and integrating the CBAM module. Notably, the incorporation of CBAM results in a substantial 3.1% accuracy improvement.

Chen and Dang [11] introduce an approach to find the defects in the printed circuit boards based on YOLOv7 [32], with FasterNet as the main network in their paper. The inclusion of CBAM in the feature fusion module notably boosts accuracy, resulting in an mAP0.5 of 97.5%. This represents a substantial accuracy improvement compared to conventional methods, alongside a speed increase from 54.3 to 83.3 frames per second, making it a more efficient PCB defect detection solution.

Liu et al.'s GhostNet-CBAM-YOLOv4 approach [12] improved detection speed by 6.83% and increased the detection rate by 42.1% in comparison with the basic YOLOv4 model [33]. This underscores the effectiveness of CBAM in enhancing foreign object detection.

Miao et al. introduce CBAM-YOLOv5 [13], an infrared ship detection model. It leverages CBAM for enhanced feature extraction and small target bounding box regression, resulting in a 1.1% mAP increase and a 2.0% recall rate increase compared to YOLOv5 on the infrared ship dataset, thus improving detection accuracy.

In their study, Munir et al. [14] evaluate the performance of MobileViT V2, highlighting its significant improvement in mobile vision technology. MobileViT V2-1.0 achieves a top-1 accuracy of 78.1% on ImageNet-1k, surpassing the 71.8% accuracy of MobileNetV2x1.0. This comparison is part of their broader research aim to explore and enhance the efficiency and accuracy of mobile vision applications.

In the study by Yang et al., "Efficient Diagnosis of Hematologic Malignancies Using Bone Marrow Microscopic Images," [15] the MobileViTv2 model achieved an impressive accuracy of 94.28% on test sets and 96.72% for patient-level diagnoses, significantly improving the precision of hematologic malignancy diagnosis.

Cao et al. introduce YOLO-SF [16], an algorithm that combines instance segmentation with YOLOv7-Tiny for more accurate fire detection. It utilizes a fire segmentation dataset, integrates MobileViTv2 for improved accuracy, and achieves a 5.9% increase in precision, 3.3% increase in recall, and a 6% increase in mAP compared to YOLOv7-Tiny, while maintaining real-time detection at 55.64 FPS.

Lv et al. [17] perform a study on MobileViTv2, which functions as a key feature extractor in the defect segmentation model, significantly contributing to accuracy improvement. Its inclusion raises the mean Intersection over Union (mIoU) to 87.01%, demonstrating its pivotal role in enhancing defect identification precision.

Zhang et al. [18] introduce an efficient model for detecting abnormal surface features in in-water fish using an improved YOLOv5s. The model achieves impressive results for precision at 99.05%, recall at 99.1%, mAP50 at 99.1%, mAP50:95 at 73.9%, and 88 FPS that surpass the baseline by 1.4, 1.2, 3.2, 8.2%, and 1 FPS. Notably, the integration of MobileViTv2 plays a significant role in enhancing the model's accuracy, enabling it to outperform other state-of-the-art models.

In the following sections, we will elaborate on our methodology, encompassing optimization procedures and architectural choices. We will also discuss the dataset, highlighting its source and preprocessing steps, present experimental results, and provide a conclusion with insights for future research.

3 Methodology

The updated approach of this study marks a significant transition in the development of neural networks, moving from TensorFlow to PyTorch while continuing to utilize Python. This strategic shift is driven by the dynamic computational landscape in neural network research, with PyTorch offering superior flexibility and dynamic computation capabilities. These features are particularly beneficial for handling complex operations in genetic algorithms and the nuanced requirements of neural network training and evolution.

The foundational design of the neural networks remains consistent with the original approach, featuring input layers dimensioned at 28 × 28 and dense output layers sized for 10 categories. The architecture of an individual within the ECNNB model is detailed in Figure 2. Key enhancements include a more efficient and adaptable crossover function, and promising optimized network architectures. Figure 1 illustrates the selection method of activation function for each hidden layer, which is critical for improving the network's efficiency.

PyTorch's adoption reflects a broader trend in machine learning and AI, where the choice of framework significantly influences model development and training efficiency. By aligning with PyTorch, the study leverages its ease of use, flexibility, and computational prowess, which are crucial in environments that prioritize rapid prototyping and experimentation.

The complete architecture of the network is illustrated in Figure 3.

3.1 Optimization procedure

The evolutionary algorithm has undergone considerable refinement. Originally implemented in TensorFlow and limited to five generations with a single-member population

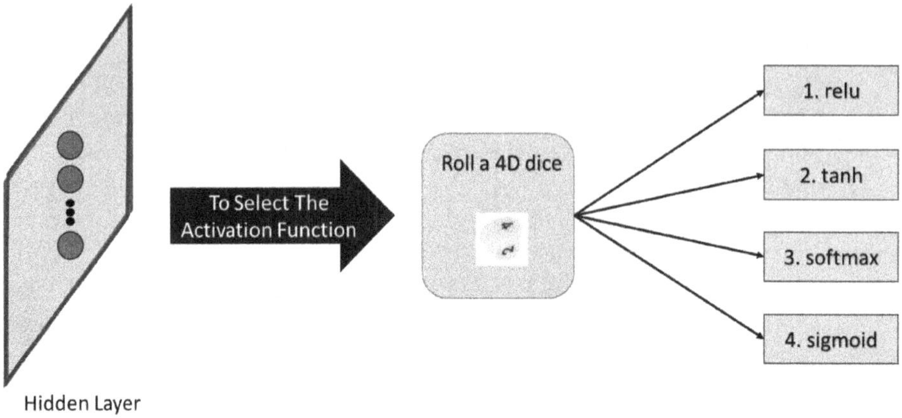

Figure 1: Illustration of how to select an activation function for each hidden layer [19].

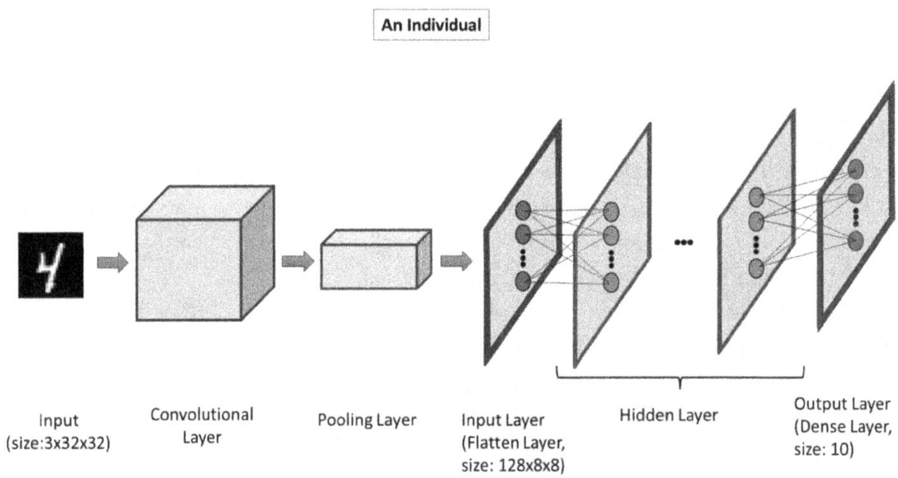

Figure 2: The complex architecture of an individual within the ECNNB model [19].

(MU = 1), the updated methodology expands the evolutionary exploration by adopting PyTorch and DEAP, extending the process to 20 generations. The increase in generational depth and expanded parent population size of 3 (up from 1) enriches genetic diversity, enhancing the potential for discovering superior network architectures.

3.1.1 Representation

The revised approach of this study enhances the representation of individuals within the genetic algorithm, aligning with the principles of evolutionary programming and

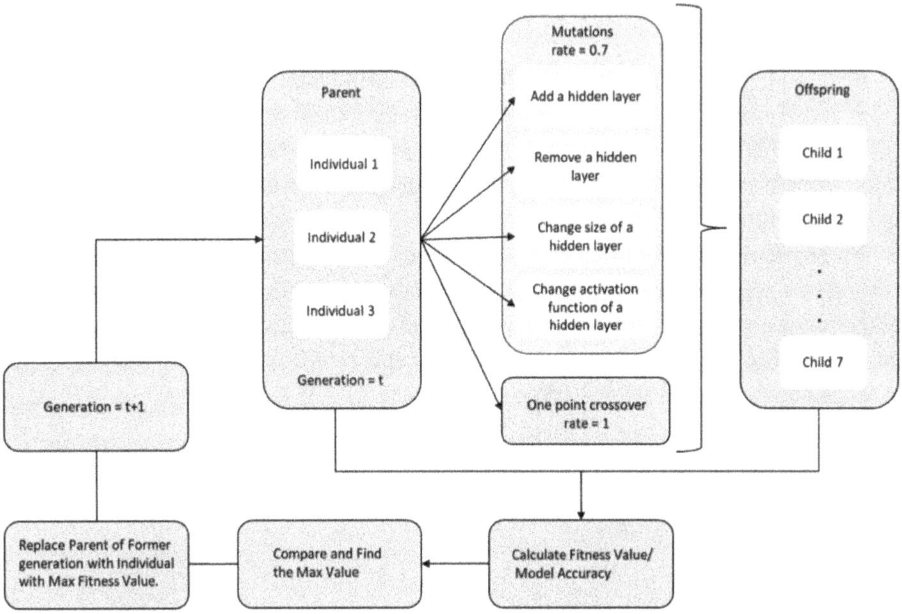

Figure 3: Evolutionary Neural Network architecture. The input to this neural network is the very first individual created and selected as the parent. It is worth mentioning that the termination criteria in this network are set to be 20 generations. The loop shown in this figure will be repeated 20 times, and the network with the highest fitness value among current children and parents will be considered as the output of the method. Note that this approach is based on the (μ+λ) concept, where μ represents the parents and λ represents the children.

leveraging the capabilities of the PyTorch framework. The representation of each individual encompasses a combination of feature extractors and classifiers, optimized for the dynamic and efficient manipulation of neural network architectures. This updated representation facilitates the creation of complex CNN networks, surpassing the basic model used previously. The transition to PyTorch enables dynamic modifications to network architectures, offering the potential for designs that are finely tuned for image classification tasks.

A critical aspect of the evolutionary process is the training and preparation of each individual. In this revised methodology, the optimization for PyTorch ensures that every neural network is primed for immediate performance assessment. This optimization is expected to accelerate training cycles and improve effectiveness in handling complex image datasets.

In contrast to the original implementation with a single-member population, the updated algorithm features a tripartite population. Each generation within this new framework produces seven offspring through a combination of mutation and crossover. This increase in population size and the generation of offspring substantially enhance the genetic diversity of the algorithm, broadening the exploration of potential network architectures and improving the chances of discovering more optimal configurations. The starting point for the population has also evolved. Unlike the original method, which began with an ECNNB-based neural network, the current approach seeds the initial population with a predefined but adaptable PyTorch-based network structure. This advanced starting point provides a robust foundation for the evolutionary process, potentially accelerating the development of highly effective neural network architectures.

In summary, the adoption of PyTorch and adjustments in population dynamics represent significant advancements in the methodology. This study underscores the potential of an evolved genetic algorithm, equipped with crossover functionality and an expanded evolutionary operation scope to optimize CNN architectures for image classification tasks effectively.

3.1.2 Fitness function

The revised genetic algorithm emphasizes the fitness function's crucial role in assessing neural network models. This function evaluates the accuracy of each model on a test dataset, representing each individual as a complex structure within a CNN, now optimized in the PyTorch environment. This optimization enhances the computational efficiency of the evaluation process.

The fitness assessment involves a dual approach, utilizing both a loss function and an accuracy metric, with a primary emphasis on test dataset accuracy. This ensures that the most performant models in terms of accuracy are identified and given precedence. In the PyTorch-enhanced algorithm, each individual in the expanded population of three undergoes this rigorous fitness evaluation. Those exhibiting the highest fitness values, indicative of superior accuracy, are selected for breeding.

The mutation methods within the algorithm, such as layer addition, activation function changes, and layer removal, are now more intricately aligned with PyTorch's capabilities. This alignment allows for nuanced alterations in network architecture, contributing to the overall effectiveness of the genetic algorithm. Additionally, the crossover operation has been notably enhanced, employing a single-point crossover method with a higher probability, indicating its increased importance in the new algorithm. This adjustment in the crossover process fosters a more robust exchange of features and traits between parent networks.

As part of the generational progression, the algorithm operates over an extended span of 20 generations. This increase in generations supports a more comprehensive evolutionary process, continuously generating offspring through a tactical application of crossover and mutation. After applying these operators, the fitness of each offspring is assessed using the updated function, determining the composition of the subsequent generation.

3.1.3 Mutation

In the updated genetic algorithm, sophisticated mutation operations have been meticulously designed to introduce targeted genetic variation within the neural network architectures. These operations are crucial for the evolution of network structures, providing a nuanced approach to architectural optimization. Each mutation operation is applied to a parent network, producing offspring with unique structural configurations, thus influencing the genetic trajectory of subsequent generations. The details of these operations are as follows:

– **Add a Hidden Layer:** This operation strategically injects a new hidden layer into the network architecture. The layer is inserted before the output layer, and its configuration includes a variable number of nodes, randomly selected from a range between 0 and 128. Additionally, the activation function for this new layer is chosen from a pool of functions, including ReLU, Tanh, Softmax, or Sigmoid, offering a diverse set of nonlinear transformation capabilities. This mutation is designed to investigate the impact of additional computational units on the network's learning capacity and performance. To ensure a balance between network complexity and computational efficiency, the total number of hidden layers in a network is restricted to a maximum of three. In scenarios where a network already comprises three hidden layers, the mutation alters either the activation function or the size of one of the existing layers, allowing for refined adjustments to the network's processing capabilities.

– **Remove a Hidden Layer:** This operation involves the selective removal of a hidden layer from the neural network. By eliminating a layer, the algorithm explores the potential of more compact and potentially more efficient network architectures. This operation challenges the assumption that larger networks are always superior, allowing for the exploration of performance in reduced complexity scenarios. Additionally, it facilitates the identification and elimination of layers that may not contribute optimally to the network's overall performance. If a network is already in its simplest form, the mutation operation adapts by introducing a new hidden layer, ensuring that every mutation leads to a tangible architectural change.

– **Change the Size of a Hidden Layer:** This mutation operation entails the random adjustment of the size of a hidden layer, with the number of nodes in the layer varying within the range of 0 to 128. Such flexibility in layer size adjustment allows the algorithm to navigate a wide spectrum of architectural configurations, fostering exploration beyond conventional layer sizing strategies. This mutation is particularly significant for examining how variations in the number of computational units within a layer influence the network's ability to process and learn from data.

– **Change the Activation Function of a Hidden Layer:** In this operation, the activation function of a hidden layer is randomly altered among choices like ReLU, Tanh,

Softmax, or Sigmoid. This mutation is grounded in the premise that different activation functions can have varied impacts on a network's learning dynamics and performance. The strategic alteration of the activation function within a hidden layer plays a crucial role in the neural network's ability to learn complex patterns and functions. This mutation probes into the heart of the network's learning mechanism, examining how different nonlinear transformations affect the overall learning process and model performance. By introducing this variety in activation functions, the algorithm delves into a realm of architectural experimentation, unshackling itself from preconceived notions about the efficacy of specific activation types. This mutation not only diversifies the network's operational dynamics but also provides empirical insights into the most effective activation functions for various tasks.

The mutation rate of 0.2 ensures a delicate balance between the introduction of new genetic variations and the preservation of overall population stability. This rate is deliberately chosen to strike a balance between evolutionary innovation and continuity. It is a pivotal aspect of the algorithm's design, aimed at fostering an environment where evolution can proceed dynamically yet coherently, leading to the discovery of efficient and effective neural network architectures.

3.1.4 Crossover

The one-point crossover method, now synergistically integrated with the PyTorch framework, is a testament to the adaptability and advanced capabilities of the genetic algorithm. This crossover operation is critical in facilitating the amalgamation of diverse genetic traits, creating a rich tapestry of neural network architectures within the population. By strategically combining segments from parent networks, the algorithm fosters the emergence of offspring with potentially superior attributes, enhancing the overall quality of the population.

This crossover technique, particularly in the context of PyTorch's dynamic environment, offers a significant advantage in the exploration of the neural network design space. The flexibility afforded by PyTorch allows for an efficient and seamless combination of disparate network elements, ensuring that each crossover event is not only a fusion of genetic material but also an opportunity for architectural innovation. The resulting offspring are a blend of the parent networks' strengths, poised to contribute novel solutions to the evolving population.

Moreover, the heightened probability of crossover in this updated algorithm underscores the increased emphasis on this operation. The higher probability reflects the algorithm's strategy to leverage crossover as a primary mechanism for introducing variety and complexity into the genetic pool. This approach is essential in avoiding stagnation and ensuring that each generation presents a fresh array of architectural possibilities, thereby enhancing the potential for discovering optimized solutions for the targeted tasks like image classification.

The one-point crossover, in essence, serves as a cornerstone of the genetic algorithm's exploration strategy. It not only facilitates the mixing of genetic traits but also propels the algorithm toward new regions of the solution space, enabling the discovery of innovative and efficient neural network models. This method, aligned with the advanced features of PyTorch, exemplifies the study's commitment to harnessing the power of genetic algorithms in the pursuit of superior neural network architectures.

3.1.5 Termination criteria

The termination criteria in the revised genetic algorithm are meticulously crafted to align with the computational sophistication of the PyTorch framework. This alignment is crucial in ensuring that the evolutionary process is both efficient and effective, culminating in the discovery of optimal neural network architectures.

The primary criterion for termination is the achievement of a predefined number of generations, set at 20 for this study. This duration marks a significant increase from previous iterations and reflects a deeper commitment to exploring a broader range of neural network configurations. The extended run time allows the algorithm to thoroughly investigate various architectural possibilities, ensuring a comprehensive optimization process.

Furthermore, the algorithm incorporates a nuanced approach to managing selection pressure and population dynamics. The ($\mu + \lambda$) tournament selection scheme, adapted to the complexities of PyTorch, plays a pivotal role in this process. The selection mechanism focuses on maintaining a robust and diverse genetic pool, essential for the continuous evolution of effective network designs. The tournament size, strategically set at 6 in the (3+7) selection model, is a critical component of the termination criteria. This size, slightly smaller than the total number of competing individuals, regulates the competition within the selection process. Such a configuration is designed to consistently favor individuals with superior fitness scores, effectively guiding the evolutionary process toward the most promising solutions.

An essential feature of the termination criteria is the elitist plus strategy. This strategy ensures the preservation of the highest-performing individual from each generation, irrespective of whether it is a parent or offspring. By retaining these top-performing models, the algorithm prevents the loss of advantageous traits, fostering a lineage of success that informs the development of future generations.

The termination criteria, as a whole, are a testament to the algorithm's refined focus on identifying and perpetuating the most effective neural network architectures. This focus is paramount for advancing the field of image classification. The combination of an extended generational span, an increased tournament size, and an elitist plus strategy underlines the algorithm's dedication to achieving a balance between exploration and optimization, ensuring the continuous evolution of increasingly sophisticated neural network models.

3.1.6 Selection scheme

In the latest iteration of the genetic algorithm, there is a conscious effort to balance computational complexity with the extensive training requirements inherent in neural networks. This balance is reflected in the nuanced approach to setting termination criteria and managing selection pressure.

The tournament size in the updated (3+7) tournament selection scheme is a calculated decision, aimed at regulating the selection pressure within the algorithm. With a population size of 3 and 7 offspring generated per generation, the tournament size of 6 is strategically chosen. Six individuals are chosen randomly from the ten parents and children. The best three among the six become the parents of the next generation. This specific size serves to enhance the competition within the selection process, ensuring that the algorithm consistently favors individuals demonstrating the highest fitness scores. Such a configuration guides the evolutionary process towards identifying and retaining the most promising neural network architectures.

Extending the evolutionary process to 20 generations signifies a commitment to a more thorough exploration and optimization of neural network designs. The duration of this extended run is crucial in allowing a comprehensive examination of various architectural solutions. The ultimate goal of this process is to identify and preserve the individual with the highest fitness score at the conclusion of the 20th generation, ensuring the selection of the most effective model.

The potential for future adjustments in the number of generations is acknowledged, as the optimal duration may vary depending on specific problem requirements and goals. Future iterations of the algorithm might experiment with varying the generational span to fine-tune the balance between runtime efficiency and model accuracy.

Figure 3 in the study visually represents this comprehensive methodology, illustrating the principles and mechanics underlying the implemented genetic algorithm. It serves as a guide to understanding the interplay of various components of the algorithm, including population dynamics, tournament selection, and termination criteria.

Overall, the updated genetic algorithm exemplifies a sophisticated enhancement of the evolutionary process. It demonstrates a deep understanding of the computational demands of training neural networks and the importance of efficient yet effective exploration within the evolutionary framework. The adjustments in tournament size and the extension of the number of generations are indicative of a refined approach to optimizing neural network architectures, aligning with the goal of achieving superior performance in image classification tasks.

3.2 Architecture

This section delves into an in-depth examination of the neural network architectures deployed in this study, meticulously outlining the strategic design choices and innovative implementations that underpin the models' effectiveness. The architecture is not merely a framework for computations but a confluence of state-of-the-art techniques and cutting-edge design principles, reflecting the latest advancements in the field of deep learning.

In parallel to adopting PyTorch, we have undertaken a strategic transition to the CIFAR-10 dataset from EMNIST. This shift is not just a change in data sources but a thoughtful realignment of our models to engage with more complex and varied image data. The CIFAR-10 dataset challenges our architectures to adapt and excel in a more demanding environment, pushing the boundaries of what is possible in image classification tasks.

The architectural designs presented in this section are the culmination of extensive research and development efforts. Each model – from the foundational CNN to the more advanced implementations featuring CBAM and MobileViTv2 attention mechanisms – is a testament to our dedication to innovation. These models are not static entities; they are dynamic, evolving structures that have been fine-tuned and optimized through rigorous testing and refinement.

We have incorporated sophisticated evolutionary algorithms and genetic programming techniques, further augmenting the capabilities of our models. These algorithms are not merely add-ons; they are integral to the architecture, and woven into the fabric of our designs to enhance adaptability and efficiency. The evolutionary algorithms, with their mutation and crossover strategies, embody a novel approach to architectural optimization, pushing the envelope in neural network design.

Furthermore, this section will showcase how the architectures have been fine-tuned and calibrated to achieve optimal performance. From the implementation of advanced data preprocessing techniques to the integration of automated GPU selection mechanisms based on memory availability, every aspect of the architecture has been crafted with precision and foresight.

In essence, the architectures delineated in this section are a confluence of innovative design, advanced technology, and strategic foresight. They stand as a testament to our commitment to pushing the boundaries of machine learning and AI, constantly seeking new horizons in the quest for excellence in neural network design and implementation.

3.2.1 AECNNB (Advanced Evolutionary CNN-Based)

The transition to PyTorch from TensorFlow Federated 0.20.0 represents a strategic shift, capitalizing on PyTorch's dynamic computation graph and enhanced usability. This shift is aligned with the latest advancements in deep learning frameworks, offering a platform that supports more flexible and efficient model development.

We have undertaken a methodical transition to the CIFAR-10 dataset from EMNIST, accompanied by a robust preprocessing pipeline. This pipeline integrates advanced techniques such as random cropping with a 4-pixel padding, horizontal flipping, and color jittering for adjustments in brightness, contrast, saturation, and hue. Additionally, data normalization is meticulously executed with specified mean values (0.4914, 0.4822, 0.4465) and standard deviations (0.247, 0.243, 0.261), ensuring data standardization and aiding in the model's generalization capabilities.

The architectural cornerstone is the introduction of the CNN class within the PyTorch framework. This class is composed of four convolutional layers, with filter counts of 32, 64, 128, and 128. It incorporates post-convolution batch normalization, dual MaxPooling layers, and strategically placed dropout layers. This architecture is designed to enhance feature extraction and reduce overfitting, forming a well-rounded model for image classification tasks. An innovative addition is the implementation of an automated GPU selection mechanism based on memory availability. This strategy maximizes computational efficiency by allocating resources where they are most available.

The training methodology is enhanced with a comprehensive fitness evaluation function and a sophisticated training protocol. This protocol integrates an Adam optimizer [34] set at a learning rate of 0.001 and a learning rate scheduler reducing this rate by a factor of 0.1 every 10 epochs, aligning with best practices in modern deep learning. The evolutionary process is meticulously managed with the introduction of two-point crossover and refined mutation logic. This approach ensures the validity of fitness measures and the efficient progression of generations, aligning with advanced evolutionary computation methods.

In summary, these comprehensive enhancements embody a forward-thinking approach to model development and training, showcasing a commitment to adopting state-of-the-art practices in machine learning, ensuring the model's robustness, adaptability, and efficacy in addressing complex computational tasks.

3.2.2 AECNNB with CBAM attention mechanism

The CBAM [24] is a sophisticated attention mechanism incorporated into the existing CNN architecture, designed to enhance the network's focus on the most informative features of input images. CBAM applies channel and spatial attention mechanisms sequentially, refining feature maps produced by convolutional layers [24].

As Woo et al. [24] have mentioned in their paper, the channel attention submodules zero in on "what" is meaningful within an image, using Adaptive Max Pooling and Adaptive Average Pooling to compress spatial information into channel descriptors. These are processed through a shared network of convolutional layers without bias and a sigmoid activation function. The element-wise sum of Max Pooling and Average Pooling outputs passes through the sigmoid function, generating a channel attention map that recalibrates the original feature maps by enhancing specific channels.

Following channel attention, the spatial attention submodule determines "where" to focus. It uses Max Pooling and Average Pooling across the channel axis, concatenating these results and passing them through a convolutional layer with a kernel size of 5. A subsequent ReLU activation and sigmoid function produce a spatial attention map, further refining the feature maps spatially [24].

CBAM's incorporation into CNN is designed to enhance feature representation by emphasizing significant features and suppressing less relevant ones. This is particularly beneficial for image classification tasks where precise focus on certain features or spatial regions is critical. The CBAM module is lightweight and generalizable, making it easy to integrate into various CNN architectures without substantial structural changes.

Key parameters within CBAM, such as the reduction ratio in Channel Attention, set to 8 in this implementation, and the kernel size in Spatial Attention, are critical for balancing computational cost and attention precision. A lower reduction ratio implies a more complex attention mechanism but potentially yields a more accurate focus on relevant features.

Empirical results demonstrate that the integration of CBAM leads to improved performance in image recognition tasks. The dual-focused attention mechanism provided by CBAM offers a more refined and focused feature set for classification, enhancing the overall capability of the CNN architecture.

3.2.3 AECNNB with Mobilevit2 attention mechanism

In the AECNNB architecture enhanced with the MobileViTv2 attention mechanism, we have developed a sophisticated model tailored for the CIFAR-10 dataset's complex image classification tasks. This section provides a focused exposition on the integration and operational specifics of the MobileViTv2 module within the existing framework. The architecture initiates with a series of convolutional layers, which are fundamental in extracting a wide range of features from the dataset images. These layers are configured with filter counts of 32, 64, 128, and 128, ensuring a comprehensive initial feature capture.

Post-convolution, batch normalization is applied consistently to stabilize the learning process. Dropout layers are strategically incorporated with rates of 0.25 and 0.5, aiming to mitigate overfitting and maintain model robustness. This setting plays an important role in improving the model generalization.

The distinctive aspect of this architecture is the MobileViTv2 attention mechanism. Positioned strategically in the network, this module employs scaled dot-product attention to concentrate on salient image features. This selective focus aids in refining the feature maps, significantly enhancing the model's discernment and classification accuracy. Max pooling layers condense the spatial dimensions of the feature maps; hence this approach can retrain the features that are dominant and also reduce the computational load [24]. The network then transitions into preparing the feature vector, flattening the 2D maps into a 1D vector for the final classification stage.

The classifier, a fully connected layer, is tasked with mapping the processed features to the CIFAR-10 dataset's output classes. The training regimen employs an Adam optimizer [34] with a learning rate of 0.001, modulated through a scheduler to ensure efficient and balanced weight optimization.

The MobileViTv2 attention mechanism's integration is a testament to our commitment to leveraging advanced techniques for improved performance in complex tasks. This approach positions the AECNNB model at the forefront of innovation in neural network design for image classification, demonstrating a harmonious blend of foundational CNN elements with cutting-edge attention mechanisms.

4 Dataset

In contrast to the MNIST dataset, which predominantly features handwritten digits and is frequently employed for training image processing systems, our research centers on the CIFAR-10 dataset [28, 29]. CIFAR-10 is well-regarded for its diversity and the intricate image classification tasks it presents. It consists of 60,000 images categorized into 10 classes, each with a resolution of 32 × 32 pixels. To ensure our CNN models' optimal performance on this demanding dataset, we implemented a tailored data augmentation and preprocessing approach aimed at improving their ability to generalize. This dataset selection allowed us to thoroughly assess our models' effectiveness in tackling a broader spectrum of image classification challenges.

Our preprocessing pipeline incorporated random crop with padding, random horizontal flipping, and color jitter to introduce variations in image positioning, orientation, and color. These augmentations are critical for training robust models capable of handling real-world image inconsistencies. Additionally, normalization of the dataset was performed using specific mean and standard deviation values, ensuring uniform data distribution for efficient model training.

For the test data, we restricted preprocessing to normalization alone to maintain a consistent evaluation framework. This strategic approach in data augmentation and preprocessing was designed to optimize the models' learning potential, making them adept at managing the varied and complex nature of real-world image datasets.

5 Experiments

This section presents an experimental analysis focused on evaluating the performance of CNN architectures across varying complexities, specifically targeting the CIFAR-10 dataset. Beginning with the ECNNB [19] previously successful on grayscale datasets like

EMNIST Digit, MNIST, and Fashion MNIST, our study extends to explore its efficacy on the more complex CIFAR-10 dataset. This progression allows us to assess the adaptability of ECNNB [19] structures when faced with the increased complexity of color images.

Subsequent experiments involve an AECNNB model without attention mechanisms and advanced iterations incorporating attention modules like CBAM and MobileViTv2. These steps are designed to discern the impact of structural enhancements and attention mechanisms on handling intricate image classification tasks. The results from these various model configurations offer insights into their relative performance, highlighting the evolution and innovation necessary in neural network design to address the challenges of sophisticated image datasets.

5.1 ECNNB [19] architecture performance on CIFAR-10

In our seminal research in "Evolving Efficient CNN-Based Model for Image Classification" [19], we embarked on a comprehensive study of two distinct models: the Evolutionary Neural Network (ENN) and the Evolutionary Convolutional Neural Network-Based (ECNNB) model. This study focused on evaluating these models across three datasets: MNIST [20, 21], EMNIST_Digits [22], and Fashion MNIST [23]. Notably, our ECNNB [19] model exhibited commendable performance, outshining its ENN [19] counterpart in various metrics, as shown in Table 1.

Building upon this foundation, our current investigation pivots towards applying the ECNNB [19] model to the CIFAR-10 dataset. This strategic decision was predicated on the superior results obtained by ECNNB [19] in our initial explorations.

The transition to CIFAR-10, however, presented a significant paradigm shift. Originally tailored for grayscale imagery, our model's architecture was confronted with CIFAR-10's color images and their inherently more complex feature distributions. This complexity was mirrored in the obtained accuracy metrics, which oscillated between 76.01% and 78.5%, culminating in an average accuracy of 77.124%. Such results, though modest in comparison to the model's performance on MNIST [20] and EMNIST Digit datasets (surpassing 99% accuracy), underscore the nuanced challenges posed by the CIFAR-10 dataset.

5.1.1 Comparative analysis across datasets

Our comparative analysis accentuates that while the ECNNB [19] model demonstrates proficiency in handling simpler datasets, its adaptability to the CIFAR-10 dataset necessitates further refinement. This revelation prompted a systematic evolution of the ECNNB's [19] architectural framework, enhancing its capability to navigate the intricate classification challenges of CIFAR-10. Table 2 compares the image classification accuracy of ECNNB [19] model on different datasets.

Table 1: Performance of the ENN and ECNNB models On Fashion _MNIST, MNIST, and EMNIST _Digits datasets [19].

Dataset	Accuracy range	Average accuracy	Std. deviation
	ENN Model [19]		
Fashion MNIST	87.76% to 89.44%	88.28%	0.39%
MNIST	97.04% to 98.05%	97.68%	0.26%
EMNIST Digits	98.53% to 98.95%	98.76%	0.1%
	ECNNB Model [19]		
Fashion MNIST	91.40% to 93.07%	92.58%	0.3%
MNIST	98.91% to 99.49%	99.32%	0.11%
EMNIST Digits	99.35% to 99.66%	99.58%	0.06%

Table 2: Detailed performance comparison of CNN models on the CIFAR-10 dataset.

Dataset	Accuracy range	Average accuracy	Std. deviation
Fashion MNIST	91.40% to 93.07%	92.58%	0.3%
MNIST	98.91% to 99.49%	99.32%	0.11%
EMNIST Digits	99.35% to 99.66%	99.58%	0.06%
CIFAR-10	76.01% to 78.50%	77.12%	0.78%

Subsequent sections of our paper will meticulously outline the incremental advancements in the ECNNB [19] model. This includes comprehensive experimentation with advanced structural components such as CBAM and MobileViTv2, aiming to further optimize the model for the multifaceted CIFAR-10 dataset.

5.2 Developed CNN with no attention architecture performance on CIFAR-10

In this study, we examined the "AECNNB" model, an evolved version of our ECNNB [19], which previously showed commendable results on grayscale image datasets. Our objective was to test the hypothesis that structural improvements alone, without attention mechanisms, can enhance model performance on the CIFAR-10 dataset, known for its complexity due to color images and intricate feature patterns.

As shown in the Table 3, over 30 trials, the AECNNB model demonstrated a significant uplift in performance when compared to the ECNNB model [19]. With an increase in accuracy, the trials for the AECNNB model achieved a minimum accuracy of 81.65%, a maximum of 83.84%, and an average accuracy of 82.95%. This suggests that the structural improvements contributed to the model's enhanced ability to discern features within the CIFAR-10 dataset.

Table 3: Performance comparison on CIFAR-10 dataset.

Model type	Min. accuracy	Max. accuracy	Average accuracy	Std. deviation
ECNNB [19]	76.01%	78.50%	77.12%	0.78%
AECNNB	81.65%	83.84%	82.95%	0.555%

5.2.1 Analysis of the AECNNB on CIFAR-10

The AECNNB model's superior performance on CIFAR-10, when contrasted with the ECNNB's earlier results, reveals that thoughtful architectural modifications can indeed yield substantial improvements. While the ECNNB [19] model had an average accuracy of 77.12%, the AECNNB's average accuracy of 82.95% marks a considerable advancement, emphasizing the efficacy of our developmental strategies.

The consistency of the AECNNB model is also notable, as evidenced by a lower standard deviation in accuracy. This reduction indicates a more reliable model performance across multiple trials, a crucial factor for real-world applications.

The subsequent exploration of our research will involve the integration of CBAM and MobileViTv2 attention mechanisms, building upon the successes of the AECNNB architecture. We aim to investigate whether the incorporation of these mechanisms can provide additional gains in accuracy and model robustness, particularly for challenging datasets such as CIFAR-10.

5.3 Empirical assessment of AECNNB with CBAM on CIFAR-10

The experiment set out to explore the impact of the CBAM on the accuracy of a CNN model when applied to the CIFAR-10 dataset. The CBAM is engineered to afford the model an enhanced focus on salient features through both spatial and channel-wise attention, theoretically enabling a more nuanced understanding of complex image data. This model is built on top of the study done in [27]. We subjected the enhanced model to 30 trials over 20 generations to thoroughly assess its performance.

In stark contrast to the base CNN model, the integration of CBAM yielded an appreciable increase in accuracy across trials. The minimum accuracy observed was 83.63%, a substantial uplift from the base model's 76.01%, highlighting the attention mechanism's capability to mitigate the performance variance due to random initialization and epoch-to-epoch fluctuations. The maximum accuracy achieved was 85.38%, exceeding the base model's 78.50% peak, signifying the potential of CBAM to amplify the model's predictive prowess.

The aggregate performance of the AECNNB with CBAM is captured by an average accuracy of 84.39%, a significant improvement over the base model's 77.12%. This improvement corroborates the hypothesis that attention mechanisms are instrumental in enhancing model performance for complex classification tasks like those presented by the CIFAR-10 dataset.

5.3.1 Statistical and practical implications

Delving into the statistical nuances of our results, we observed the following outputs, illustrated in Table 4:

Table 4: Enhanced CNN performance with CBAM on CIFAR-10.

Model type	Min. accuracy	Max. accuracy	Average accuracy	Std. deviation
ECNNB [19]	76.01%	78.50%	77.12%	0.78%
AECNNB with CBAM	83.63%	85.38%	84.39%	0.508%

The decrease in standard deviation from the ECNNB's [19] 0.78% to the CBAM-enhanced CNN's 0.508% illuminates a reduction in performance dispersion, underscoring a higher reliability in the model's operation across varying conditions. The narrower spread of accuracies indicates a model that is not only capable of higher peak performance but also less prone to the vicissitudes of stochastic gradient descent and initialization variability.

The results offer compelling evidence of the CBAM's value in complex image classification tasks. The data suggest that the attention-enhanced model is not only capable of achieving superior accuracy but does so with a consistency that is vital for practical applications where predictability and reliability are paramount.

5.4 Examination of AECNNB with MobileVitv2 on CIFAR-10

Leveraging the MobileViTv2 attention mechanism, the CNN model's capabilities were extended to improve its focus on informative features, expected to result in higher classification accuracy on CIFAR-10. This model is built on top of the study done in [27]. Throughout 30 trials, each consisting of 20 generations, the modified CNN's performance was meticulously recorded.

The results indicate that the MobileViTv2 attention mechanism provides a robust improvement over the base model. The minimum accuracy achieved throughout the trials was 86.23%, which represents an advanced baseline when compared to the 76.01% minimum of the original CNN model. This minimum threshold's elevation reflects the MobileViTv2's ability to consistently enhance the model's baseline performance.

The maximum accuracy recorded was an impressive 87.69%, which notably exceeds the maximum accuracy of both the base CNN and the AECNNB with CBAM standing at 78.50% and 85.38% respectively. This peak is indicative of the MobileViTv2 architecture's potential to reach superior performance levels.

With an average accuracy of 86.89%, the CNN model equipped with MobileViTv2 outperforms the previous iterations, showcasing the significant impact of this attention mechanism. This average, alongside a relatively low standard deviation of 0.348%, suggests a strong and stable performance across different trials and conditions.

5.4.1 Performance metrics and analysis

Table 5 summarizes the enhanced performance brought about by the MobileViTv2 attention mechanism:

Table 5: Performance enhancement with MobileViTv2 on CIFAR-10.

Model type	Min. accuracy	Max. accuracy	Average accuracy	Std. deviation
ECNNB [19]	76.01%	78.50%	77.12%	0.78%
AECNNB with MobileVitv2	86.23%	87.69%	86.89%	0.348%

The addition of MobileViTv2 has not only raised the floor and ceiling of the model's accuracy but also centered the model's performance on a higher mean. The standard deviation reduction is indicative of the model's consistent performance across the trials, further affirming the reliability and effectiveness of the MobileViTv2 attention mechanism.

These results provide empirical evidence of MobileViTv2's ability to significantly bolster a CNN's accuracy in image classification tasks. This enhancement can be attributed to the mechanism's focus on optimizing the representational capacity of the network, allowing for a more profound learning of complex features within the CIFAR-10 dataset.

The statistical improvement observed with the MobileViTv2 attention mechanism emphasizes its potential for practical applications where high accuracy and consistency are crucial. Its ability to consistently output high-quality predictions makes it a valuable addition to the CNN framework for complex image classification challenges.

5.5 Discussion

In our experimental analysis conducted on the CIFAR-10 dataset, we evaluated a range of CNN architectures, each representing a distinct phase of advancement and complexity. These models included the ECNNB [19], an AECNNB without attention mechanisms, and more sophisticated versions integrating CBAM and MobileViTv2 attention mechanisms. The performance metrics of each model are presented in Table 6.

Table 6: Performance comparison of CNN models on CIFAR-10.

Model type	Min. accuracy	Max. accuracy	Average accuracy	Std. deviation
ECNNB [19]	76.01%	78.50%	77.12%	0.78%
AECNNB	81.65%	83.84%	82.95%	0.555%
AECNNB with CBAM	83.63%	85.38%	84.39%	0.508%
AECNNB with MobileVitv2	86.23%	87.69%	86.89%	0.348%

The ECNNB [19], originally designed for simpler, grayscale datasets, demonstrated modest performance on the CIFAR-10 dataset. In contrast, the AECNNB without attention mechanisms exhibited a significant improvement in accuracy, marking a 5.83% increase over the ECNNB. This enhancement highlights the impact of structural advancements in the model's design.

Further improvements were achieved with the integration of attention mechanisms. The AECNNB with CBAM outperformed the AECNNB by 1.44% and the ECNNB by 7.27%, indicating the effectiveness of the CBAM attention mechanism in enhancing the model's focus and feature extraction capabilities.

Table 7 illustrates the improvement in average accuracy of each CNN model compared to the other models:

Table 7: Output schema of the component repository.

Model type	Improvement over ECNNB (%)	Improvement over AECNNB (%)	Improvement over AECNNB with CBAM (%)
ECNNB [19]	0.00	N/A	N/A
AECNNB	5.83	0.00	N/A
AECNNB with CBAM	7.27	1.44	0.00
AECNNB with MobileVitv2	9.77	3.94	2.50

The most substantial performance enhancement was observed with the integration of the MobileViTv2 attention mechanism. The AECNNB with MobileVitv2 surpassed all other models, achieving an average accuracy of 86.89%, which is a 2.5% improvement over the AECNNB with CBAM and a 9.77% improvement over the ECNNB. This performance underscores the potential of MobileViTv2's design in optimizing the network's representational capacity for complex image classification tasks, offering greater efficiency and consistency.

In conclusion, our comparative analysis highlights a progressive improvement in model performance with each architectural enhancement, particularly with the integration of advanced attention mechanisms. The model equipped with MobileViTv2 emerged as the most efficient and accurate, underscoring its suitability for challenging image classification tasks that demand high performance and operational efficiency. This progression from basic architectures to sophisticated designs emphasizes the importance of continuous innovation in neural network development to effectively address the complexities of evolving image datasets.

6 Conclusion and future work

In this extended study focusing exclusively on the CIFAR-10 dataset, we have demonstrated the progressive enhancement of convolutional neural network (CNN) architectures, culminating in significant improvements in image classification performance. The evolution from ECNNB [19] structures to advanced models incorporating attention mechanisms like CBAM and MobileViTv2 highlights the critical role of architectural sophistication in handling complex datasets. Among these, the AECNNB with Mobile-Vitv2 attention mechanism emerges as the most effective model, showcasing superior accuracy and consistency in the CIFAR-10 classification task.

Further research should explore the potential of novel neural network architectures and attention mechanisms beyond the conventional CNN framework. This includes investigating architectures such as Transformer networks or advanced RNNs and experimenting with hybrid models combining different neural network types. The exploration of new attention mechanisms, inspired by the latest advancements in AI, could also yield significant improvements.

References

[1] "Calculus in Action: Neural Networks," https://machinelearningmastery.com/calculus-in-action-neuralnetworks/, Mar. 16 2022.

[2] W. M. Spears, K. A. De Jong, T. Back, D. B. Fogel and H. De Garis, "An Overview of Evolutionary Computation," in *Proc. Eur. Conf. Mach. Learning (Lecture Notes in Computer Science)*, vol. 667. Berlin, Germany: Springer-Verlag, pp. 442–459, Apr. 1993.

[3] J. Shapiro, "Genetic Algorithms in Machine Learning," in *Machine Learning and Its Applications*, G. Paliouras, V. Karkaletsis and C. D. Spyropoulos, Ed., vol. 2049. Berlin, Heidelberg: Springer, pp. 146–168, 2001, doi: 10.1007/35404467377.

[4] H. Fu, G. Song and Y. Wang, "Improved YOLOv4 Marine Target Detection Combined with CBAM," *Symmetry*, vol. 13, no. 4, p. 623, 2021.

[5] Y. Chen, X. Zhang, W. Chen, Y. Li and J. Wang, "Research on Recognition of Fly Species Based on Improved RetinaNet and CBAM," *IEEE Access*, vol. 8, no. 2020, pp. 102907–102919.

[6] Y. Luo and Z. Wang. "An Improved Resnet Algorithm Based on CBAM." In 2021 International Conference on Computer Network, Electronic and Automation (ICCNEA), pp. 121–125. IEEE, 2021.

[7] Y. Zhang, H. Wang, J. Liu, X. Zhao, Y. Lu, T. Qu, H. Tian, J. Su, D. Luo and Y. Yang, "A Lightweight Winter Wheat Planting Area Extraction Model Based on Improved DeepLabv3+ and CBAM," *Remote Sensing*, vol. 15, no. 17, p. 4156, 2023.

[8] R. Ma, J. Wang, W. Zhao, H. Guo, D. Dai, Y. Yun, L. Li, F. Hao, J. Bai and D. Ma, "Identification of Maize Seed Varieties Using MobileNetV2 with Improved Attention Mechanism CBAM," *Agriculture*, vol. 13, no. 1, p. 11, 2022.

[9] C. Yuan, T. Liu, F. Gao, R. Zhang and X. Seng, "YOLOv5s-CBAM-DMLHead: A Lightweight Identification Algorithm for Weedy Rice (Oryza Sativa F. Spontanea) Based on Improved YOLOv5," *Crop Protection*, vol. 172, no. 2023, pp. 106342.

[10] L. Lin, J. Zhang, X. Gao, J. Shi, C. Chen and N. Huang, "Power Fingerprint Identification Based on the Improved VI Trajectory with Color Encoding and Transferred CBAM-ResNet," *Plos One*, vol. 18, no. 2, p. e0281482, 2023.

[11] B. Chen and Z. Dang. "Fast PCB Defect Detection Method Based on FasterNet Backbone Network and CBAM Attention Mechanism Integrated with Feature Fusion Module in Improved YOLOv7." IEEE Access, 2023.

[12] J. Liu, H. Qiao, L. Yang and J. Guo, "Improved Lightweight YOLOv4 Foreign Object Detection Method for Conveyor Belts Combined with CBAM," *Applied Sciences*, vol. 13, no. 14, p. 8465, 2023.

[13] L. Miao, N. Li, M. Zhou and H. Zhou. "CBAM-Yolov5: Improved Yolov5 Based on Attention Model for Infrared Ship Detection." In International conference on computer graphics, artificial intelligence, and data processing (ICCAID 2021), vol. 12168, pp. 564–571. SPIE, 2022.

[14] M. Munir, W. Avery and R. Marculescu. "MobileViG: Graph-Based Sparse Attention for Mobile Vision Applications." In Proceedings of the IEEE/CVF Conference on Computer Vision and Pattern Recognition, pp. 2210–2218, 2023.

[15] G. Yang, Z. Qin, J. Mu, H. Mao, H. Mao and M. Han, "Efficient Diagnosis of Hematologic Malignancies Using Bone Marrow Microscopic Images: A Method Based on MultiPathGAN and MobileViTv2," *Computer Methods and Programs in Biomedicine*, vol. 237, p. 107583, 2023.

[16] X. Cao, Y. Su, X. Geng and Y. Wang. "YOLO-SF: YOLO for Fire Segmentation Detection." IEEE Access, 2023.

[17] Z. Lv, Y. Li, S. Qian and L. Wu, "Online Surface Defect Segmentation on Aluminum Strip Production Line Using a Lightweight and Efficient Model.," *Engineering Applications of Artificial Intelligence*, vol. 126, no. 2023, pp. 107023.

[18] Z. Zhang, L. Xiang and S. Cao, "An Efficient Detection Model Based on Improved YOLOv5s for Abnormal Surface Features of Fish.," *Mathematical Biosciences and Engineering*, vol. 21, no. 2, p. 1765–1790, 2024.

[19] A. Shams, D. Becker, K. Becker, S. Amirian and K. Rasheed, "Evolving Efficient CNN Based Model for Image Classification." In Proceedings of the 2023 International Conference on the World Congress in Computer Science, Computer Engineering, and Applied Computing (CSCE'23), July 24-27, 2023, Luxor (MGM), Las Vegas, Nevada, USA. IEEE Computer Society, 2023. Hamid R. Arabnia, Leonidas Deligiannidis, Fernando G. Tinetti, and Quoc-Nam Tran (Eds.). ISBN: 979-8-3503- 2759-5, IEEE Catalog Number: CFP23UB2-USB, BMS Part #: CFP23UB2-USB. DOI: 10.1109/CSCE60160.2023.00041. p. 228.

[20] L. Deng, "The MNIST Database of Handwritten Digit Images for Machine Learning Research [Best of the Web]," *IEEE Signal Process Magazine*, vol. 29, no. 6, pp. 141–142, Nov. 2012.

[21] Y. LeCun, C. Cortes and C. J. Burgess, "MNIST Handwritten Digit Database," ATT Labs Online, https://www.tensorflow.org/datasets/catalog/mnist, 2, 2010.

[22] G. Cohen, S. Afshar, J. Tapson and A. Van Schaik, "EMNIST: An Extension of MNIST to Handwritten Letters," 2017. [Online]. Available: arXiv:1702.05373.

[23] L. Deng, "The MNIST Database of Handwritten Digit Images for Machine Learning Research [Best of the Web]," *IEEE Signal Process Magazine*, vol. 29, no. 6, pp. 141–142, Nov. 2012.

[24] S. Woo, J. Park, J.-Y. Lee and I. So Kweon. "Cbam: Convolutional Block Attention Module." In Proceedings of the European conference on computer vision (ECCV), pp. 3–19, 2018.

[25] A. Vaswani, N. Shazeer, N. Parmar, J. Uszkoreit, L. Jones, A. N. Gomez, Ł. Kaiser and I. Polosukhin, "Attention Is All You Need.," *Advances in Neural Information Processing Systems*, vol. 30, 2017.
[26] S. Mehta and M. Rastegari. "Separable Self-attention for Mobile Vision Transformers." arXiv preprint arXiv:2206.02680, 2022.
[27] https://github.com/xmu-xiaoma666/External-Attention-pytorch
[28] Nagadomi, "Kaggle CIFAR-10," 2014. [Online]. Available: https://github.com/nagadomi/kaggle-cifar10-torch7.
[29] A. Krizhevsky and G. Hinton, ""Learning Multiple Layers of Features from Tiny Images," p. 7, 2009.
[30] D. Zhang, Y. Ding, P. Chen, X. Zhang, Z. Pan and D. Liang, "Automatic Extraction of Wheat Lodging Area Based on Transfer Learning Method and Deeplabv3+ Network," *Computers & Electronics in Agriculture*, vol. 179, no. 2020, pp. 105845.
[31] G. Jocher, K. Nishimura, T. Mineeva and R. Vilariño. "YOLOv5." GitHub repository, 2020. https://github.com/ultralytics/yolov5.
[32] C.-Y. Wang, A. Bochkovskiy and H.-Y. Mark Liao. "YOLOv7: Trainable Bag-of-freebies Sets New State-of-the-art for Real-time Object Detectors." In Proceedings of the IEEE/CVF Conference on Computer Vision and Pattern Recognition, pp. 7464–7475, 2023.
[33] A. Bochkovskiy, C.-Y. Wang and H.-Y. Mark Liao, "Yolov4: Optimal Speed and Accuracy of Object Detection.," *arXiv Preprint arXiv:2004.10934*, 2020.
[34] D. P. Kingma and B. Jimmy, ""Adam: A Method for Stochastic Optimization," *arXiv Preprint arXiv:1412.6980*, 2014.

Convolutional neural network (CNN)

Md Mahmudur Rahman and Bikesh Regmi

Multi-label concept detection in imaging entities of biomedical literature leveraging deep learning-based classification and object detection

Abstract: Scientific articles in the biomedical domain convey important information for clinical purpose, research, and education, using both text and image modalities. Numerous illustrations in the form of charts, graphs, diagrams, and photos are frequently used along with text in those articles. Authors also often use overlaid annotation markers using different arrows, letters, or symbols to highlight Region of Interests (ROIs) or specific medical concepts. There are many cases, where a single image depicts multiple ROIs or concepts based on assignment of multiple markers/pointers. This work presents such a proof-of-concept (POC) experiment leveraging deep learning for classifying chest CT images with multi-concept labels and detecting ROIs as concepts appeared in biomedical articles. This study conducts multi-class experiments in a dataset of annotated concepts with Convolutional Neural Networks (CNNs) and Vision Transformers (ViTs) as deep learning (DL) models. Three different DL-based object detection techniques are also applied in the annotated dataset with the goal to generate a bounding box around each ROI in the image and identify it with a concept label. We achieved around 70% micro average precision and recall accuracies in a test set for multi-label classification. Our results also demonstrate that all three object detection techniques achieved high accuracies in recognizing and localizing the ROIs in biomedical images, whereas YOLOv7 exhibited the highest precision of 92.5%, indicating its ability to accurately identify ROIs. Overall, this study demonstrates the effectiveness of DL models in concept detection in biomedical images and establishes the feasibility and rationale of the POC.

Keywords: software component, retrieving, storing component, repository, retrieving techniques, component reuse

---

Acknowledgment: The work is supported by an NSF grant (#2131207), entitled, "CISE-MSI: DP: IIS: III: Deep Learning-Based Automated Concept and Caption Generation of Medical Images Towards Developing an Effective Decision Support System"

Md Mahmudur Rahman, Computer Science Department, Morgan State University, Maryland, USA, e-mail: md.rahman@morgan.edu
Bikesh Regmi, Computer Science Department, Morgan State University, Maryland, USA, e-mail: bireg1@morgan.edu

https://doi.org/10.1515/9783111344126-007

1 Introduction

A wide variety of users, such as medical residents and clinicians as well as patients, use tools to search for relevant information from biomedical literature (e.g., PubMed). However, due to the rapidly evolving and increasing volume of literature in the healthcare domain, it is challenging to search for information in the right place at the right time [1–4]. Scientific articles in the biomedical domain are multimodal (text and image) in nature where authors frequently use figures (e.g., images, graphs, diagrams, etc.) to clarify or elucidate the text. Moreover, authors also use different arrows or symbols as overlaid markers in images to highlight specific portion of images as Region-of-Interest (ROI) [2].

However, in the search process little attention is devoted to the use of images in the articles due to the difficulty of image understanding and comprehension [5–7]. A majority of the existing search tools retrieves images related to the query topic by looking at the associated image captions only, while completely ignoring image contents. However, the importance of illustrations and figures in improving retrieval of literature is well-established, where investigators examined the possibility of integrating information derived directly from image data [8–10]. As a result, content-based visual image retrieval has gained significant popularity during the past three decades [11, 12].

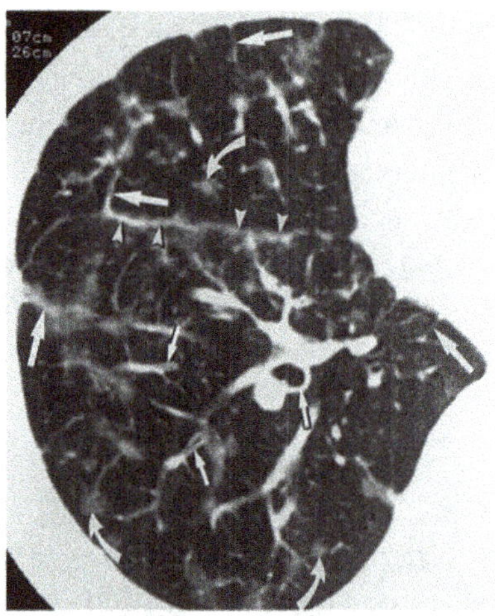

Caption: Hypereosinophilic syndrome in a 59-year-woman. Transverse thin-section CT scan (1-mm collimation) through the right lower lung demonstrates bronchial wall thickening (small straight arrows), interlobular septal thickening (large straight arrows), faint centrilobular nodules (curved arrows), patchy areas with ground-glass attenuation, and thickening of the interlobar fissure (arrowheads)

Figure 1: Example chest CT image with caption in a biomedical journal article.

Moreover, it is very common practice in biomedical literature to use overlaid arrows, pointers, and text labels as annotation markers on figures to highlight ROIs, which are associated and correlated with concepts (such as UMLS) in the caption text or figure mentions in the article text. For example, Figure 1 shows a chest CT image along with caption (highlighted with different colors for better visualization) where different arrows (e.g., small straight arrows, large straight arrows, curved arrows, arrowhead) point to different visual patterns as concepts, such as bronchial wall thickening, septal thickening, ground-glass attenuation, etc. [13]. For example, an area of a computed tomography (CT) scan having a mass of hexagonal wax cells-like structure can be mapped to the pattern "***honeycombing***" and slightly bright and hazy appearances can be mapped to the pattern "***ground-glass opacity.***" Hence, identifying and localizing the ROIs with different concepts (such as the Unified Medical Language System (UMLS) [14] or RadLex [15]) categories might be an important step towards reducing the semantic gap in the search process.

This work presents a proof-of-concept experiment by annotating ROIs and classifying 200 Computed tomography (CT) images of the chest with multi-label concepts, which appear in biomedical articles obtained through a large ImageCLEFmed benchmark collection [17]. The images are classified into 11 different concept categories where each image may contain one or more concept categories suitable for multi-class and multi-label task. To demonstrate the utility, we trained several CNNs [18–20] and Visual transformer (ViT) [21–23]-based classifiers to automatically assign thoracic imaging concepts to image regions based solely on their appearance. In addition, we extended our previous work [24] by incorporating DL-based object detection by manually annotating ROIs with several concept categories such as ground-glass, bronchi, honeycomb, cyst, etc. For this purpose, an easy to use graphical image annotation tool (LabelImg [25]) is utilized, which is written in Qt library of Python. The annotations are saved as XML files in PASCAL VOC format and in the format to support the YOLO-based implementation. The experiments were performed using three DL-based object detection techniques, Mask R-CNN [26], YOLOv5 [27], and YOLOv7 [28] in the annotated dataset with the goal of drawing a bounding box around each ROI in the image and assigning them a class (concept) label. The methods are evaluated in a small gold standard set of annotated image regions and descriptions with encouraging results, which are the first steps towards the creation of a visual ontology of biomedical imaging entities. The goal here is to demonstrate the feasibility and potential utility of using DL-based multi-label classification and object detection models for assisting medical professionals in extracting relevant information from biomedical images.

In this chapter, DL-based multi-label classification and object (ROI) detection are described. The related works are described in Section 2. The proposed multi-label classification and concept detection are explained in Sections 3 and 4. The experiments and result analysis are described in Sections 5 and 6 respectively. At the end, we provide the conclusion and future work in Section 7.

2 Related work

In the biomedical domain, Deep Neural Networks (DNN)-based models are predominant with state-of-the art performances in various medical image processing jobs, such as classification, retrieval, segmentation, and object detection [29]. Several studies showed that the use of DL can significantly improve the performance of Computer Aided Detection (CAD) systems, such as COVID-19 detection in X-ray and CT images [30, 31], interstitial lung disease (ILD) classification [32], skin cancer classification of dermoscopic images [33], breast cancer detection in mammograms [34]. For example, CheXNet [35] makes use of a Dense Convolutional Network (DenseNet) architecture with 121 layers and trains in ChestX-ray14 dataset, which consists of 112,120 frontal-view X-ray images from 30,805 distinct patients. Based on radiology reports, each patient's X-ray image is labeled with one of 14 thoracic diseases. The Medical Vision Transformer (MVT) model [36] was developed to address skin cancer classification tasks. It leverages the Vision Transformer (ViT) architecture with a multi-layer perceptron for the top classification layer. With 10,015 images in seven different classes from the Human Against Machine (HAM10000) dataset, the suggested model showed excellent results in metrics. In addition, several methods have been proposed during the past decade, where DL-based techniques are used for image feature extraction and retrieval. For example, content-based medical image retrieval (CBMIR) systems using deep CNN-based high-level and rich features are proposed by Cai et al. [37] and Qayyum et al. [38], whereas a multimodal image retrieval system is proposed by Vikram et al. [39] based on early fusion of deep autoencoder and modified VGG-16 based deep features as well as late fusion using ensemble technique.

Consequently, there is a considerable need for automatic methods, which enable physicians to focus on interesting image regions or to describe findings as condensed textual descriptions from visual information. Due to the recent success of DL-based methods on automatic image captioning in natural images, researchers are also being motivated to use similar techniques for medical image interpretation or caption generation for the past few years [40–42]. In general, a standard encoder-decoder recurrent neural network (RNN) and more recently transformers-based architectures are used to address the image caption generation problem. DL-based models also require a large amount of training data to avoid overfitting problems and to improve the generalizability of the model. Although training the model from scratch using domain dataset can improve the performance, existing publicly available datasets for medical image captioning are limited in number and rather noisy. For example, the yearly ImageCLEFCaption [43] benchmark campaign contains a concept detection task with an aim of automatically detecting concepts by leveraging the clinical concept vocabulary (e.g., UMLS Concept Unique Identifiers) from the images and a caption prediction task to predict a precise and coherent text caption for the images in the test data set [44]. We have participated in this campaign for the last few years in different capacities where our methods were evaluated and compared with other research groups around the globe [45, 46].

3 Multi-label classification of concepts

In the domain of scientific literature, AI/ML techniques are often used to classify the contents of articles to categories from a set of a priori known categories based on binary classification, multi-class classification, or multi-label classification. In general, binary and multi-class classification approaches are widely used in several downstream Natural Language Processing (NLP) and Computer Vision tasks, such as document categorization, sentiment analysis, image classification, object detection, etc. However, many of real-life data sets (such as the image in Figure 1) are too complex to impose the restriction of only one category or label for each data instance. The main difference between multi-class and multi-label classification is that the classes are mutually exclusive for the first one, whereas multiple non-mutually exclusive labels may be assigned to each instance for the later one. The multi-label classification problem can be implemented in two ways using transformation methods or algorithm adaptation methods [47]. The transformation methods transform the multi-label problem into a set of binary classification problems, which are mainly handled using single-class classifiers, whereas the algorithm adaptation methods try to address the problem in its full form for directly performing multi-label classification by adapting the algorithms [48].

This work focuses on the first approach of multi-label classification (e.g., detection task is divided into a series of multiple binary classification problems) methods using both deep CNNs and ViTs. Since, it is not feasible to use the standard ***LabelBinarizer*** class for multi-class classification, the scikit-learn library's ***MultiLabelBinarizer*** class is used, which transforms the labels of each image to a vector with a total 11 categories where one-hot encoding transforms categorical labels from a single integer to a vector. The concept labels of images in the dataset are extracted from a manually annotated CSV file and added in a list.

3.1 Convolutional Neural Networks (CNNs) and Vision Transformers (ViTs)

When examining CNNs and transformers, or more specifically ViTs, in medical image processing, noticeable architectural differences are evident. CNNs use convolutional layers to analyze spatial structures in images, extracting local details and combining them to build a holistic understanding of the image [16]. ViTs use self-attention mechanisms to analyze connections between different parts of an image at the same time, catching distant interconnections and contextual details [21].

Over the past decade, CNNs have shown state-of-the-art performances in different medical image analysis tasks, such as disease classification, tumor segmentation, and lesion detection [29]. CNNs are architecturally designed with multiple types of layers, each serving a unique purpose in the feature extraction and classification process. The convolutional layer, a core component, utilizes filters to perform convolution operations that capture spatial features from the input images [16]. Following the convolutional layer, the pooling layer reduces the dimensionality of the data, helping to decrease computational complexity and overfitting by summarizing the features. Normalization layers, such as batch normalization, are employed to stabilize and accelerate the training process by normalizing the inputs of each layer. Finally, fully connected layers integrate the learned features from previous layers to perform classification [16]. These layers collectively enable the network to learn from data in both supervised and unsupervised settings, efficiently mapping input images to their corresponding categories by progressively learning more complex feature representations. Through this hierarchical learning process, CNNs can effectively detect and utilize discriminative features for accurate image classification, reconstruction, object-detection, and text generation.

Three well-known and popular CNN architectures, such as Xception [18], DenseNet-121 [19], and ResNet-50 [20] are experimented in this study by training the dataset both from scratch and using transfer learning (TL) with fine-tuning approach. The training of the CNNs is performed by minimizing a loss function using gradient descent-based methods and backpropagation of the error with the following configuration:

- Number of nodes in the output layer matches the number of labels.
- Keep activation function of the classification (output) layer in our models to sigmoid, which enables to perform multi-label classification with *Keras*.
- Use ***binary cross-entropy*** loss function instead of the commonly used ***categorical cross-entropy*** to treat each output label as an independent Bernoulli distribution and to penalize each output node independently.

Transformers are a type of DL architecture, based primarily upon the self-attention module, which were originally proposed for language translation task in NLP. Thanks to their superior ability to model long-distance relationships between sequence elements and their parallel processing capabilities, which set them apart from recurrent networks like Long Short-Term Memory, Transformer models have had a significant impact on the field of computer vision [21]. Their architecture enables flexible handling of various data types – images, videos, text, and speech – using uniform components, free from the predefined assumptions inherent in convolutional networks [21–24]. With its straightforward structure and versatility, transformer networks have enabled effective scaling for big datasets and high network capacities, leading to notable advancements in vision-related applications.

Recent works have shown that transformers can fully replace the standard convolutions in DL networks by operating on a sequence of image patches, giving rise to ViTs [22]. These ViT models continue the long-lasting trend of removing handcrafted visual features and inductive biases from models to leverage the availability of larger datasets coupled with increased computational capacity. Being inspired by the success of ViTs in computer vision and its application in medical imaging fields in recent years [24], this work also experimented with different ViT models trained from both scratch and fine-tuned for classification using transfer learning.

The core architecture of the Vision Transformer is based on a sequence of encoder layers. The two primary components of each encoder layer are feedforward neural networks and multi-head self-attention [21, 22]. The multi-head self-attention module computes a weighted sum of all patch embedding, where the weights are assigned based on how relevant each patch is to the others. This allows the module to capture the complex interactions between various input patches. In addition to highlighting the important patches, this method connects local and global contexts by using several attention heads that reveal different kinds of relationships. The output of each patch's self-attention phase is then processed via a feedforward neural network, which adds nonlinearity to the system by combining activation functions such as ReLU with fully connected layers. This allows the model to identify intricate relationships between patches. In addition, residual connections and layer normalization are added to each encoder layer to improve learning and model performance by stabilizing training, promoting gradient flow, and reducing the vanishing gradient issue [21].

For this work, the set-up of the ViT encoder consists of a patch encoder that receives an input of 224 × 224 images and produces a dense projection of the patches and a positional embedding representative of each patch's locality. Overlapping patches of 14 × 14 x 3 are obtained from the image (e.g., 256 patches per image) and are flattened into a 588-D array. The patch encoder layer linearly transforms a patch by projecting it into a vector of size projection dim, 64-D in the experiment.

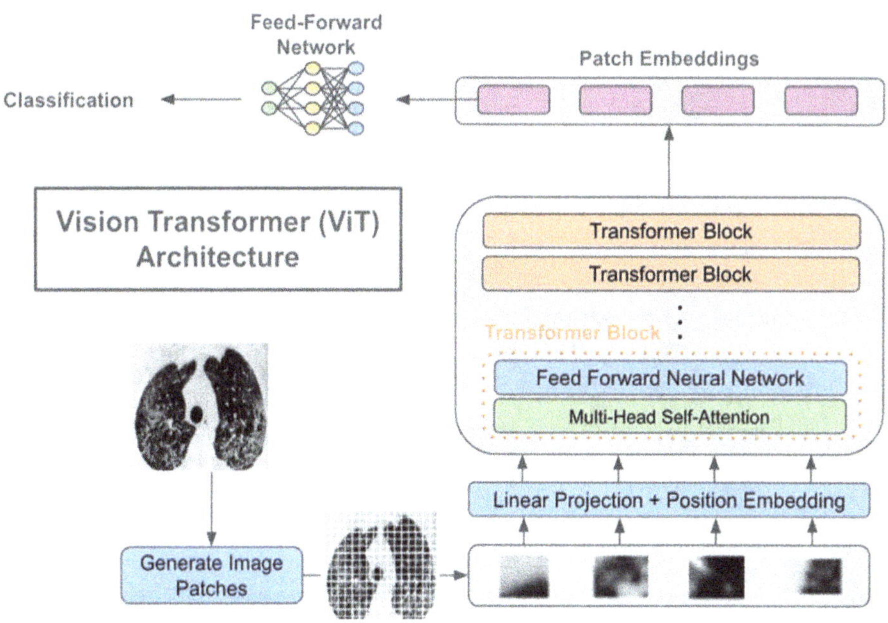

Figure 2: Process flow diagram for ViT-based classification.

As shown in Figure 2, the patch projections and positional embedding are fed into the encoder stack with multi-head attention layer that provides the attended representation of the features, a skip connection, and an intermediate dense layer that projects the visual feature representation into the specified dimension size. The transformer blocks produce a [batch_size = 8, num_patches = 256, projection_dim = 64] tensor, which is processed via the classifier head with sigmoid (like CNN classification) to produce the final class probabilities output.

4 Concept detection as region-of-interests (ROIs)

Identifying medical concepts as ROIs is similar to problem of object detection in computer vision where object localization involves drawing a bounding box around one or more objects (e.g., ROIs) in an image. Image classification discussed in the previous section is usually the straightforward application of computer vision but identifying concepts as ROIs in the image unravels so many endless possibilities. Object detection is more challenging compared to image classification where the detection model typically outputs the bounding box, confidence score, and class. The confidence value is the model's confidence in the detection, which ranges between 0 and 1 [49].

For the past several years, DL-based algorithms have shown great promise in improving the accuracy of object detection in recent years [49–51]. One of the pioneer methods is the Region-based Convolutional Neural Networks (R-CNNs) first introduced around 2012, which are a family of techniques (e.g., fast R-CNN, faster R-CNN, Mask R-CNN) for addressing object localization and recognition tasks, which operate in two stages [52]. In the first stage, potential object proposals are generated using a selective search algorithm. In the second stage, a deep CNN is trained to classify the object proposals and refine their bounding boxes. R-CNN has been widely used in computer vision tasks and has been shown to achieve high accuracy.

Mask R-CNN: extends faster R-CNN by adding a branch for predicting an object mask in parallel with the existing branch for bounding box recognition [52]. On the other hand, YOLO (You Only Look Once), is a second family of techniques designed for speed and real-time use with state-of-the-art accuracy in recent years [53, 54]. In these techniques, there is no region creation and then again processing on top of that, there is one convolution network that creates boxes and class predictions for each box as one stage detector.

YOLOv5 (You Only Look Once version 5) is a popular real-time object detection algorithm that uses a single neural network to perform both object detection and classification [53]. YOLOv5 operates in a single stage, where the network directly predicts bounding boxes and class probabilities for all objects in the image. YOLOv5 has shown impressive performance on a variety of object detection tasks and has been widely used in industry and research.

YOLOv7 is an extension of the YOLOv5 algorithm, developed by the authors of the study [54]. YOLOv7 uses a larger and more complex neural network architecture than YOLOv5, with more layers and greater capacity. YOLOv7 also incorporates additional training techniques, such as data augmentation and progressive resizing, to improve the performance of the model. YOLOv7 is a state-of-the-art object detection algorithm that has shown promising results in various benchmarks.

Along with multi-label classification, this study aims to experiment with the above three different DL-based object detection techniques in our proof-of-concept experiment, where ROIs of 200 Chest CT images were annotated using an online tool, LabelImg, and several patterns such as ground-glass, bronchi, honeycomb, cyst, etc. were categorized based on their associated captions.

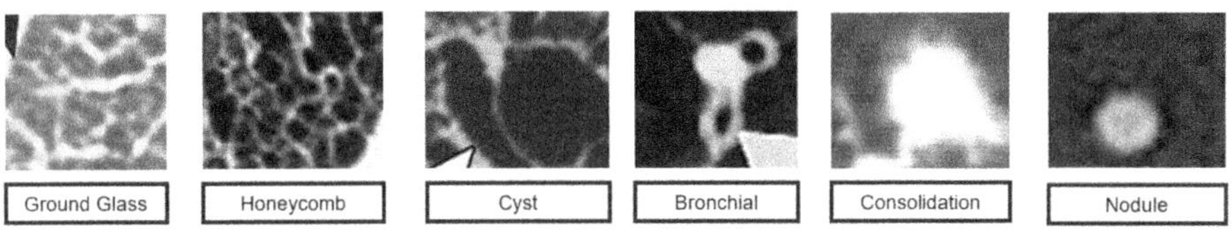

Figure 3: Sample concepts in lung CT images.

```
ct_labels

[['bronchiectasis', 'tree-in-bud'],
 ['bronchiectasis', 'nodules'],
 ['bronchiectasis', 'tree-in-bud'],
 ['cyst', 'honeycombing', 'bronchiectasis'],
 ['bronchiectasis'],
 ['bronchiectasis', 'mosaic'],
 ['bronchiectasis', 'tree-in-bud'],
 ['ground-glass'],
 ['ground-glass'],
 ['bronchiectasis', 'ground-glass'],
 ['bronchiectasis', 'cyst', 'consolidation'],
 ['bronchiectasis'],
 ['bronchiectasis'],
 ['bronchiectasis'],
 ['bronchiectasis'],
 ['bronchiectasis'],
 ['nodules', 'bronchiectasis', 'septal-thickening'],
 ['bronchiectasis'],
 ['bronchiectasis', 'mosaic'],
 ['bronchiectasis', 'ground-glass'],
 ['bronchiectasis', 'ground-glass', 'nodules'],
 ['bronchiectasis', 'ground-glass'],
 ['ground-glass', 'bronchiectasis', 'septal-thickening', 'consolidation'],
 ['bronchiectasis', 'septal-thickening', 'nodules', 'ground-glass'],
 ['bronchiectasis', 'septal-thickening', 'consolidation'],
 ['bronchiectasis', 'ground-glass'],
 ['bronchiectasis', 'ground-glass'],
```

Figure 4: Sample multi-label annotation of images in the CSV file.

5 Experiments

To validate the assumptions of the proof-of-concept, we experimented with a manually annotated ground truth dataset of 200 lung CT images with 11 different concept labels (Figure 4), which is a subset of images under a much larger ImageCLEFmed benchmark dataset [17]. However, these images are always low-resolution compared to their clinical counterparts, in varying sizes and lightning conditions; moreover the dataset is highly imbalanced where a few concept categories (patterns) occur more frequently compared to other less frequent ones. Hence, although currently smaller, the dataset still might be considered as a realistic set for evaluating medical image classification and retrieval techniques for images in biomedical articles.

The multi-label annotation of different concept categories is saved in a CSV file (where each image is associated with one or more labels (Figures 3 and 4)).

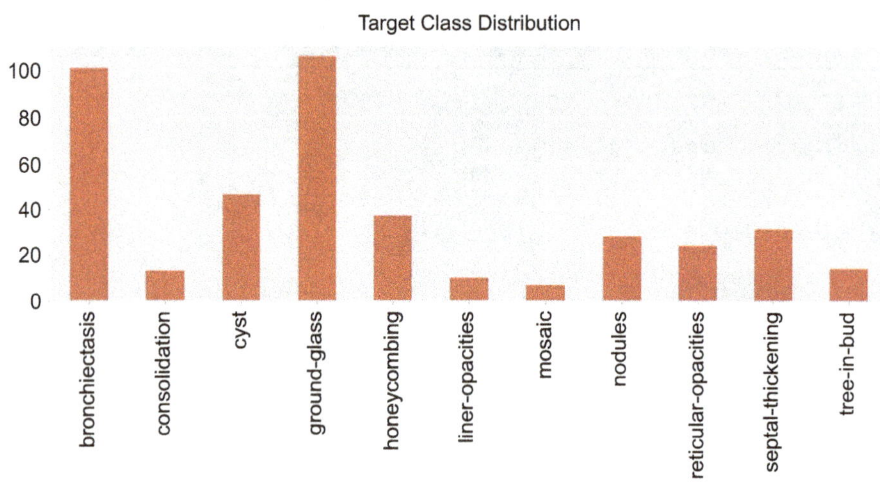

Figure 5: Distribution (frequency) of 11 concept categories.

Figure 5 shows the frequency distribution of 11 different concept categories in the CSV file. It is observed that around one-fourth images (out of 200) exist in this dataset, which contain only a single label (category), such as 18 images with a label "cyst," 16 images with label "bronchiectasis," and 14 images with "ground-glass" label. Since, it is a small dataset currently, almost half of the multi-labels occurred only once in the dataset, thus making the training and model generation very difficult.

An annotation tool, LabelImg [25] is also used to annotate the ROIs (with coordinates information) and annotations are saved as XML files in PASCAL VOC format and YOLO text file format for object (ROI) detection, such as different versions of R-CNN and YOLO algorithms. Figure 6 shows the interface of LabelImg tool while annotating a sample image with different ROIs with concept labels in the dataset.

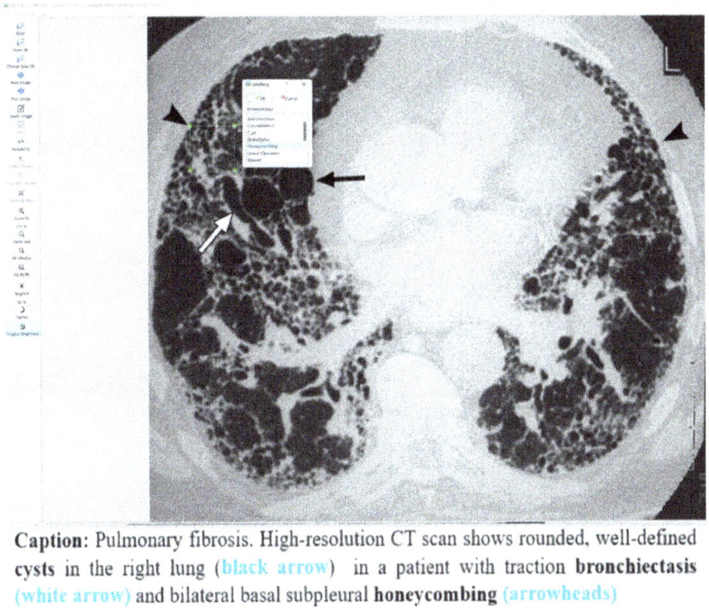

Figure 6: Sample annotation of ROIs of an image with associated caption using LabelImg tool.

5.1 Preprocessing with image augmentation

All dataset images are resized to 224 × 224 (except 299 × 299 for Xception model) pixels scaling the raw pixel intensities to the range [0, 1] and stored as NumPy arrays. After that, labels are binarized for multi-class classification by utilizing the scikit-learn library's *MultiLabelBinarizer* class, which actually transforms the concept labels into a vector that encodes those concepts that are present in the image. The high imbalance in the label frequency results in a huge bias towards the multi-label classification problem. Hence, data augmentation (scaling, rotation, flipping, etc.) is also applied while training as we have only a handful of images per concept class. The images are randomly rotated (25 degrees), horizontally and vertically shifted by a factor of 0.2, sheared by 0.2, and randomly horizontally flipped.

The goal of applying data augmentation is to increase the generalizability of the model. Applying a (small) amount of these transformations to an input image will change its appearance slightly, but it does not change the class label – thereby making data augmentation a very natural and easy method to apply to deep learning for computer vision tasks. The dataset is divided into random training (80%) and testing (20%) subsets where different accuracies are measured in the testing sets to compare different models and feasibility of the classification.

5.2 Model training

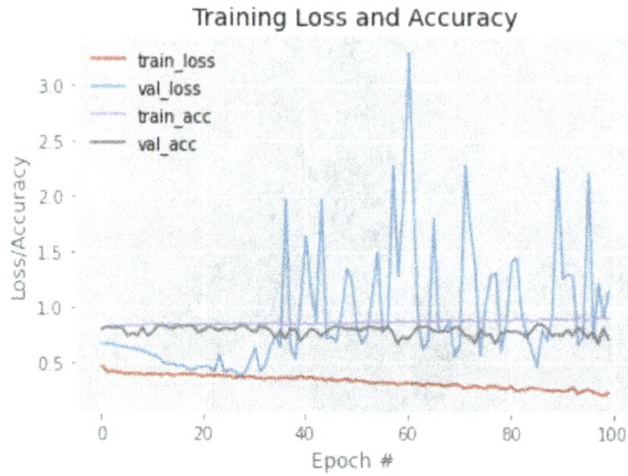

Figure 7: Training/validation loss and accuracy curves for the Xception model.

All the models (CNNs and ViTs) are built by initializing the Adam optimizer and compiled using binary cross-entropy rather than categorical cross-entropy to treat each output label as an independent Bernoulli distribution where the labels are not disjoint. After training is complete, the models and label binaries are saved to disk and loaded later during prediction in the test set. For training of the models from scratch, a learning rate = 0.001 and for pre-trained models a learning rate = 0.0001 is used and all the models are trained with 100 epochs with batch size = 8. Figure 7 shows the loss and accuracy curves for train and validation sets while training a CNN model from scratch.

6 Result analysis

For evaluating the classification performances of different models, measuring simple accuracy is not sufficient when working with a class-imbalanced data set, like this one, where there is a significant disparity between the class labels. Hence, aggregate metrics like ***macro, micro, weighted,*** and ***sampled avg.*** are calculated as they give us a high-level view of how the models are performing.

The low avg. accuracies (in the range of 45–65%) are because the dataset size is currently small and there is not simply enough representation of different concept labels in these low-resolution and highly varied images. As can be observed in Figure 7, the classifier even obtained zero (0) ***precision, recall,*** and ***F1-scores*** for three concept labels (e.g., consolidation, linear opacities, and mosaic patterns).

Table 1: Accuracy in test set for different CNN and ViT models configurations.

Method	Micro_avg. precision	Micro_avg. recall	Micro_avg. F1-score	Weighted_ avg._F1-score
DenseNet-121 (Scratch)	0.70	0.44	0.54	0.52
DenseNet-121 (pre-trained)	0.53	0.48	0.50	0.34
Xception (Scratch)	**0.71**	0.52	**0.60**	**0.57**
Xception (Pre-trained)	0.56	0.31	0.40	0.37
ResNet-50 (Scratch)	0.59	0.37	0.46	0.33
ResNet-50 (Pre-trained)	0.59	0.47	0.52	0.41
ViT (Scratch)	0.65	0.37	0.47	0.40
ViT_B_16 (pre-trained)	0.30	**0.70**	0.42	0.52
ViT_l_32 (pre-trained)	0.35	0.66	0.46	0.53

Table 1 shows the aggregate metrics, such as micro avg. precision, recall, and F-scores and weighted avg., F1-scores for different classifiers based on using CNN and ViT models and training both from scratch and fine tuning with TL. For pre-trained ViTs, both the ***ViT-Small*** model (ViT-B/16) and ***ViT-Large*** model (ViT-L/32) from original paper [20] are used. It is observed from Table 1 that Xception model (scratch) performed better compared to other models in terms of micro avg. precision, and micro- and weighted avg. F1-scores. In addition, it seems the pre-trained ViTs achieved good avg. recalls; however their precisions are very low (30–35%) compared to other models. The quality of the model is affected by several factors, such as architecture choices, learning rate schedule, optimizer, weight decay, etc. [20]. In practice, fine-tuning a ViT model that was pre-trained using a large, high-resolution dataset is always recommended. Overall, the accuracy (precision, recall, and F1-scores) in the range of 60–70% are satisfactory considering all other facts related to the problem domain, types of images, and current small dataset size.

The class-wise multi-label confusion matrix is also generated (Figure 8) using scikit-learn library to evaluate the accuracy of the classification and output confusion matrices for each concept class. The output of the confusion matrices in Figure 8 also confirmed the reason of low accuracies (Figure 9) for certain class labels, such as linear opacities, mosaic etc.

Figure 10 shows the classification probabilities of different class labels, and top two labels with associated probabilities are overlaid in a sample test image in the original ImageCLEFmed dataset with associated caption "CT scan at the level of the upper lobes in a 26 year-old woman demonstrates mild to moderate signs of bronchiectasis and peribronchial wall thickening. Mosaic perfusion, bullae (straight arrows), emphysema (*), and an area of consolidation (curved arrow) are also seen" [17]. From the output we can figure out that it correctly predicted *"bronchiectasis"* and *"mosaic"* patterns and confused probably *"consolidation"* with *"tree-in-bud"* pattern.

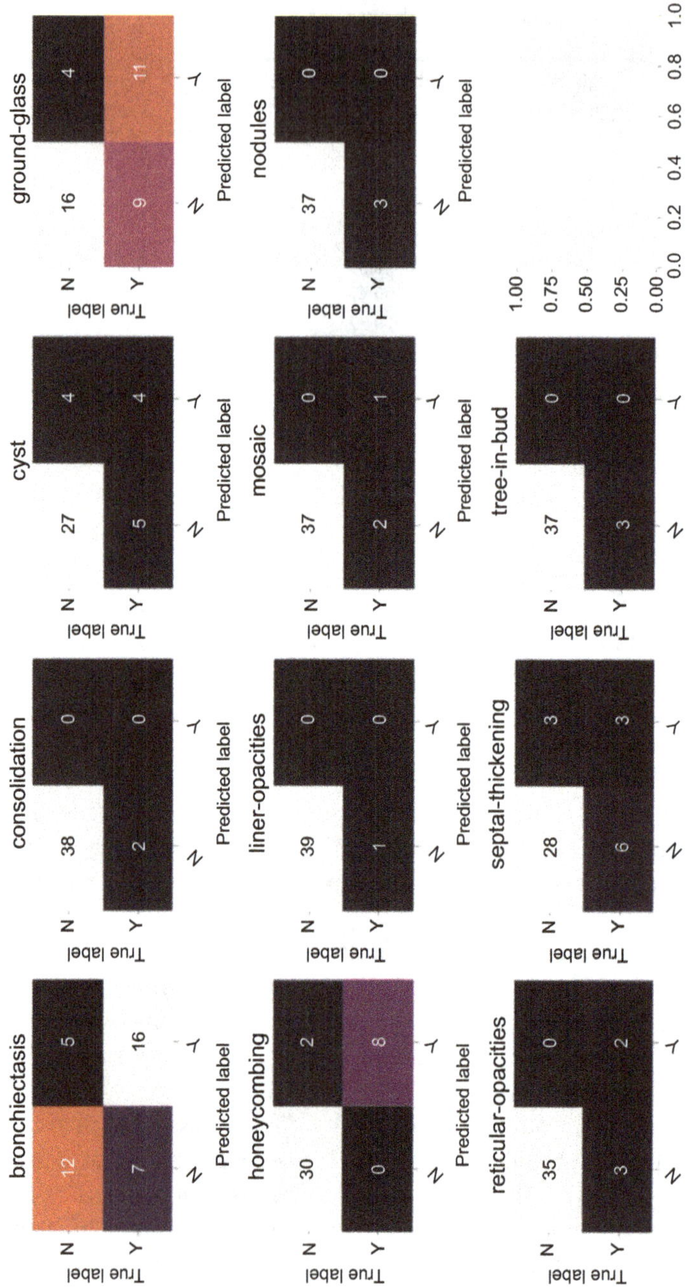

Figure 8: Multi-label confusion matrix (test set) for the Xception model.

	precision	recall	f1-score	support
bronchiectasis	0.71	0.65	0.68	23
consolidation	0.00	0.00	0.00	2
cyst	0.50	0.44	0.47	9
ground-glass	0.76	0.65	0.70	20
honeycombing	0.80	0.50	0.62	8
liner-opacities	0.00	0.00	0.00	1
mosaic	0.00	0.00	0.00	3
nodules	0.33	0.33	0.33	3
reticular-opacities	0.50	0.20	0.29	5
septal-thickening	0.20	0.11	0.14	9
tree-in-bud	1.00	0.67	0.80	3
micro avg	0.64	0.48	0.55	86
macro avg	0.44	0.32	0.37	86
weighted avg	0.59	0.48	0.52	86
samples avg	0.62	0.47	0.51	86

Figure 9: Classification accuracy (test set) report for the Xception model.

```
bronchiectasis: 99.58%
consolidation: 0.01%
cyst: 2.66%
ground-glass: 0.50%
honeycombing: 0.01%
liner-opacities: 0.00%
mosaic: 20.46%
nodules: 1.46%
reticular-opacities: 0.00%
septal-thickening: 0.60%
tree-in-bud: 15.89%
```

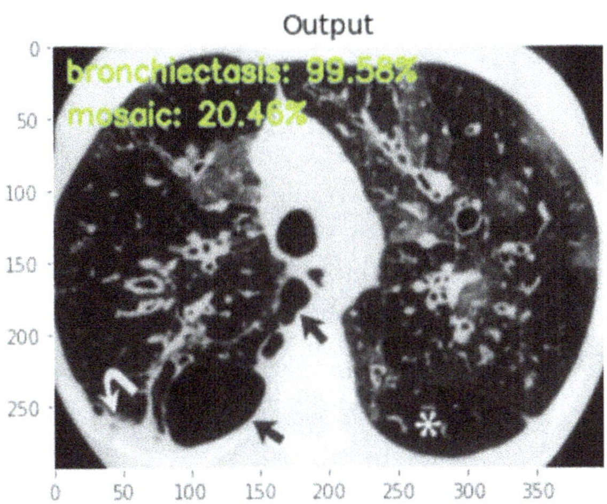

Figure 10: Output result of the test image (31927.jpg).

```
bronchiectasis: 0.57%
consolidation: 36.71%
cyst: 0.21%
ground-glass: 96.34%
honeycombing: 13.83%
liner-opacities: 0.04%
mosaic: 0.00%
nodules: 0.01%
reticular-opacities: 0.01%
septal-thickening: 48.75%
tree-in-bud: 0.00%
```

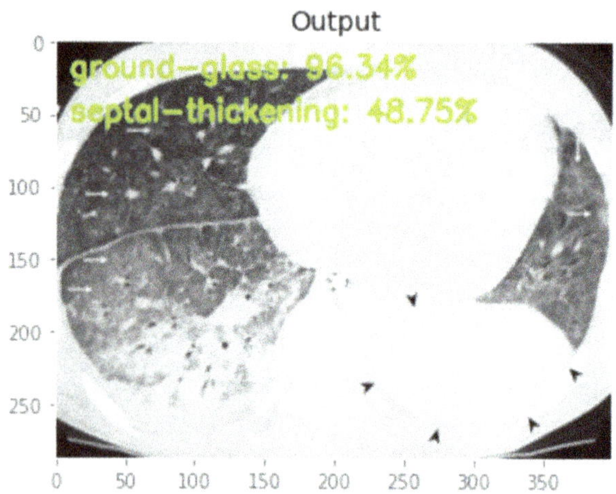

Figure 11: Output result of the test image (62167.jpg).

Figure 11 shows the classification probabilities of another test image (62167.jpg) in the ImageCLEFmed dataset [14] with associated caption as "Acute systemic lupus erythematosus pneumonitis. CT scan reveals extensive ground-glass attenuation throughout both lungs (arrows), interlobular septal thickening, bilateral lower lobe consolidations (complete on the left side [arrowheads]), and minimal pleural effusion" [17]. The output shows that this time it correctly predicted the *"ground-glass," "septal-thickening,"* and *"consolidation"* class labels with higher probabilities.

6.1 ROI detection results

We evaluated the performance of Mask R-CNN, YOLOv5, and YOLOv7 by measuring their accuracy in detecting and localizing the annotated markers within the chest CT images. The evaluation metrics used were ***precision, recall***, and ***mean average pre-***

cision (mAP). Additionally, we compared the computational efficiency of the techniques to assess their practical variability.

Table 2: Accuracy comparison of different DL-based object detection methods.

Method	Precision (%)	Recall (%)	mAP (%)
Mask R-CNN	85.6	81.8	0.83
YOLOv5	87.2	84.6	0.85
YOLOv7	92.5	87.3	0.89

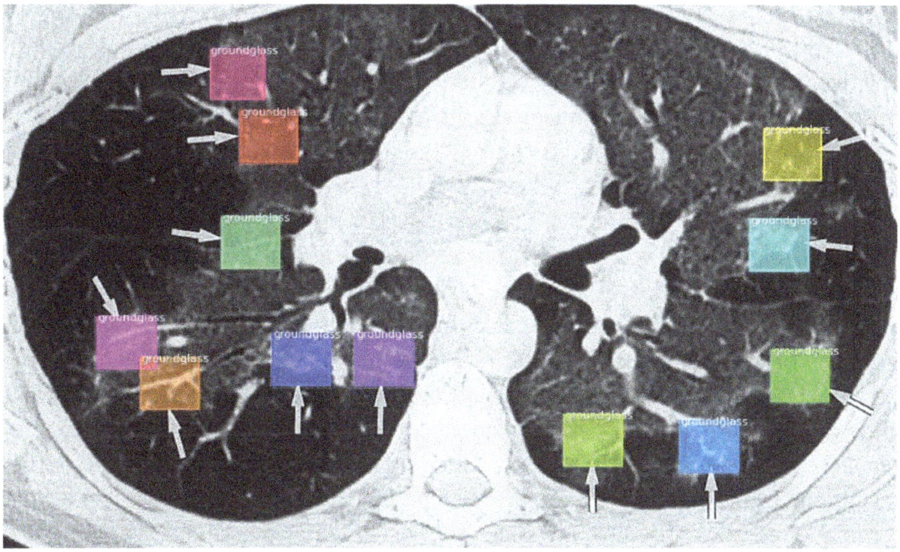

Figure 12: Sample result of ROI detection for the "ground-glass" concept.

The results in Table 2 demonstrate that all three DL-based object detection techniques achieved high accuracies in recognizing and localizing (annotation markers) the ROIs in biomedical images. YOLOv7 exhibited the highest precision of 92.5%, indicating its ability to accurately identify ROIs. YOLOv5 and Mask R-CNN also demonstrated respectable precision scores of 87.2% and 85.6%, respectively. In terms of recall, Mask R-CNN achieved 87.3%, indicating its effectiveness in capturing a high percentage of the annotated markers. YOLO v5 and YOLOv7 showed slightly lower recall rates of 84.6% and 81.8%, respectively, but still performed well in detecting the markers. Figure 12 shows an example image where the ROIs are detected correctly as "ground-glass" patterns, which are pointed by white arrows.

The study found that YOLOv7 marginally outperformed the other two models, achieving the highest average precision, recall, and F1 scores. Furthermore, adjusting the hyperparameters significantly improved the models' performance. Regarding computational efficiency, YOLOv5 and YOLOv7 exhibited faster inference times compared to Mask R-CNN. This makes them more suitable for real-time applications where speed is crucial, although with a slight compromise on precision and recall rates. In conclusion, the results of this study demonstrate that DL-based object detection techniques, specifically Mask R-CNN, YOLOv5, and YOLOv7, can achieve high accuracy in detecting and classifying different types of ROIs in low-resolution chest CT images as has often appeared in biomedical journal articles.

7 Conclusion

This work presents a proof-of-concept study to demonstrate the effectiveness of images that have appeared in biomedical articles as a valuable resource for ML and information retrieval tasks, such as concept-based classification and image search. In this study, we demonstrated the efficacy of CNN and ViTs for multi-label classification of concepts and Mask R-CNN, YOLO v5, and YOLO v7 in detecting annotation markers and extracting relevant image concepts from chest CT images. The results indicate the potential of DL-based classification and object detection techniques in improving the accuracy and effectiveness of biomedical image analysis. For classification, Xception model performed comparatively better regarding micro average precision and F1-score. For object (ROI) detection, YOLO v7 excelled in precision and recall, while YOLO v5 and Mask R-CNN offered faster inference times. Future research could explore the integration of transfer learning techniques to leverage pre-trained DL models on large-scale medical imaging datasets. Furthermore, combining multiple DL-based object detection techniques or using ensemble methods could potentially enhance the overall performance of biomedical image analysis systems.

It is expected that this work can be extended further to generate more data (training ground truth), which would offer building blocks for the development of advanced information retrieval systems aided by a visual ontology. The main limitation of this study is that the models (networks) are unable to predict on data they were never trained on using Keras networks for multi-label classification. In future, we plan to expand our work further in concept detection based on a larger annotated dataset, which is currently under construction. Overall, the impact of this work is substantial as many applications such as digital libraries and image search engines for teaching and training purposes require effective and efficient techniques to categorize and access images.

References

[1] D. Demner-Fushman, S. K. Antani, M. S. Simpson and G. R. Thoma, "Annotation and Retrieval of Clinically Relevant Images," *International Journal of Medical Informatics*, vol. 78, no. 12, pp. e59–e67, 2009.

[2] Z. Lu, "Pubmed and Beyond: A Survey of Web Tools for Searching Biomedical Literature," vol. baq036, 2011.

[3] M. S. Simpson, D. You, M. M. Rahman, Z. Xue, D. Demner-Fushman, S. K. Antani and G. R. Thoma G, "Literature-based Biomedical Image Classification and Retrieval," *Computerized Medical Imaging and Graphics*, vol. 39, pp. 3–13, 2015.

[4] M. S. Simpson, D. You, M. M. Rahman, S. K. Antani, G. R. Thoma and D. Demner-Fushman, "Towards the Creation of a Visual Ontology of Biomedical Imaging Entities," *AMIA Annual Symposium Proceedings*, pp. 866–875, 2012.

[5] E. K. Charles and T. Cheng, "GoldMiner: A Radiology Image Search Engine," *The Practice of Radiology*, vol. 188, no. 6, pp. 1475–1478, 2007.

[6] M. A. Hearst, A. Divoli *et al.*, "Biotext Search Engine: Beyond Abstract Search," *Bioinformatics*, vol. 23, no. 16, pp. 2196–2197, 2007.

[7] S. Xu, J. McCusker and M. Krauthammer, "Yale Image Finder (YIF): A New Search Engine for Retrieving Biomedical Images," *Bioinformatics*, vol. 24, no. 17, pp. 1968–1970, 2008.

[8] R. J. Sandusky and C. Tenopir, "Finding and Using Journal Article Components: Impacts of Disaggregation on Teaching and Research Practice," *Journal of the American Society for Information Science and Technology*, vol. 59, no. 6, pp. 970–982, 2008.

[9] A. Divoli, M. A. Wooldridge and M. A. Hearst, "Full Text and Figure Display Improves Bioscience Literature Search," *PLoS One*, vol. 5, no. 4, p. e9619, 2010.

[10] H. Shatkay, N. Chen and D. Blostein, "Integrating Image Data into Biomedical Text Categorization," *Bioinformatics*, vol. 22, no. 14, pp. e446–53, 2006.

[11] H. Müller, N. Michoux, D. Bandon and A. Geissbuhler, "A Review of Content Based Image Retrieval Systems in Medical Applications Clinical Benefits and Future Directions," *International Journal of Medical Informatics*, vol. 73, pp. 1–23, 2014.

[12] A. Smeulders, M. Worring, S. Santini, A. Gupta and R. Jain, "Content-based Image Retrieval at the End of the Early Years," *IEEE Transactions on Pattern Analysis Machine Intelligence*, vol. 22, no. 12, pp. 1349–1380, 2000.

[13] E. B. Meltzer and P. W. Noble, "Idiopathic Pulmonary Fibrosis," *Orphanet Journal of Rare Diseases*, vol. 3, no. 8, pp. 1–15, 2008.

[14] D. Lindberg, B. Humphreys and A. McCray, "The Unified Medical Language System," *Methods of Information in Medicine*, vol. 32, no. 4, pp. 281–291, 1993.

[15] C. P. Langlotz, "RadLex: A New Method for Indexing Online Educational Materials," *Radiographics*, vol. 26, no. 6, pp. 1595–1597, 2006.

[16] R. Venkatesan and B. Li, *Convolutional Neural Networks in Visual Computing: A Concise Guide*. CRC Press, 2017.

[17] H. Müller, A. Herrera, J. Kalpathy-Cramer, D. Demner-Fushman, S. K. Antani and E. Ivan, "Overview of the ImageCLEF2012 Medical Image Retrieval and Classification Tasks," in *The Working Notes for the CLEF 2012 Labs and Workshop*. Rome, Italy, 17–20 Sept. 2012.

[18] F. Chollet, "Xception: Deep Learning with Depthwise Separable Convolutions," *Proceedings of the IEEE Conference on Computer Vision and Pattern Recognition*, pp. 1251–1258, 2017.

[19] G. Huang, Z. Liu, L. Van Der Maaten and K. Q. Weinberger, "Densely Connected Convolutional Networks," in *2017 IEEE Conference on Computer Vision and Pattern Recognition (CVPR)*. Honolulu, HI, USA, 2017, pp. 2261–2269. doi: 10.1109/CVPR.2017.243.

[20] K. He, X. Zhang, S. Ren and J. Sun, "Deep Residual Learning for Image Recognition," in *2016 IEEE Conference on Computer Vision and Pattern Recognition (CVPR)*. Las Vegas, NV, USA, 2016, pp. 770–778. doi: 10.1109/CVPR.2016.90.
[21] S. Khan, M. Naseer, M. Hayat, S. W. Zamir, F. S. Khan and M. Shah, "Transformers in Vision: A Survey," *ACM Computing Surveys (CSUR)*, vol. 54, no. 10s, pp. 1–41, 2022.
[22] D. Alexey, B. Lucas, K. Alexander *et al.*, "An Image Is Worth 16x16 Words: Transformers for Image Recognition at Scale," *arXiv Preprint arXiv:2010.11929*.
[23] K. Han, Y. Wang, H. Chen *et al.*, "A Survey on Visual Transformer," *ArXiv*, abs/2012.12556, 2020.
[24] S. Fahad, K. Salman, W. Z. Syed *et al.*, "Transformers in Medical Imaging: A Survey," *arXiv:2201.09873*, 2022.
[25] M. M. Rahman and B. Regmi, "Multi-label Concept Classification in Imaging Entities of Biomedical Literature Using CNN and Vision Transformers," in *The 9th International Conference on Health Informatics & Medical Systems, (HIMS'23)*. Las Vegas, USA, 24–27 July 2023.
[26] LabelImg: https://pypi.org/project/labelImg/.
[27] K. He, G. Gkioxari, P. Dollar and R. Girshick, "Mask R-CNN," *IEEE Transactions on Pattern Analysis and Machine Intelligence*, vol. 42, no. 2, pp. 386–397, Feb. 2020, doi: 10.1109/TPAMI.2018.2844175. Epub 2018 Jun 5. PMID: 29994331.
[28] Object Detection using YOLOv5 OpenCV DNN in C++ and Python: https://learnopencv.com/object-detection-using-yolov5-and-opencv-dnn-in-c-and-python/.
[29] C. Wang, A. Bochkovskiy and H. M. Liao, 2022. YOLOv7: Trainable bag-of-freebies sets new state-of-the-art for real-time object detectors. 10.48550/arXiv.2207.026
[30] G. Litjens, T. Kooi, B. E. Bejnordi *et al.*, "A Survey on Deep Learning in Medical Image Analysis," *Medical Image Analysis*, vol. 42, pp. 60–88, 2017, doi: 10.1016/j.media.2017.07.005.
[31] T. Ozturk, M. Talo, E. A. Yildirim, U. B. Baloglu, O. Yildirim and U. Rajendra Acharya, "Automated Detection of COVID-19 Cases Using Deep Neural Networks with X-ray Images," *Computers in Biology and Medicine*, p. 103792, 28 Apr. 2020, doi: 10.1016/j.compbiomed.2020.103792. Epub ahead of print. PMCID: PMC7187882.
[32] S. Wang, B. Kang, J. Ma, X. Zeng, M. Xiao, J. Guo and B. XuA, "Deep Learning Algorithm Using CT Images to Screen for Corona Virus Disease (COVID-19)," *medRxiv*, 2020.
[33] L. Quin, C. Weidong and D. F. David, "Lung Image Patch Classification with Automatic Feature Learning," *Conference Proceedings IEEE Engineering in Medicine and Biology Society*, pp. 6079–6082, 2013.
[34] N. Codella, J. Cai, M. Abedini, R. Garnavi, A. Halpern and J. R. Smith, "Deep Learning, Sparse Coding, and SVM for Melanoma Recognition in Dermoscopy Images," *MICCAI MLMI*, vol. 9352, pp. 118–126, 2015.
[35] W. Lotter, A. R. Diab, B. Haslam, J. G. Kim, G. Grisot, E. Wu, K. Wu, J. O. Onieva, Y. Boyer, J. L. Boxerman, M. Wang, M. Bandler, G. R. Vijayaraghavan and A. Gregory Sorensen, "Robust Breast Cancer Detection in Mammography and Digital Breast Tomosynthesis Using an Annotation-efficient Deep Learning Approach," *Nature Medicine*, vol. 27, no. 2, pp. 244–249, Feb. 2021, doi: 10.1038/s41591-020-01174-9. Epub 2021 Jan 11. PMID: 33432172.
[36] P. Rajpurkar, J. Irvin, K. Zhu, B. Yang, H. Mehta, T. Duan, D. Ding, A. Bagul, C. Langlotz, K. Shpanskaya *et al.*, "Chexnet: Radiologist-level Pneumonia Detection on Chest X-rays with Deep Learning," *arXiv Preprint arXiv:1711.05225*, 2017.
[37] S. Aladhadh, M. Alsanea, M. Aloraini, T. Khan, S. Habib and M. Islam, "An Effective Skin Cancer Classification Mechanism via Medical Vision Transformer," *Sensors*, vol. 22, no. 11, p. 4008, 2022.
[38] Y. Cai, Y. Li, C. Qiu, J. Ma and X. Gao, "Medical Image Retrieval Based on Convolutional Neural Network and Supervised Hashing," *IEEE Access*, vol. 7, pp. 51877–51885, 2019, doi: 10.1109/ACCESS.2019.2911630.

[39] A. Qayyum, S. M. Anwar, M. Awais and M. Majid, "Medical Image Retrieval Using Deep Convolutional Neural Network," *Neurocomputing*, vol. 266, pp. 8–20, 2017, doi: 10.1016/j.neucom.2017.05.025.

[40] M. Vikram, A. Anantharaman and S. BS, "An Approach for Multimodal Medical Image Retrieval Using Latent Dirichlet Allocation," *Proceedings of the ACM India Joint International Conference on Data Science and Management of Data*, pp. 44–51, 2019.

[41] H. Ayesha, S. Iqbal, M. Tariq, M. Abrar, M. Sanaullah, I. Abbas, A. Rehman, M. Farooq Khan Niazi and S. Hussain, "Automatic Medical Image Interpretation: State of the Art and Future Directions," *Pattern Recognition*, vol. 114, no. 2021, p. 107856, ISSN 0031-3203, https://doi.org/10.1016/j.patcog.2021.107856.

[42] J. Pavlopoulos, V. Kougia and I. Androutsopoulos, *A Survey on Biomedical Image Captioning*. 2019, pp. 26–36, doi: 10.18653/v1/w19-1803.

[43] B. Jing, P. Xie and E. P. Xing, On the Automatic Generation of Medical Imaging Report.

[44] https://www.imageclef.org/2020/medical/caption last accessed on April 14th, 2021.

[45] C. Eickhoff, I. Schwall, A. Garc´ıa, S. de Herrera and H. Müller, "Overview of ImageCLEFcaption 2017 – Image Caption Prediction and Concept Extraction Tasks to Understand Biomedical Images," in *CLEF CEUR Workshop*. Dublin, Ireland, 2017.

[46] M. M. Rahman, "A Cross Modal Deep Learning Based Approach for Caption Prediction and Concept Detection by CS Morgan State," in *CLEF CEUR Workshop*. Avignon, France, 2018, http://ceur-ws.org/Vol-2125/paper_138.pdf.

[47] O. Lyode and M. M. Rahman, "Multi-Label and Cross-Modal Based Concept Detection in Biomedical Images. ImageCLEF2020," in *CLEF CEUR Workshop, CEUR-WS.org Proceedings* (http://ceur-ws.org/). 2020, http://www.dei.unipd.it/~ferro/CLEF-WN-Drafts/CLEF2020/.

[48] J. Bogatinovski, L. Todorovski, S. Džeroski and D. Kocev, "Comprehensive Comparative Study of Multi-label Classification Methods," *Expert Systems with Applications*, vol. 203, no. 2022, p. 117215, ISSN 0957-4174.

[49] S. H. A. García, J. Kalpathy-Cramer, D. Demner-Fushman, S. Antani and H. Müller, "Overview of the Imageclef 2013 Medical Tasks," in *CLEF 2013 Online Working Notes/Labs/Workshop*. 2013, pp. 1–15.

[50] S. S. Abbas Zaidi, M. Samar Ansari, A. Aslam, N. Kanwal, M. Asghar and B. Lee, "A Survey of Modern Deep Learning Based Object Detection Models," *Digital Signal Processing*, vol. 126, p. 103514, 2022, ISSN 1051-2004 https://doi.org/10.1016/j.dsp.2022.103514.

[51] Z.-Q. Zhao, P. Zheng, S.-T. Xu and X. Wu, "Object Detection With Deep Learning: A Review," *IEEE Transactions on Neural Networks and Learning Systems*, vol. 30, no. 11, pp. 3212–3232, Nov. 2019, doi: 10.1109/TNNLS.2018.2876865.

[52] R. Girshick, J. Donahue, T. Darrell and J. Malik, "Region-Based Convolutional Networks for Accurate Object Detection and Segmentation," *IEEE Transactions on Pattern Analysis and Machine Intelligence*, vol. 38, no. 1, pp. 142–158, 1 Jan. 2016, doi: 10.1109/TPAMI.2015.2437384.

[53] Github: Yolov5. https://github.com/ultralytics/yolov5

[54] Wang, Chien-Yao & Bochkovskiy, Alexey & Liao, Hong-yuan. (2022). YOLOv7: Trainable bag-of-freebies sets new state-of-the-art for real-time object detectors. 10.48550/arXiv.2207.02696.

Beilei Zhu and Chandrasekar Vuppalapati

Revolutionizing supply chain dynamics: deep meta-learning and multi-task learning for enhanced predictive insights

Abstract: Supply chain enterprises in the global trade domain seek methods to achieve operational excellence, enhance profitability, reduce costs, and boost customer satisfaction. To this end, they adopt Artificial Intelligence (AI) and Machine Learning (ML), which provide unparalleled benefits by using large datasets for task automation and strategic decision-making. However, these technologies are not without challenges. The main ones are the frequent changes in business environments and the need for deep industry-specific knowledge to ensure effective implementation.

Conventional machine learning approaches often struggle, as they lack inherent abilities to deeply understand and associate fundamental knowledge with new data, which affects predictive accuracy. This paper explores advanced ML methods –deep meta-learning and multi-task learning – as potential solutions to these complexities. Through a thorough analysis of regression and classification models in realistic scenarios, we demonstrate the skill of sophisticated algorithms in overcoming traditional limitations.

We focus on the novel application of meta-learning techniques to build highly flexible supply chain models. These models, guided by common domain-specific knowledge, enable improved learning from new data and tasks, offering a remarkable improvement in operational adaptability. By highlighting the transformative potential of deep meta-learning, this paper supports its role in optimizing supply chain processes – creating a paradigm shift in business performance and competitive advantage in the fast-changing marketplace.

Keywords: deep meta-learning, multitask learning, supply chain operations, COVID-19 market disruptions

1 Introduction

Over recent years, the supply chain has experienced substantial disruptions. The recent U.S. Census Small Business Pulse survey, carried out between May 31 and June 6, 2021, indicates that 36% of small businesses are facing domestic supplier delays [1], primarily in manufacturing, construction, and trade, as illustrated in Figure 1. Though

Beilei Zhu, Global Supply Chain, Intel Corp., Hillsboro, Oregon, USA, e-mail: beilei.zhu@intel.com
Chandrasekar Vuppalapati, Computer Engineering, San Jose State University, San Jose, USA, e-mail: chandrasekar.vuppalapati@sjsu.edu

https://doi.org/10.1515/9783111344126-008

pre-pandemic comparable survey data is lacking, industry-specific surveys reveal that input shortages are unusually high. While the pandemic, natural disasters, and freight complications have made headlines, they are not entirely responsible for the global component shortage. Instead, supply chain experts must introspectively scrutinize their longstanding procedures contributing to the current predicament [2].

A critical issue in the electronic supply chain is multi-level demand forecasting. Unclear signals in the semiconductor and component market cause confusion, impacting organizations reliant on these products and complicating growth prediction and investment planning.

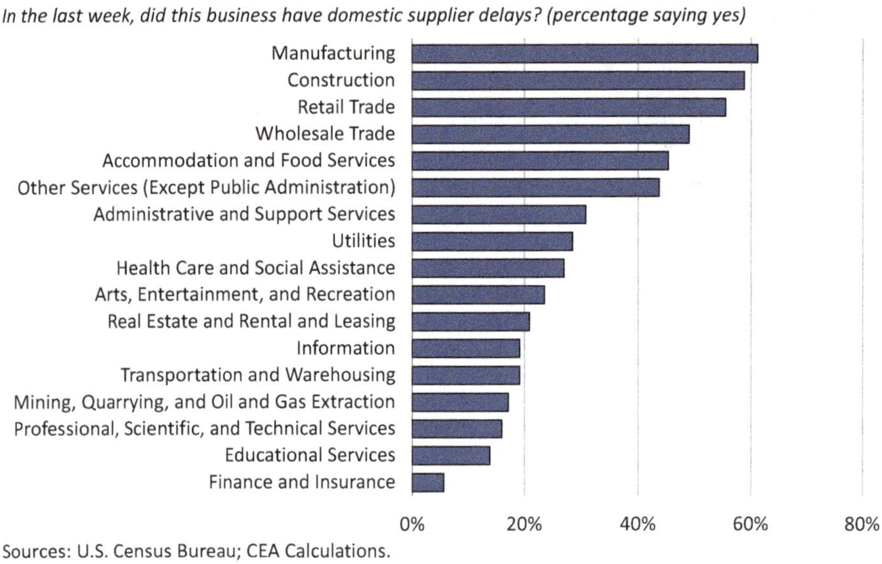

Figure 1: Supply chain disruptions.

The ramifications extend beyond the market's current shortages and price hikes, potentially causing a demand downturn as companies halt orders to deplete excess inventories [3].

Addressing supply chain issues and assessing demand are intricate tasks [4, 5], leading to an increased reliance on data analytics in supply chain management. Yet, a prevalent issue is ***data scarcity***, especially within high-precision electronics and chip manufacturing, where data collection is often hindered by privacy concerns or proprietary systems.

Meta-learning and multi-task learning techniques can address these challenges, enabling companies to leverage *smaller datasets* and learn efficiently in new business areas. These techniques can be utilized in supply chain management areas like demand forecasting, inventory management, and logistics planning.

We will proceed with a structured exposition. Our study initiates by articulating the foundational methodologies to be employed, setting a rigorous academic context. We shall systematically unpack the theoretical underpinnings of meta-learning and multi-task learning, highlighting their relevance and applicability in addressing challenges within supply chain dynamics. This theoretical exploration is paramount in elucidating the reasons these advanced techniques have been selected for our investigation.

Subsequent to this theoretical grounding, we delve into innovative methodologies to address persistent challenges in supply chain management, specifically focusing on demand inventory and planning forecast issues. Traditional models often fall short in these areas due to *the complexity and dynamic nature of supply chain data*. To navigate these challenges, our paper introduces a sophisticated approach employing deep meta- and multi-task learning techniques within a generic regression framework. This approach is designed to enhance predictive accuracy and adaptability by leveraging shared information and insights across various tasks in the supply chain domain. By utilizing a generic data structure, this method not only maintains versatility in handling diverse data types but also ensures robustness in predictions, making it particularly effective for complex supply chain forecasting. Although our paper analyzes supply chain data of the electronic component industry, the proposed framework could be equally applicable to any industrial vertical [6–8].

Furthermore, this paper will expand its scope to include supply chain classification problems, which are crucial in master data management and risk identification. We propose a novel solution that integrates optimized meta-learning models to improve the classification tasks inherent in supply chain contexts. This optimization is pivotal, as it allows the model to quickly adapt to new, yet similar tasks by understanding the underlying meta-information shared across different learning scenarios. We demonstrate the efficacy of this approach through rigorous experiments, the flow and results of which are detailed comprehensively in subsequent sections. These results are anticipated to showcase not only the theoretical validity of using deep learning and meta-learning in tandem but also their practical utility in revolutionizing supply chain analytics.

2 Enhancing adaptability in supply chain management: Integrating mechanistic and probabilistic meta-learning approaches

In tackling various predictive challenges within supply chain management, employing mechanistic and probabilistic meta-learning perspectives can be instrumental in creating models that adapt quickly and efficiently. These methods hold significant relevance due to their unique approaches.

Under the "mechanistic view," a meta-learning model operates to make swift, efficient predictions across a range of products, regions, or logistical scenarios. For instance, Model-Agnostic Meta Learning (MAML) stands out for its versatility in quickly adapting to new circumstances or data. This *characteristic is particularly beneficial in supply chain management, where conditions and requirements can shift rapidly*, requiring models that can keep pace with minimal additional training or data [9].

Conversely, the "probabilistic view" harnesses Bayesian probability theory and hierarchical modeling to deepen the understanding of the relationships between various prediction tasks in the supply chain. Techniques such as ***Gaussian*** [10] process-based meta-learning or ***hierarchical Bayesian*** models draw on previous task knowledge to make informed inferences, employing a robust structure that integrates existing information and dynamically updates based on new data.

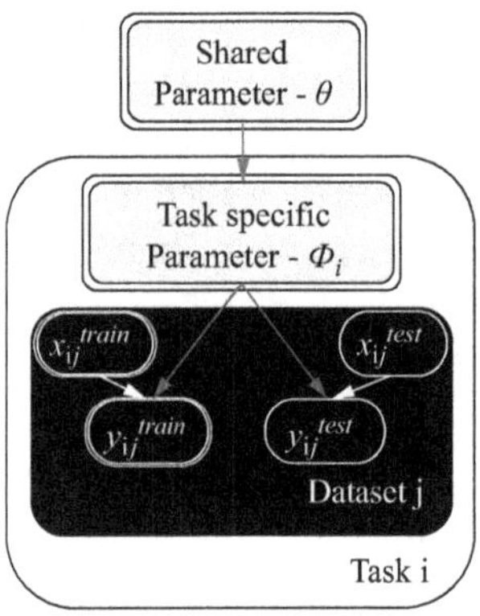

Figure 2: General meta-learning and multi-task process.

Integrating multi-task learning with these meta-learning strategies further strengthens the model. By training on multiple tasks simultaneously, the model grasps underlying patterns and insights, beneficial for tackling the common issue of data scarcity in different segments of the supply chain. This integration facilitates a more nuanced approach, particularly when the tasks share structural similarities or related data points [11].

The design of meta-learning algorithms, informed by both mechanistic and probabilistic insights, aims to bolster performance across various supply chain tasks. Their effectiveness hinges [12] on the ability to leverage shared knowledge, adapt based on context, data specificity, and structural assumptions of the tasks at hand. In essence, these strategies (as depicted in Figure 2) rely on shared latent information, recognizing hidden factors that interlink multiple tasks, thereby enhancing overall learning efficiency.

In the model outlined in Figure 2, the principle is clear: both multi-task learning and meta-learning capitalize on shared information (θ) across different tasks to bolster performance (ϕ) in each specific task (i). By recognizing and exploiting the commonalities between tasks, these methodologies facilitate effective knowledge transfer, thus fostering a more interconnected and holistic learning process within the realm of supply chain management [13].

3 Meta-learning algorithms for addressing data dynamics in supply chain management

Meta-learning algorithms for addressing supply chain use cases focus on enabling accurate predictions for new instances of task **i**, input $\mathbf{x}_i^{test}$ and producing corresponding output, $\mathbf{y}_i^{test}$. The *goal is to learn a mapping between a supply chain dataset and a test example to the appropriate learning output.* To do this, we use training datasets with more than **k** examples. At least one extra example is reserved for testing, so we can sample **k** instances for the training dataset. It becomes necessary to design a function, **f**, capable of generalizing from an input dataset related to our use cases' predictions by conceptualizing the learning process this way (eq. 1) [14]:

$$\mathrm{Y}^{test} = f_\theta\left(\mathrm{D}^{train},\ \mathrm{X}^{test}\right) \tag{1}$$

In order to optimize this function, a set of datasets is used, essentially representing a learning procedure that is optimized [15] for learning from small amounts of data for the new tasks. The process is illustrated in Figure 3.

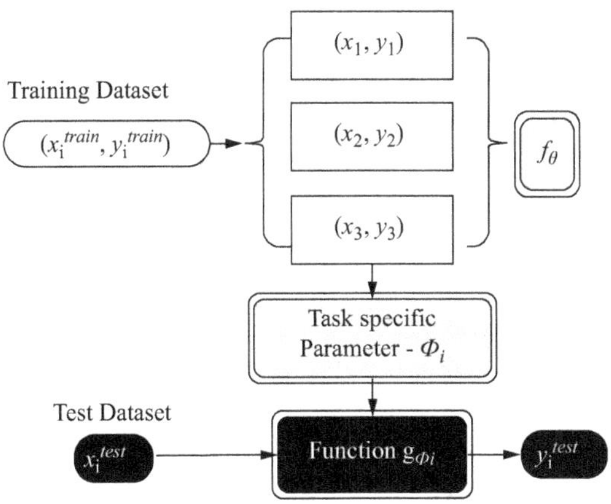

Figure 3: Model – multi-task and meta-learning.

Let us combine meta-learning and multi-task learning for our use cases. Following these steps will help us apply meta-learning and multi-task learning to the supply chain problem [16]:

Step 1: Define the tasks.
Identify the multiple related tasks for multi-task learning, like predicting regression or classification outcome for different tasks we defined in our shared datasets. Define a set of tasks that can be used to teach the model how to learn so it can adapt quickly to new tasks.

Step 2: Develop shared model architecture.
Create a model architecture that incorporates both multi-task learning and meta-learning. We will explain two approaches in detail by utilizing shared layers in a neural network or learning a shared latent space for multi-task learning.

Step 3: Train the model in multiple tasks and meta-learning scenarios.
Using both shared and task-specific knowledge, train the combined multi-task and meta-learning model on multiple tasks *concurrently*. Meta-learning techniques can be used to train the model on a set of tasks, allowing it to adapt quickly to new tasks with *limited data*.

Step 4: Evaluate the model.
Evaluate the combined multi-task and meta-learning model's performance on each task using relevant evaluation metrics and assess its adaptability to new tasks.

We will define generic data structure for this kind of supply chain numeric prediction use cases:

We will have Task column, for example, for each different data type if predicting for multiple signals, or different product grouping as tasks if we want to fine-tune a single signal. We will have a Quantity column for numeric prediction base. We will have Feature columns to retain all the attributes or descriptors of the data. Finally, we will have Predicted quantity column for the predicted numeric value.

We are navigating the landscape of supply chain management use cases with a spotlight on two contrasting methodologies: *the black box approach and the optimized approach*. The nuances and depths of these techniques will be illuminated in subsequent sections. After critically analyzing both strategies, our objective will be to discern which method proves most adept at addressing our designated use cases.

Starting with the black box approach, let us dive deep into its application in a regression use case specific to supply chain management. The intricate web of supply chain management comprises multifaceted signals such as demand, supply, capacity, inventory, pricing, and logistics planning. These indicators, predominantly represented in numerical values, serve as potential goldmines for machine learning applications.

While these data points are interconnected through a master data relationship, they each exhibit unique trends. However, the dynamic nature of business, characterized by the advent of new business areas or shifts in existing ones, often poses challenges. Traditional machine learning methodologies might falter under these circumstances, unable to achieve the desired accuracy. Herein lies the *golden opportunity* for meta-learning and multi-task learning to step in. These data-centric scenarios, with their inherent intricacies, become ideal contenders for the application of advanced learning strategies.

Now, for a structured representation of such supply chain numeric prediction use cases, let us conceptualize generic data architecture:

Task column: This serves as an identifier. If the goal is to predict multiple signals, each distinct data type (e.g., demand, supply, capacity) would be labeled as a separate task. Conversely, if the aim is to fine-tune predictions for a singular signal, product groupings might be used as tasks.

Quantity column: This is the bedrock of our numeric predictions. All actual values against which predictions will be measured will be housed here.

Feature columns: These columns capture the essence of our data. They retain all attributes or descriptors, encapsulating the finer details that might influence the numeric predictions.

Predicted quantity column: Post the application of our machine learning model, this column will store the predicted numeric values. It acts as the culmination of our efforts, presenting a tangible metric to gauge the efficiency and accuracy of our models.

In the following sections, we will delve into the intricacies of implementing the black box approach, using this data structure as our foundation. By juxtaposing its results with those of the optimized approach, we aim to identify the most effective strategy for our supply chain use cases.

4 Black box approach

In our supply chain regression use case, black box approaches involve parameterizing the learning procedure with a neural network. Based on historical data, this network outputs model parameters. Following this, the supply chain signals for new tasks are predicted using a network parameterized by these parameters, ϕ. This usually involves a recurrent neural network (RNN) that processes data from the supply chain training dataset and outputs a set of parameters (please see Figure 7). Another neural network then employs these parameters to make predictions for new logistics. In this process, we are going to use both the training set ($\boldsymbol{D}_i^{train}$) and the test set ($\boldsymbol{D}_i^{test}$). The neural networks are trained using standard supervised learning across all tasks. The overall goal is to measure the model's performance, with parameters ϕ, in predicting supply chain signals for new examples across all supply chain tasks [17].

***Equation 2:** Loss functions for black box meta-learning*

$$\max_{\theta} \Sigma_{\tau_i} \Sigma_{(x,y)\sim D_i^{test}} \text{Log}\, g_{\phi i}(y|x)$$

$$\textit{where}\ \Sigma_{(x,y)\sim D_i^{test}} \text{Log}\, g_{\phi i}(y|x) = -L(\phi_i,\ D_i^{test})$$

$$\max_{\theta} \sum_{\tau_i} \sum_{(x,y)\sim D_i^{test}} \text{Log}\, g_{\phi i}(y|x) = \max_{\theta} \sum_{\tau_i} -L(\phi_i,\ D_i^{test}) = \max_{\theta} \sum_{\tau_i} -L(f_\theta(D_i^{train}), D_i^{test})$$

$$= \min_{\theta} \sum_{\tau_i} L(f_\theta(D_i^{train}), D_i^{test}) \tag{2}$$

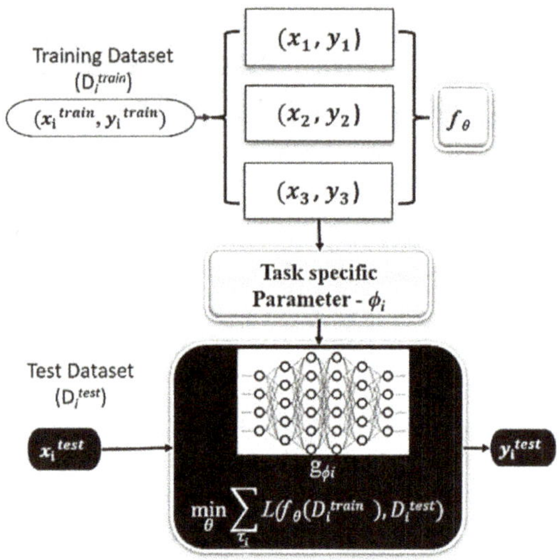

Figure 4: Black box approach.

$$Y_i^{test} = f_{\theta\text{-black-box}}\left(D_i^{train},\ X_i^{test}\right) \tag{3}$$

In the Figure 4 and eq. (3), the function f_θ predicts the shared parameters ϕ, which represent various aspects of supply chain prediction, such as carrier rates, regional factors, and product-specific requirements as different tasks. The parameter's effectiveness in generalizing test examples is then evaluated. Can we train the parameters separately for each task, or is there an alternative?

In this context, the index *i* represents different tasks, and ϕ_i can be considered more as an activation of the network rather than actual parameters or weights. Using the output of the neural network, another network $g_{\varphi i}$ can use the weights. The *separate representation* of the networks is key, since the neural network produces weights for another network that predicts supply chain indicators. Optimizing the weights of the network is more important than optimizing ϕ_i. As a result, the log-likelihood is maximized.

In essence, ϕ_i acts as an ***activation*** instead of being trained directly in the context of supply chain numeric indicators prediction. The goal is to make sure the secondary neural network predicts the indicators well. As a result, ϕ is not explicitly optimized, but only the primary network's parameters θ_i are optimized.

As observed, this approach resembles standard supervised learning in the sense that it aims to maximize the log-likelihood of $g_{\phi i}$, as demonstrated below Figure 5. This is equivalent to minimizing the loss function.

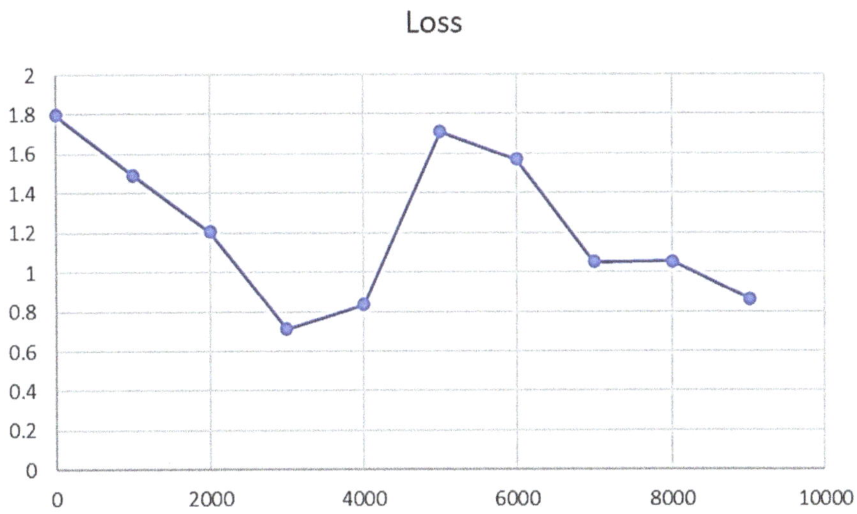

Figure 5: Loss for black box approach.

The utilization of black box meta-learning approach in prediction of supply chain indicators involves training a neural network to quickly adapt to new tasks, leveraging knowledge from previous ones. This approach utilizes the entire training dataset to capture complex relationships in the data and offers expressiveness. However, the use of black box approaches in prediction of supply chain indicators can also introduce challenges related to optimization, particularly when the training data is limited, unrepresentative, or contains noise or outliers. The quality and quantity of training data significantly impact the stability of the model, as demonstrated by Figure 8, and the eqs. (2) and (3).

In order to adapt the model to new tasks, black box meta-learning usually requires a large number of training iterations. Every time a new task is encountered, the model must be updated from scratch or fine-tuned, which can be computationally expensive and time-consuming. As a result, the model may have difficulty adapting quickly to new tasks with limited data. This can lead to instability in the model's performance.

Considering our specific use case, the black box approach may not be the most suitable solution. However, for use cases that meet certain conditions, we recommend considering black box meta-learning. These conditions include dealing with complex models or tasks with ***undefined distributions***, where the ability to learn and solve a wide range of problems without explicitly modeling the task structure is ***valuable***. Additionally, the expressiveness of black box methods, as they do not rely on specific optimization procedures or task distribution assumptions, can be advantageous. Lastly, the availability of sufficient data is important for black box methods to learn accurate representations of underlying task distributions.

5 Optimization-based meta-learning

In our regression-based use case for prediction of supply chain indicators, we will continue exploring an optimized meta-learning approach. This approach incorporates optimization within the meta-learning process, allowing for the adjustment of free parameters during both *inner-and-outer learning* phases. The aim is to achieve better generalization and address challenges associated with calculating second-order derivatives.

To address these challenges, we propose treating meta-learning as an optimization procedure similar to *gradient descent* on training data. This approach involves incorporating optimization within the meta-learning process, which allows for the adjustment of free parameters during both *inner-and-outer-learning phases*, as depicted below:

$$\phi_i = f_\Theta\left(D_i^{train}\right) \tag{4}$$

The fundamental concept behind optimization-based meta-learning is to integrate an optimization algorithm (see eq. (4) and Figure 9) into the meta-learning process. This enables the fine-tuning of pre-trained parameters to effectively adapt to new tasks.

1) Fine-tuning procedure for small datasets (inner loop for task-specific parameters, Figure 6)

Given the limitation of small datasets commonly found in supply chain business use cases, our objective is to optimize the fine-tuning procedure. This involves adjusting pre-trained parameters and other components of the procedure, as illustrated in Figure 9. By performing an optimization process during test time with the small dataset, we can ***enhance*** the model's generalization to test data points.

2) Optimizing pre-trained parameters θ across meta-training tasks (outer loop)

Our approach involves optimizing pre-trained parameters across all meta-training tasks to achieve better performance, even when fine-tuning with a minimal dataset.

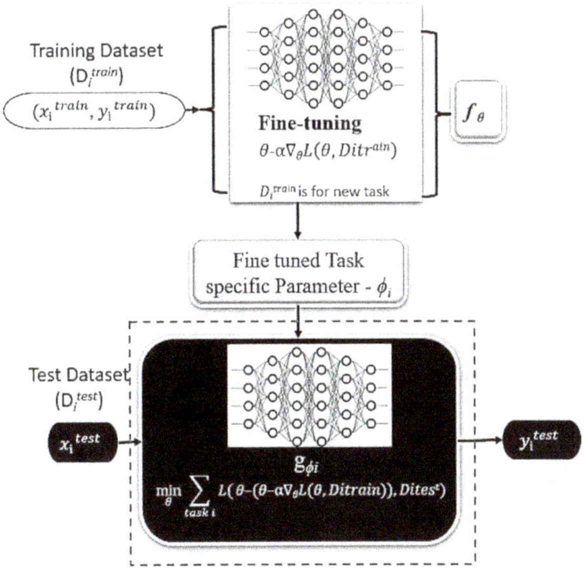

Figure 6: Meta optimization – outer loop.

$$\min_{\theta} \sum_{task\ i} L(\phi_i, D_i^{test}) = \min_{\theta} \sum_{task\ i} L(\theta - (\theta - \alpha\nabla_\theta L(\theta, D_i^{training})), D_i^{test}) \tag{5}$$

The objective of the outer loop optimization process is to minimize the loss function for the shared parameter θ with respect to all the task-specific parameters ϕ_i, and test data set D_i^{test}. This objective is depicted in eq. (5) (with corresponding colors in the equation to illustrate the equivalent mathematical formula throughout the steps), and Figure 6 (where the content enclosed in a blue dashed frame represents the trajectory of the optimization process).

The optimization aims to ensure that a single gradient step with respect to a specific task brings the model close to the task's optimum, *even with limited data.*

3) Computational complexities in optimized meta-learning

One potential challenge with optimization-based meta-learning is the computational complexity associated with calculating *second-order derivatives* or *Hessians* as shown in Figure 7 (dashed line framed) [14–17].

For instance, a neural network with n parameters requires a full Hessian of *size* $n \times n$, which is quadratic in terms of parameters. This complexity arises when computing ϕ, as it requires taking the gradient of L with respect to θ on the training dataset, followed by taking the gradient at the previously computed gradient value.

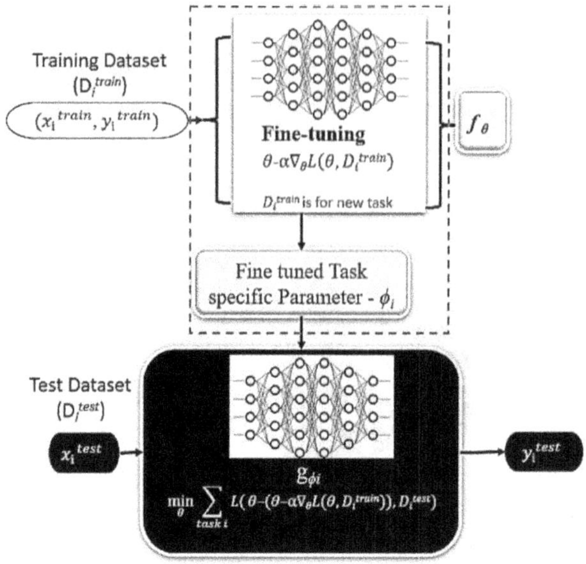

Figure 7: Second-order derivatives or Hessians -inner loop.

Optimization-based meta-learning proves to be a highly suitable approach for tasks characterized by similar structures or well-defined task distributions. Its effectiveness is particularly prominent in *few-shot learning* scenarios where data per task is limited. By leveraging the inherent structure of the tasks and optimizing shared parameters, these methods demonstrate enhanced sample efficiency, facilitating faster adaptation to new tasks.

Moreover, optimization-based approaches provide the added benefit of improved ***interpretability*** as they optimize shared parameters θ across multiple tasks, enabling a more comprehensive understanding of the learning process. Given these reasons, *optimization-based meta-learning aligns exceptionally well with our supply chain indicators use case.*

Having examined the applicability of black box and optimization-based meta-learning to our specific use case, we will now transition to the subsequent section, where we will delve into the implementation details and outcomes of our experimental investigations.

6 Optimize based meta-learning implementation and results for supply chain regression use case

Based on our comprehensive analysis, we have determined that the Model-Agnostic Meta-Learning (MAML) framework is the most suitable approach for addressing our case study problem. In the following steps, we will elucidate the seamless integration of MAML's shared parameter learning and task-specific parameter learning components, which enable us to effectively predict supply chain numeric indicators for different tasks. Our proposed methodology consists of two essential components: *the inner loop function, responsible for learning task-specific parameters, and the outer loop function, responsible for learning shared parameters across tasks.* We will provide detailed insights into the implementation of these core functions within the MAML framework.

The ***inner loop function*** includes the following steps:

1) It receives the model, task data, inner learning rate, and the number of inner steps as input parameters.
2) An ***SGD (Stochastic Gradient Descent)*** optimizer is instantiated for the inner loop.
3) The task-specific loss function, in this case, ***Mean Squared Error (MSE)*** loss for the price prediction problem, is defined.
4) The inner loop is executed for the predetermined number of steps. Each iteration performs the following tasks:
   a. Extracts input features (x) and target price (y) from task data.
   b. Transforms x and y into tensor formats.
   c. Executes a forward pass through the model to obtain predictions.
   d. Computes the loss by comparing the predicted and true prices.
   e. Clears any previous gradients in the model.
   f. Executes backpropagation to compute the gradients relative to the model parameters.
   g. Updates the model parameters employing the SGD optimizer.
5) Finally, the updated model is returned.

The ***outer loop function*** operates according to the following steps:

1) It takes as input the model, dataset, outer learning rate, number of tasks, inner learning rate, inner steps, number of epochs, log interval, and a TensorBoard writer.
2) An ***Adam optimizer*** is instantiated for the outer loop.
3) The task-specific loss function, again ***MSE loss***, is defined for the price prediction problem.
4) The outer loop is run for the specified number of epochs. *Each epoch* undertakes the following tasks:
   a. Samples a batch of tasks from the dataset.
   b. For each task in the batch:

   - Retrieves task-specific data.
   - Calls the inner loop function to adapt the model for the current task.
   - Retrieves input features (x) and target price (y) from task-specific data.
   - Transforms x and y into tensor formats.
   - Executes a forward pass through the adapted model to obtain predictions.
   - Computes the loss between the predicted and true prices.
   - Clears any previous gradients in the model.
   - Performs ***backpropagation*** to compute the gradients relative to the model parameters.
   - Updates the model parameters utilizing the Adam optimizer.

c. Logs the loss and accuracy (if applicable) to TensorBoard.

Figure 8 presents the MAML loss performance: There are several reasons for the lower loss observed in MAML (Figure 8) compared to black box meta-learning (Figure 4).

Firstly, MAML is designed specifically for few-shot learning scenarios where a limited amount of data is available per task. Using meta-learning, it trains a model to adapt quickly to new tasks by learning from previous ones. The adaptive nature of MAML allows it to generalize across tasks and improve overall performance.

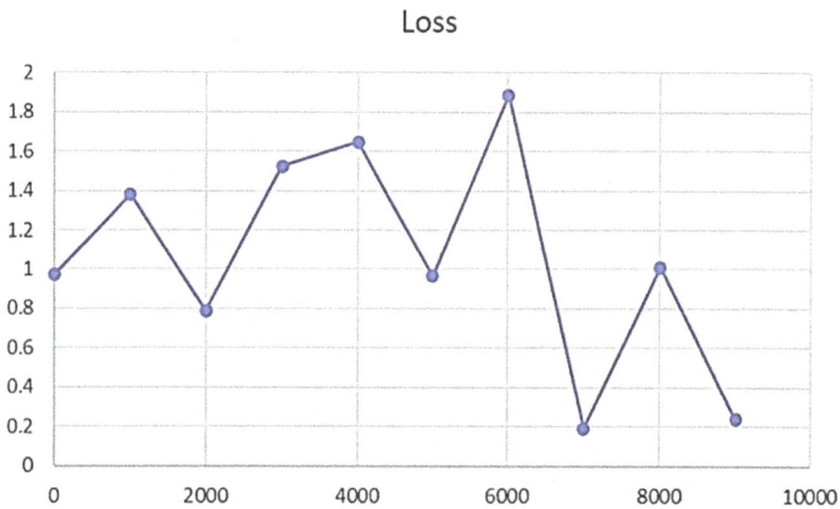

Figure 8: Loss for MAML.

Secondly, MAML employs an optimization-based approach to fine-tune the parameters of the model to meet the needs of a particular task. Through this fine-tuning process, the model can better capture the nuances and specific characteristics of each task, resulting in more *accurate predictions and lower loss rates.*

Furthermore, MAML's architecture and training methodology optimize the shared parameters across tasks, allowing the model to learn more efficiently. Through this shared parameterization, information is shared and transferred between tasks, resulting in improved performance and reduced loss.

There is no doubt that the performance and loss outcomes will differ based on the specific dataset and the specific task at hand. However, MAML's superior performance in terms of lower loss suggests that it has the ability to adapt to new tasks and leverage previous knowledge in order to make better predictions.

Our research extended beyond the adoption of MAML to include the evaluation of various other machine learning models in comparison to MAML (see Table 1). This comparative analysis involved the utilization of classical machine learning models to gain a comprehensive understanding of their potential and their performance in relation to each other.

Table 1 showcases a comparison of the performance metrics of four distinct machine learning models: Linear Regression, Random Forest Regressor, Logistic Regression, and MAML. The R2 metric, representing the proportion of variance in the dependent variable that can be predicted from independent variables, suggests the fit of the model. Here, the Linear Regression has a relatively low R2 of 17%, indicating a potential misfit for the dataset. In contrast, the Random Forest Regressor impressively explains 85% of the variance. Logistic Regression moderately explains 50%, while the R2 for MAML is not provided.

Focusing on error metrics like MSE and RMSE, MAML shines brightly. Although Linear Regression and Logistic Regression show higher errors with MSE values of 2.735 and 765.115, respectively, and RMSE values of 1.65394 and 27.66, the Random Forest Regressor demonstrates better performance with an MSE of 0.5034 and an RMSE of 0.7095. Yet, it is MAML that stands out with the most impressive error metrics. Despite a typographical error in its RMSE value, which is likely 0.616, and an MSE of 0.37966, MAML not only outperforms the other models but showcases its potential benefits in machine learning tasks, particularly in terms of minimized errors.

In the context of supply chain management, particularly in the domain of supply chain indicators, the adaptability of machine learning models to limited or small-scale datasets is a significant concern. This is commonly encountered during business transitions or when undertaking new tasks where sufficient historical data may be lacking.

In such scenarios, the distinct advantages of MAML became evident (see Table 1). Unlike traditional machine learning models that typically require large-scale datasets for effective training and optimization, MAML showcased superior ability in leveraging prior learning experiences and adapting quickly to new tasks, even with smaller datasets.

On the other hand, the other models examined in our study, while individually powerful, fell short in this particular aspect. They struggled to match the efficiency and flexibility demonstrated by MAML when faced with new tasks and smaller datasets, resulting in their underperformance compared to MAML.

Table 1: Classical MAML performance metrics.

	Linear regression	Random forest regressor	Logistic regression	MAML
R^2	17%	85%	50%	–
MSE	2.735	0.5034	765.115	0.37966
RMSE	1.65394	0.7095	27.66	o.616

This performance contrast highlights the limitations of classical machine learning models in dealing with smaller datasets or dynamic business scenarios. In contrast, MAML, with its meta-learning approach, offers a robust solution to address these challenges, making it a highly valuable tool in such contexts. The superiority of MAML, as observed in our evaluation, justifies its selection as the preferred approach for our case study problem.

The successful deployment of an optimized meta-learning approach has illuminated its potential. Consequently, we have embarked on assessing its efficacy within the realm of supply chain classification use cases. Let us delve deeper into this exploration.

7 A deep dive into supply chain product classification: meta-learning in master data management

In the intricate tapestry of today's business ecosystem, enterprises continually confront the arduous task of preserving coherent and precise master data. At its core, master data encapsulates the quintessential information integral to an organization's operations, encompassing pivotal data elements such as products, suppliers, and customers. This data serves as the bedrock, universally employed across a spectrum of functions and systems within a corporate structure. Given the diverse array of systems, data sources, and stakeholders involved, attaining consistency emerges as a formidable challenge.

One significant challenge in the realm of supply chain management is the classification of products. Products from different systems are frequently mapped to product families. However, due to the variability in product family naming conventions across systems, there can be inaccuracies. These inconsistencies can, in turn, lead to erroneous calculations in demand forecasting, inventory management, and expenditure reports.

Enter meta-learning – a promising solution to this conundrum. Meta-learning, often termed "learning to learn," stands apart from traditional machine learning paradigms. Rather than training models for specific tasks, meta-learning emphasizes training on multiple tasks. This allows models to swiftly acclimatize to new tasks, even with minimal data. This adaptive capability of meta-learning makes it an exemplary fit for managing the varied and often inconsistent data in supply chains.

To elucidate further, consider a multinational corporation with products sourced from a plethora of global vendors. Each vendor employs its own system, leading to diverse naming conventions. For instance, while "laptop" might be the terminology used by one vendor, another might prefer "portable computer." Such disparities, although superficially minor, can culminate in substantial inconsistencies across various business facets.

With meta-learning at its disposal, such a corporation could train a model using diverse vendor data. This model wouldn't merely assimilate individual vendor classifications but would grasp the broader principles underlying product categorization. When a new vendor joins the fray, the model can rapidly adjust to the unique naming schema of this vendor, ensuring homogeneity.

While our discussion has predominantly revolved around product categorization in the context of supply chain, the principles of meta-learning and its benefits extend to other spheres of master data as well. Whether it is assimilating customer data from a myriad of CRM systems or parsing supplier data from an assortment of procurement platforms, the versatile and adaptive nature of meta-learning ensures that master data remains congruous, precise, and of premium quality. This congruence translates into enhanced business decisions, streamlined operations, and favorable fiscal outcomes.

The pivotal role of master data in shaping business operations necessitates the employment of advanced tools and methodologies for its management. Meta-learning, endowed with the prowess to generalize across diverse tasks and swiftly adapt to novel data scenarios, offers a beacon of hope in navigating the intricate maze of master data management. The insights provided herein underscore the potential of meta-learning, spotlighting its efficacy not only in the realm of supply chain product classification but as a holistic solution to multifaceted master data challenges.

7.1 Structuring meta-learning data for master data supply chain classification

In the realm of meta-learning, particularly in the context of supply chain master data, it is imperative to establish a shared, consistent structure to facilitate multi-task learning. This not only ensures uniformity in the process but also enhances the model's ability to generalize across different tasks. The proposed structure aims to make this process streamlined, flexible, and broadly applicable.

Task column

Purpose: To categorize and segregate data into distinct tasks, especially crucial for multi-task learning. Each task represents a different classification challenge, often emanating from different systems or vendors in the supply chain.

Application: For instance, if products come from three distinct vendors or systems, each with their unique naming conventions, the "Task" column might be populated with labels such as “Vendor A”, “Vendor B”, and “Vendor C.” This delineation allows the meta-learning model to learn the idiosyncrasies of each vendor’s classification and thereby achieve specialization in each task.

Label column
Purpose: Represents the ground truth or the actual categorization for each data entry. This serves as the benchmark against which the model’s predictions are validated and compared.

Application: If a product from “Vendor A” is truly a “Laptop,” the “Label” column for that entry would reflect “Laptop.”

Feature columns
Purpose: To retain all the necessary attributes or descriptors of the data. These features provide the vital information that the meta-learning model uses to make its predictions.

Application: For a product dataset, this might include attributes such as “Product ID,” “Product Description,” “Vendor Price,” among others.

Predicted label column
Purpose: This column captures the model’s predictions for each data entry. After training, when the model is exposed to new or test data, it endeavors to predict the category or class of each product based on the given features.

Application: Using the earlier example, if a product from “Vendor B” has features indicating it is a portable computing device, the “Predicted label” might display “Laptop” if that is the categorization the model deduces.

The beauty of this structured approach is its *broad applicability.* While we have outlined its relevance to the master data supply chain context, its adaptability means it can be harnessed for other generic master data scenarios. Furthermore, this approach is not limited to master data alone. It can be seamlessly adapted to other classification use cases across diverse domains, providing a versatile framework for implementing multi-task learning. In essence, the proposed structure offers a standardized yet flexible blueprint for leveraging meta-learning across a spectrum of classification challenges, ensuring optimal utility and consistency in predictive endeavors.

7.2 Deep meta- and multi-task learning in master data supply chain classification: process breakdown

Having structured our data appropriately for deep meta- and multi-task learning, the stage was set for us to proceed with the actual implementation. This journey can be best described in the following steps:

1. Pre-trained Nature Language Processing (NLP) tokenizer and model are loaded to tokenize input text and extract embedding from the NLP model.
2. Data loading and preprocessing:
   a. The dataset is loaded.
   b. For each unique task in the dataset:
      i. The dataset is filtered to focus on the current task.
3. Setting up MAML:
   a. Hyperparameters such as meta-learning rate and adaptation learning rate are set.
   b. The MAML model is initialized using defined hyperparameters.
   c. An *optimizer and a loss function* are also defined for the training process.
4. MAML training loop:
   a. Inner loop (Figure 7): Fast adaptation: For each batch of data (representing a specific task) from the training set, the model quickly adapts its parameters based on the error in predictions. This is done by performing a few gradient descent steps on the loss of the current task. This adapted model is often termed as the "learner."
   b. Outer loop(Figure 6): Meta-update: After the inner loop completes for all tasks (or batches), a meta-update is performed. This is done using the losses computed from the adapted models (learners) of each task. The main idea is to update the model's parameters in a way that it becomes easier to adapt to new tasks in the future.
5. TensorBoard logging: During training, various metrics (like training loss, validation loss, training accuracy, and validation accuracy) are logged for visualization in TensorBoard.
6. Model evaluation and saving results:
   a. After training, the model evaluates its performance on the entire dataset for the current task.
   b. The predicted labels are decoded back to their original form and added to the dataset.
   c. The dataset, now with predicted labels, is saved as an output dataset file.
7. Final completion:
   a. The process repeats each unique task in the dataset.

The MAML approach ensures that the model learns to quickly adapt to new tasks using minimal data, making it suitable for scenarios like supply chain classification where new tasks might arise frequently and the data for each task might be limited.

There are some interesting comparisons when compared to results of not using MAML and having all data together.

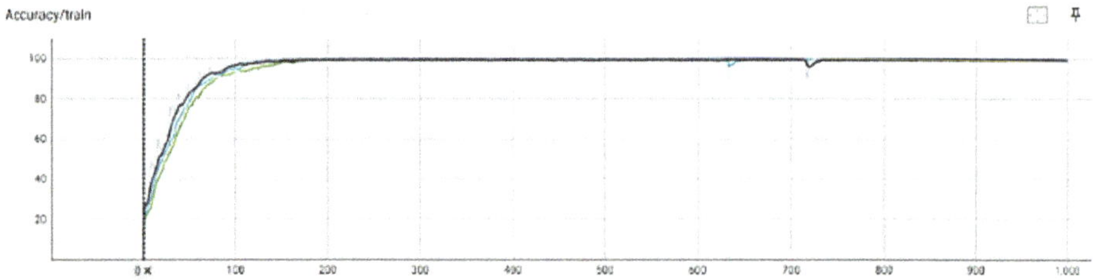

Figure 9: Accuracy/train graph for classification with MAML.

From Figure 9, we can see that on the *x*-axis, we have the number of epochs or iterations, while the *y*-axis showcases the accuracy percentage. As we observe the graph, a notable uptick in accuracy is evident in the initial epochs. By the time we approach approximately 100 epochs, this accuracy levels out, consistently hovering close to 100%. Such a trend signifies that the model has not only attained an exemplary performance level but also confidently predicts outcomes on the validation dataset. Furthermore, the presence of multiple lines for various tasks indicates a uniform performance, suggesting consistent results irrespective of different settings or initial conditions.

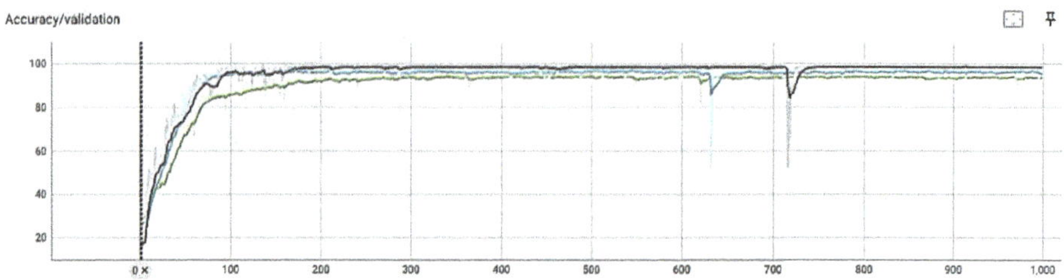

Figure 10: Accuracy/validation graph for classification with MAML.

The Accuracy/validation graph (Figure 10) illustrates the validation accuracy across epochs, with the *x*-axis representing the number of epochs and the *y*-axis showing accuracy percentage. Initially, there is a sharp increase in accuracy, moving from below 20% to around 80% in early epochs, signifying quick learning. After this ascent, the accuracy plateaus between 80% and 100%, indicating the model's performance on validation data stabilizes. However, abrupt accuracy drops, particularly around epochs between 600 and 800, reveal moments of instability, potentially due to challenging data points, *overfitting*, or a high learning rate. While the model recovers and maintains high accuracy post these dips, such fluctuations highlight areas warranting further investigation.

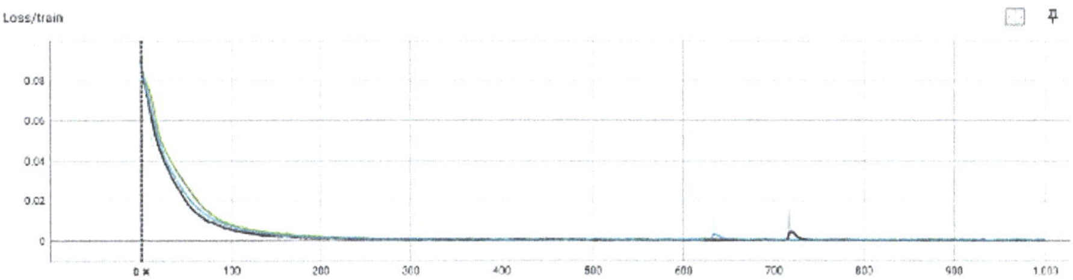

Figure 11: Loss/train graph for classification with MAML.

The Loss/train graph (Figure 11) illustrates the training loss as the model progresses through epochs, with the *x*-axis denoting the epoch count and the *y*-axis representing the loss magnitude. Initially, the model exhibits a high loss, which sharply drops, indicating the model's adaptability and learning phase to cater to the training data. As the process unfolds, the loss tends to stabilize, showing minor fluctuations, a sign of the model nearing its optimal performance on the training set. Notably, minor spikes towards the conclusion may point to possible noise in the data or an elevated learning rate that might induce these variations.

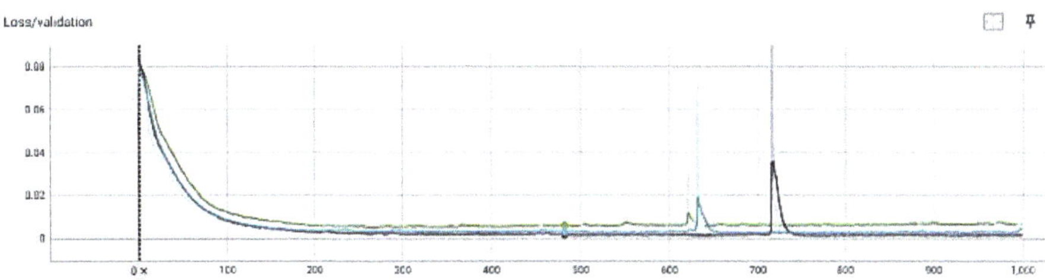

Figure 12: Loss/Validation Graph for Classification with MAML.

In Figure 12, the *x*-axis delineates the number of epochs or iterations, while the *y*-axis reflects the loss value. Early on, there is a pronounced decline in loss during the initial epochs, signaling that the model is effectively learning and refining its predictions. By the time we near the 100-epoch mark, this decline plateaus, settling around a minimal value. This plateau suggests that the model has largely converged, with further training yielding minimal reductions in error.

Much like the prior graph, multiple lines represent different tasks. Notably, all these lines gravitate towards a similar loss value, underlining the model's stability and robustness across diverse configurations.

Taking a step back to look at the overarching trend in both graphs, there is a clear pattern: swift adaptability in the early stages followed by a period of stability. The confluence of high accuracy and minimal loss post-convergence is indicative of adeptly trained models, poised to excel on both training and validation datasets. Furthermore, the consistent performance depicted by the multiple lines across both graphs suggests uniformity in results, irrespective of the model or configuration in play. Let us compare the MAML results to a simple neural network learning all the tasks without the MAML (Figure 11):

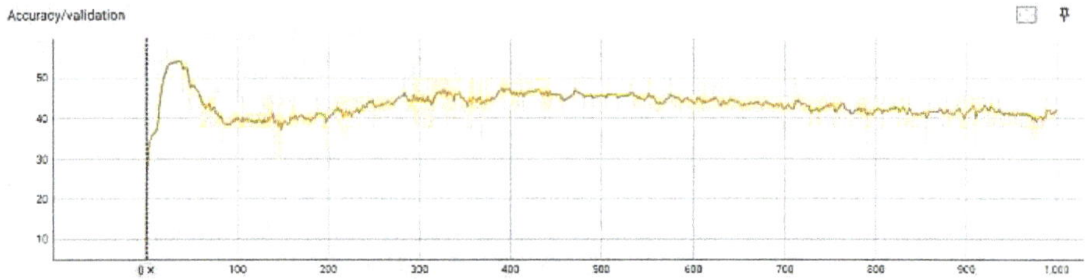

Figure 13: Accuracy/validation graph for classification deep learning without MAML.

The Accuracy/validation graph (Figure 13) offers insights into the model's predictive prowess on the validation set as it progresses through the epoch. Early on, there is a striking surge in accuracy during the initial epochs, indicating the model's capacity to swiftly identify and learn patterns present in the training data.

However, as training continues, the graph showcases an oscillatory behavior, particularly evident between the 300–700 epoch range. This wavering can be indicative of learning instability, potentially stemming from a heightened learning rate or the presence of noisy data. Beyond this phase, the accuracy levels off, plateauing in the vicinity of the 40–50% mark. This stabilization hints that the model, given its current architecture and hyperparameters, might have maximized its learning potential from the available data.

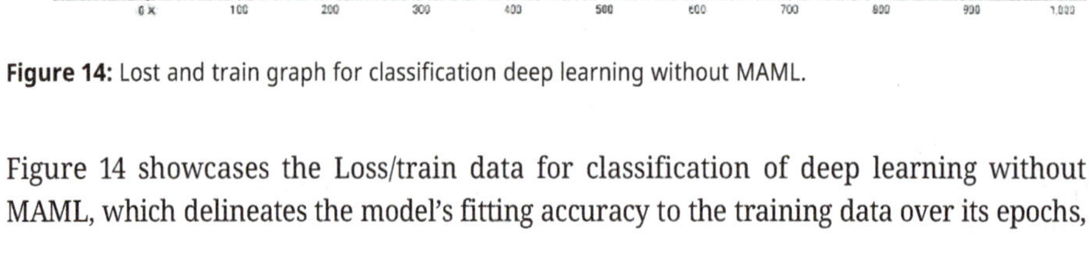

Figure 14: Lost and train graph for classification deep learning without MAML.

Figure 14 showcases the Loss/train data for classification of deep learning without MAML, which delineates the model's fitting accuracy to the training data over its epochs, as gauged by a decreasing loss function – often MSE, cross-entropy, among others. Throughout the epoch, we observe a consistent decline in the loss, a typical characteristic of a neural network refining its weight parameters to better align with the training data. However, as training progresses, the loss starts to flatten, signaling convergence. This allows us to infer that any additional training might not yield a notable reduction in training loss.

Shifting our focus to the MAML, tailored for multi-task learning, it stands out in its ability to generalize effectively to new tasks with minimal training. When juxtaposing the performance metrics of the standalone neural network against a meta-learning model like MAML, we note several distinctions. For starters, the MAML could potentially employ a more adaptable learning rate, shedding light on why the single network displays greater oscillation in accuracy. Further, while the standalone network's accuracy plateaus, suggesting possible challenges in generalizing across an amalgamation of tasks, MAML's design inherently gears towards enhanced task generalization.

The single neural network seems to learn some patterns, but it is less skilled at handling complex tasks compared to a meta-learning framework like MAML. These differences might stem from various factors, such as the inherent limitations of architecture, tendencies of overfitting, or challenges in achieving a wide-based generalization across diverse tasks. The modern business environment is full of challenges regarding the maintenance of accurate and consistent master data. With master data being the basis of organizational operations, its effective management becomes essential. However, complexities arise due to varied systems, naming conventions, and stakeholder inputs. One such domain that illustrates this struggle is the realm of supply chain management, where product classifications often show discrepancies due to different system conventions. This leads to a domino effect of errors in demand forecasting and financial reports.

To tackle these inconsistencies, the advent of meta-learning offers a promising avenue. Meta-learning, with its "learning to learn" paradigm, is uniquely equipped to handle varied data, ensuring rapid adaptability to new tasks. Its ability to generalize based on multiple tasks provides organizations the ability to manage diverse vendor data and ensures uniformity in product categorizations. Beyond the supply chain, the versatility of meta-learning is evident in other master data spheres, underpinning its value in maintaining data quality and congruence. As we explore the granular details of structuring meta-learning for master data classification, the emphasis lies on creating a coherent structure to facilitate multi-task learning. Such structuring includes designating columns for tasks, labels, features, and predicted labels, making it suitable for diverse master data scenarios and beyond. When examining the process breakdown of implementing deep meta- and multi-task learning, the journey is systematic. It starts with tokenizing input text, loading the data, initializing the MAML framework, followed by a two-step training loop involving fast adaptation and meta-update. Model performance evaluation and iterations for unique tasks complete the process. Comparative assessments between MAML and standalone neural networks highlight the former's prowess. Where traditional models might exhibit learning instability or plateauing accuracy, MAML's ability to quickly adapt and generalize shines through. In conclusion, in the maze of master data management, meta-learning emerges as a beacon of promise. Its capacity to traverse varied tasks and adapt rapidly to new data scenarios emphasizes its potential as a holistic solution to the multifaceted challenges of master data management.

8 Conclusion

The global trade domain, influenced by the adoption of AI and ML in supply chain operations, requires more than just traditional machine learning methods. The joint potential of meta-learning and multi-task learning arises as a solution specially designed for this challenge. These methods, using shared knowledge and adaptability, offer unmatched advantages: from improving predictive performance and generalization to ensuring effective data utilization and quick adaptability to new scenarios. However, the choice of tasks, models, and training strategies depends on the specific context, data availability, and underlying task structures. While we acknowledge the significant improvements enabled by meta-learning and multi-task learning, it is clear that there is scope for further research and in-depth experimentation. As supply chain management moves towards a more data-driven approach, the problem of data scarcity becomes more prominent. Meta-learning and multi-task learning stand out as solutions, allowing organizations to make informed predictions even with limited datasets. By adopting these methods, there is a clear opportunity for enterprises to enhance customer experiences, optimize costs, and streamline operations. Our research emphasizes the need for continued investigation in this area, with the aim of creating algorithms that fully exploit domain knowledge to address data limitations.

References

[1] S. Helper, and E. Soltas, Why the Pandemic Has Disrupted Supply Chains, June 17, 2021, https://www.whitehouse.gov/cea/written-materials/2021/06/17/why-the-pandemic-has-disrupted-supply-chains/, Access Date: April 10, 2023.

[2] The White House, FACT SHEET: Biden-Harris Administration Announces Supply Chain Disruptions Task Force to Address Short-Term Supply Chain Discontinuities, June 08, 2021, https://www.whitehouse.gov/briefing-room/statements-releases/2021/06/08/fact-sheet-biden-harris-administration-announces-supply-chain-disruptions-task-force-to-address-short-term-supply-chain-discontinuities/, Access Date: April 05, 2023.

[3] The US Bank, How do supply chain issues contribute to inflation?, February 21, 2023, https://www.usbank.com/investing/financial-perspectives/market-news/supply-chain-issues-contribution-to-inflation.html, Access Date: April 05, 2023.

[4] D. Ardeshana, Supply Chain Shipment Price Data Analysis, 2018, https://www.kaggle.com/code/divyeshardeshana/supply-chain-shipment-price-data-analysis, Access Date: April 05, 2023.
[5] Trade.gov, Know Your Incoterms, https://www.trade.gov/know-your-incoterms, Access Date: April 05, 2023.
[6] J. Tartal, Process Validation, November 4, 2015, https://www.fda.gov/media/109539/download, Access Date: April 05, 2023.
[7] J. Tartal, Quality System Regulation Process Validation, September 30, 2015, https://www.fda.gov/media/94074/download, Access Date: April 20, 2023.
[8] S. Bengio *et al.*, "On the Optimization of a Synaptic Learning Rule," 2007, https://www.semanticscholar.org/paper/On-the-Optimization-of-a-Synaptic-Learning-Rule-Bengio-Bengio/8784f905f4f9fb6fa4a3cc9b0faa5b5479c687ec.
[9] A. Krizhevsky, I. Sutskever, and G. E. Hinton, ImageNet Classification with Deep Convolutional Neural Networks, January 2012, https://www.researchgate.net/publication/267960550_ImageNet_Classification_with_Deep_Convolutional_Neural_Networks, Access Date: April 20, 2023.
[10] A. Ghaddar, and P. Langlais, SEDAR: A Large Scale French-English Financial Domain Parallel Corpus, 2020, https://aclanthology.org/2020.lrec-1.442.pdf, Access Date: April 20, 2023
[11] W.-N. Hsu, A. Sriram, A. Baevski, T. Likhomanenko, X. Qiantong, V. Pratap, J. Kahn, A. Lee, R. Collobert, G. Synnaeve, and M. Auli, Wav2Vec2-Large-Robust fine-tuned on Switchboard, 2014, https://huggingface.co/facebook/wav2vec2-large-robust-ft-swbd-300h, Access Date: March 10, 2023.
[12] C. Finn, P. Abbeel, and S. Levine, Model-Agnostic Meta-Learning for Fast Adaptation of Deep Networks, 2017, https://arxiv.org/pdf/1703.03400.pdf, Access Date: January 08, 2023.
[13] M. Andrychowicz, M. Denil, S. Gomez, M. W. Hoffman, D. Pfau, T. Schaul, B. Shillingford, and N. De Freitas, "Learning to Learn by Gradient Descent by Gradient Descent," in *Advances in Neural Information Processing Systems*. 2016, pp. 3981–3989, https://arxiv.org/abs/1606.04474.
[14] C. Finn, P. Abbeel, and S. Levine, Model-agnostic Meta-learning for Fast Adaptation of Deep Networks, *Proceedings of the 34th International Conference on Machine Learning-Volume 70*, 2017, pp. 1126–1135. JMLR. org.
[15] Use Case Data Source, https://data.world/usaid/supply-chain-shipment-pricing.
[16] Y. Yorozu, M. Hirano, K. Oka, and Y. Tagawa, "Electron Spectroscopy Studies on Magneto-optical Media and Plastic Substrate Interface," *IEEE Translation Journal on Magnetics in Japan*, vol. 2, pp. 740–741, August 1987 [Digests 9th Annual Conf. Magnetics Japan, p. 301, 1982].
[17] M. Young, *The Technical Writer's Handbook*. Mill Valley, CA: University Science, 1989.

Cory Davis, Patrick Stockton, Eugene B. John, Zachary Susskind, and Lizy K. John

Characterization of Neuro-Symbolic AI and Graph Convolutional Network workloads

Abstract: The explosive growth of artificial intelligence has created new domains of AI models. These domains include Neuro-Symbolic AI (NSAI) and Graph Neural Networks (GNN). NSAI and GNN models have already demonstrated the capability to significantly outperform deep learning models in domains such as image and video reasoning, and network classification, respectively. They have also been shown to obtain high accuracy with significantly less training data than traditional neural network models. However, the recent emergence of the field, and relative sparsity of published results, leads to a meager understanding of the performance characteristics of these models. In this work, we describe and analyze four models in the NSAI and GNN domains. We find that the NSAI models have less potential for parallelism than traditional neural models due to complex control flow, low compute-to-byte operations, and high cost of data movement. Additionally, in the graph network, we find an abundance of sparse matrix multiplication and similar low compute-to-byte operations. These operations have low potential for parallelism, and instead will focus on improved techniques for element-wise operations.

Keywords: neuro-symbolic, graph networks, machine learning, performance, workload profiling

Acknowledgment: This research was supported in part by Semiconductor Research Corporation (SRC) Task 3015.001/3016.001 and National Science Foundation grant number 1763848. Any opinions, findings, conclusions, or recommendations are those of the authors and not of the funding agencies.

Cory Davis, The University of Texas at San Antonio, San Antonio, Texas, USA, e-mail: cory.davis@utsa.edu
Patrick Stockton, The University of Texas at San Antonio, San Antonio, Texas, USA, e-mail: patrick.stockton@utsa.edu
Eugene B. John, The University of Texas at San Antonio, San Antonio, Texas, USA, e-mail: eugene.john@utsa.edu
Zachary Susskind, The University of Texas at Austin, Austin, Texas, USA, e-mail: zsusskind@utexas.edu
Lizy K. John, The University of Texas at Austin, Austin, Texas, USA, e-mail: ljohn@utexas.edu

https://doi.org/10.1515/9783111344126-009

1 Introduction

Traditional neural networks have been effective in solving problems in many areas through the use of Deep Learning (DL). These traditional models require well-defined hyperparameters, which need to be readable and understandable to users, a difficult feat due to the tremendous amount of hyperparameters that some models use. The topology of a model is defined during the selection of hyperparameters. This makes it a trivial observation for a user; however the abstract features that the neural network manipulates is a difficult concept. Evaluating the performance of neural network models can provide additional insight into the internal workings of the models.

Neuro-Symbolic AI (NSAI) is an emerging AI domain that combines deep learning for feature extraction and rules-based "intuition" for manipulating those features. Rules-based, or symbolic, approaches dominated the field of AI until the 1980s [1]. Symbolic models had several advantages: they required only a few input samples, generalized well to new problems, and their internal functionality was conceptually simple when compared to DL models. At the same time, they required substantial hand-tuning, which made them difficult to create for complex problems. A far larger issue was that they simply were not very accurate: in 1973, the entire field of AI was summarized as "increasingly disappointing" [2], and by the 1980s, research had spiraled into what became known as the "AI winter" [3].

The nascent field of NSAI blends traditional symbolic methodologies with modern DL to leverage the strengths of both domains. For instance, the Neuro-Symbolic Concept Learner (NSCL) [4] and Neuro-Symbolic Dynamic Reasoning (NS-DR) [5] models use DL to extract features from images or videos and generate symbolic tokens from accompanying natural language questions. Tokens correspond to attributes of objects ("red", "metallic"), relations between objects ("behind", "left of"), and verbs ("filter", "find"), and thus form a restricted language. "Sentences" in this language can be directly evaluated by a fixed-function symbolic model, which uses them to filter and manipulate the extracted image or video features. By explicitly representing logical relations between abstract concepts, neuro-symbolic models such as the NSCL and NS-DR have a degree of inherent explainability. For instance, while understanding the internal behaviors of DL models requires sophisticated analysis, the token "blue" unambiguously represents the color blue. NSAI can also be used to bring human intuition into the model creation process, by for instance defining the lexicon of tokens and the relational rules that govern how they should be processed.

The neuro-symbolic Neural Logic Machine (NLM) research was conducted by collaboration between Google Inc., ByteDance Inc., and Tsinghua University [6]. The resulting NLM architecture provides a state-of-the-art method for solving general application tasks such as array sorting, critical path finding, and more complex tasks such as Blocks World [7]. Blocks World is a classic symbolic reasoning problem where the model is given a set number of blocks and logical rules. Using the provided generalized rules, the model will need to perform the available logical actions to achieve the desired target result from the randomized starting layout.

Symbolic relational reasoning has relations with the NLM as an application in processing discrete data structures that can be represented as knowledge and social graphs [8]. This allows for learning from increasing complexity of logical rules.

The problem of classifying nodes in a graph network is one that took off at the start of the 2000s. Large graph networks can be highly complex and the need for large labeled datasets is costly. Seeger [9] delves into using semi-supervised learning and proposed several novel approaches (at the time) for learning. Zhu et al. [10] applies some of the proposed concepts to large graphical datasets. While not using neural networks, the Gaussian model in [10] greatly improved efficiency of training graphical datasets. Sen et al. [11] discusses different methods of traditional machine learning algorithms that aim to classify networks. Weston et al. [12] showed how NNs can be utilized for graph networks. Xu et al. [13] details popular GNN variants to show how powerful the models are at identifying graph structures. The Graph Convolutional Network (GCN), as described by Kipf and Welling [14], is one of the models Xu et al. [13] details as it provides a strong avenue for classification of nodes in a network.

As these models become increasingly computationally complex and data intensive, the computational cost increases greatly. Machine learning researchers must balance these costs to create efficient models. Amazon [15] implements SageMaker, an in-house profiler, to capture performance metrics. DeepProf [16] is a tool used for processing GPU traces for generating performance analyses.

In this paper, we analyze the performance characteristics of two separate neural models, using NVidia's DLProf. The NLM model uses an object's relations, properties, quantifiers, and logic connectives in order to accomplish the task of generalization. The GCN is a semi-supervised classification model designed for graph or nodal structured data. Using the Cora dataset, GCN demonstrates capabilities of highly accurate classification while only using 5% labeled data during training.

The remainder of this paper is organized as follows: Section 2 describes the models that we are analyzing in this paper in detail. Section 3 describes our methodology for analyzing the performance of these models, based on classifying activity into distinct categories. Section 4 provides the results of our research with breakdowns for each model component. Section 5 provides our takeaways on the behavior and potential opportunities for acceleration of both models. Finally, in Section 6, we summarize our findings. We also provide direct links to the repositories of the models cited in this paper as an appendix.

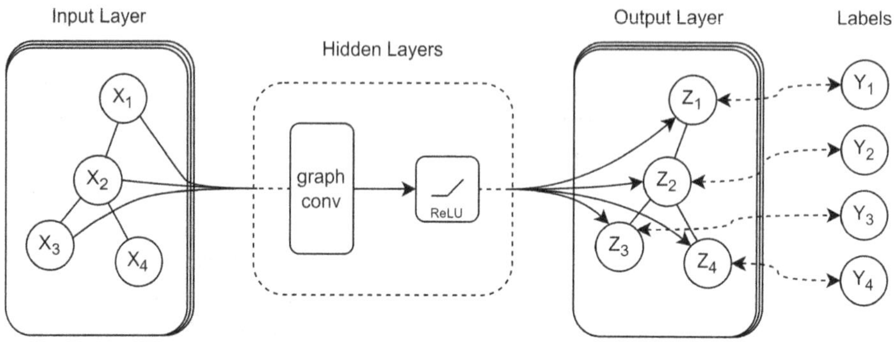

Figure 1: Illustration of the GCN framework, where X_i are the input nodes and Y_i are the predicted labels for the processed nodes. Based on Figure 1(a) in [14].

2 Model overview

2.1 Graph Convolutional Network

The GCN was developed through work between the University of Amsterdam and the Canadian Institute for Advanced Research (CIFAR), and published at the 2017 International Conference on Learning Representations (ICLR). Many real-world datasets are in the form of graphs or networks: social networks and knowledge graphs are some examples. Prior to 2015 very little focus was devoted to these forms of datasets. In [14] the authors describe the difficulties of implementing classic CNN and RNN models to work with arbitrarily structured graphs. CNNs required the simplification of graph networks into vectors; however this preprocessing caused information loss. RNNs have a higher ability to directly process graph data. Even so, RNNs could only process directed and acyclic graph networks [17].

The first Graph Neural Network [17] (GNN) was an extension of RNNs for usage with graph and node networks. In [18] the authors continued to expand on their revised RNN concept, in order to apply RNNs to cyclic and undirected graphs in addition to the acyclic and directed graphs RNNs were already capable of processing. This first GNN model is based around information diffusion and relaxation mechanisms [18], but also requires multiple applications of contraction maps until the nodes reach a point of stability; [19] attempts to solve this problem by way of gated recurrent units in conjunction with more modern RNN optimization techniques. This resulted in a more useful and generalized class of graph networks.

Duvenaud et al. [20] introduced circular fingerprints that generate feature layers using a fixed hash function. This method pools features of the prior layer's neighborhood (all nodes and edges connected to a specific node). Circular fingerprints resemble convolutional networks, but take separate mathematical approaches. This approach performs well on small graphs, but does not scale well into large networks [20].

The GCN network defined in [14] introduces a semi-supervised classification model designed for graph-structured data. Unlike other graph-structured models [14], GCN uses an efficient layer-wise propagation rule based on first-order approximations of spectral convolutions on graphs. The GCN model requires an input feature matrix, an adjacency matrix that describes the relationship between nodes, and the binary label matrix. Additionally, the model will create a diagonal node degree matrix that is necessary for the normalization of the adjacency matrix. The required adjacency matrices used in this model contain undirected edges, and, because not all graph nodes are connected, these matrices contain large amounts of zeros. When the diagonal node degree matrix is generated, only the principal diagonal contains pertinent values, while all other diagonals contain zeros. To increase processing efficiency, the GCN implementation utilizes sparse MM algorithms, available in PyTorch.

Whereas the previously mentioned prior work [20] requires node degree-specific weight matrices, GCN implements single weight matrix per layer and deals with varying node degrees through an appropriate normalization of the adjacency matrix. This implementation change allows GCN to scale well to large graph networks [14].

This categorization uses the PyTorch implementation of GCN. The corresponding dataset is a subset of the Cora dataset. This Cora subset consists of 2,708 Machine Learning papers classified into one of seven different categories: case-based, genetic algorithms, neural networks, probabilistic methods, reinforcement learning, rule learning, and theory. Every paper in the dataset is cited by at least one other paper in the dataset. This dataset is designed for a classification task. The model uses the absence or presence of words in a dictionary, in conjunction with the citation network, to infer which ML category a paper belongs.

2.2 Neural Logic Machines

The LM [6] is a neuro-symbolic architecture that uses inductive logic programming (ILP) to learn logic rules from examples [21] and logic reasoning to perform various logical tasks. The NLM was published in the International Conference on Learning Representations (ICLR) in 2019 as the realized work between Tsinghua University, Google Inc., and ByteDance Inc. As a neuro-symbolic AI architecture, the NLM can learn a set of logic rules using ILP from logic-based task examples by combining both symbols and probabilities [6]. The NLM applies both inductive learning and logic reasoning to an object's relations, properties, quantifiers, and logic connectives in order to accomplish the task of generalization. Unlike most traditional networks, NLMs are able to achieve perfect generalization in various tasks such as family tree decisions, sorting of arrays, finding the shortest path between points, or playing the Blocks World task. The NLM architecture provides the ability to solve historically challenging

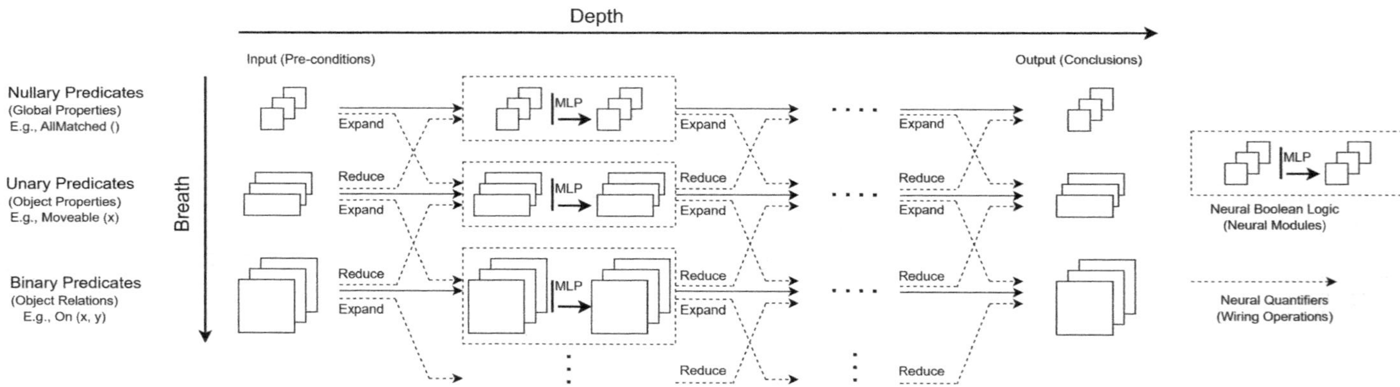

Figure 2: An illustration of the NLM framework showing object properties and object relations as inputs, the concluding outputs of the objects properties and relations, and the internal logical structure. Based on Figure 2 in [6].

tasks that traditional neural network architectures struggle to complete. The challenge of generalization of tasks from small scale to large scale has been proven to be overcome using NLM. The NLM proves to accomplish complex tasks by overcoming major challenges that traditional neural networks and inductive logic reasoning systems cannot solve alone. The tasks that were used involved the graph-based path task, the general sort application task, and the more complex Blocks World task. These different problems presented variations in the system's performance, as each task involved different levels of logic rule set complexity. In addition, the NLM architecture also solves the problem of scalability with respect to the complexity of rules given to the system. As the rules of the task scale up, the complexity of the logic rules to be learned will also scale up exponentially. This allows the NLM to adjust its trained rules based on uses of a minimal set of prior examples. Using a minimum set of prior data illustrates the ability for the NLM to effectively improve as it learns. As these different tasks use a variation of input data and parameters, the performance evaluation of these exercises gives many opportunities for improvement.

Figure 2 shows an illustration for the NLM framework. The framework of the NLM is represented by the breath and the depth of the model. The breath of the NLM model represents the number of inputs, or predicates, to determine the rule set complexity of the task being performed. As the rule set complexity increases, additional object properties and relationships are considered. The depth of the NLM represents the number of hidden layers present in the model. Each hidden layer consists of a multi-layer perceptron (MLP) to operate on the input data.

2.3 Neuro-Symbolic Concept Learner

The NSCL was designed for the CLEVR dataset [22]. CLEVR is a dataset for "image reasoning": images are presented to the model, along with a set of related questions, and the model's outputs are the answers to these questions. Image samples in CLEVR contain cubes, cylinders, and spheres with different sizes, colors, and materials. The NSCL is composed of the three submodels described below.

1) *Image parser:* The objective of the image parser is to generate object "masks": pixel-accurate regions with annotated colors, shapes, and materials. This is accomplished using a Mask R-CNN model, which constructs object segmentation masks and classifications in parallel [23]. While the addition of object mask generation makes its structure somewhat more complex than a traditional convolutional neural network, both branches of computation are internally convolutional.

The implementation of Mask R-CNN originally used for the NSCL proved challenging to run on modern hardware. Not only was no pretrained model provided, but the software libraries required were also obsolete and not compatible with modern CUDA versions. Therefore, we decided to use a more recent, pretrained Mask R-CNN model provided by Facebook's Detectron2 [24]. This also had the advantage of giving us access to the profiling tools built into more recent versions of the PyTorch framework.

2) *Question parser:* Questions in the CLEVR dataset are in the form of natural language, which presents the challenge of translating them into a form usable by the model. The authors of the NSCL accomplished this by defining a domain-specific language, including verbs, such as "filter" or "intersect", and concepts, such as "blue" or "left". This effectively converts the problem of question parsing into neural machine translation (NMT). A bidirectional GRU [25] was used to accomplish this task for the original NS-DR. However, source code, a pretrained model, or any other information is not available for this model. As such, we chose to profile a small, modern pretrained transformer-based NMT model provided by Harvard's OpenNMT toolkit [26]. This provides similar advantages to the Detectron2 image model: more recent library versions and support for modern DL profiling tools. It also provides a more realistic insight into what deployment of this model would look like in a modern datacenter environment.

3) *Symbolic program executor:* The symbolic program executor converts extracted image features and question tokenizations into predictions. Formally, this model is "quasisymbolic": unlike a true symbolic AI model, which performs logical and Boolean operations on data, the executor approximates non-differentiable functions in a probabilistic manner. The output of the NSCL is a vector in which the ith element, $\text{Mask}_i \in [0, 1]$, represents the probability of the corresponding object in the scene being in the output set. This representation is useful for two reasons. First, it allows for multiple correct answers, which increases the complexity of the questions the model can answer. Second, it makes the symbolic step fully reversible, which allows backpropagation to occur. This approach avoids the limitations of true Boolean circuits, which are in general irreversible. In turn, this reversibility allows for the learning of concept embedding vectors: representations of object features, which allow concepts (such as color, shape, and material) to be learned without explicit labels [4].

2.4 Neuro-Symbolic Dynamic Reasoning

The Neuro-Symbolic Dynamic Reasoning (NS-DR) model is a neuro-symbolic model for the CLEVRER video reasoning dataset. CLEVRER builds on CLEVR by replacing images with videos. Objects in CLEVRER may enter or exit the scene throughout the course of a (5 s, 25 frame) video, potentially colliding and rebounding with other objects. Using videos instead of images adds a new domain to the questions in CLEVRER: that of causality. Causal questions may range from the descriptive ("Which objects collided?") to more complex causal relations ("What caused the objects to collide? Would the objects still have collided if a given counterfactual was true?"). Existing neural models were shown to perform poorly on all but the simple descriptive questions in CLEVRER [5].

In order to account for causal relations, the NS-DR introduces a new submodel: a neural dynamics predictor, which is essentially a learned physics engine. The introduction of the dynamics predictor brings the model up to a total of four independent submodels. We discuss the structure below.

1) *Video frame parser:* The video frame parser treats each frame of a video separately, using the same Mask R-CNN approach as the NSCL. Thus, for each input video, inference with this model must be run 25 times.
2) *Question parser:* The original NS-DR model used a more modern NMT model than the NSCL: Seq2Seq [27], which was demonstrated to be more accurate on long inputs than prior models. Once again, we opted to replace this model with the OpenNMT transformer model.
3) *Dynamics predictor:* The dynamics predictor, PropNet [28], is a learned physics engine that can represent complex collisions between objects. PropNet improves on prior work in the domain by accurately modeling propagation of force through multiple objects (such as in a Newton's cradle), and operating correctly in the presence of partial information (where not all objects are visible). Functional correctness with partial information is crucial, since all videos in CLEVRER are taken from a fixed camera angle, where objects are allowed to enter and leave the scene. Dynamics prediction provides the positions, trajectories, and collisions between objects for the NS-DR model. The results of the dynamics predictor are augmented with the properties identified by the video frame parser to provide a complete record of what occurred during the input video.
4) *Symbolic program executor:* The program executor of the NS-DR is a true symbolic model: unlike the NSCL, it uses non-differentiable operations to make predictions. The disadvantage of non-differentiable operations is that backpropagation of error is not possible; therefore, this model can not learn concept embedding, so concepts must be learned directly by the frame parser via supervised training. This is an entirely fixed-function model with no learned component; internally, it behaves much like a programming language interpreter, using tokens to apply filter and reduction operations to the extracted video features.

The NS-DR program executor is single-threaded and CPU-only. The sequential nature of its processing does not expose any obvious opportunities for the sorts of coarse-grain parallelism typical of DL workloads: in general, processing the nth token of a sequence will require the result of processing the $(n-1)$th token. An example symbolic program, shown in Figure 3, demonstrates how the symbolic program executor uses tokens to filter and perform basic arithmetic on the features extracted by the other submodels.

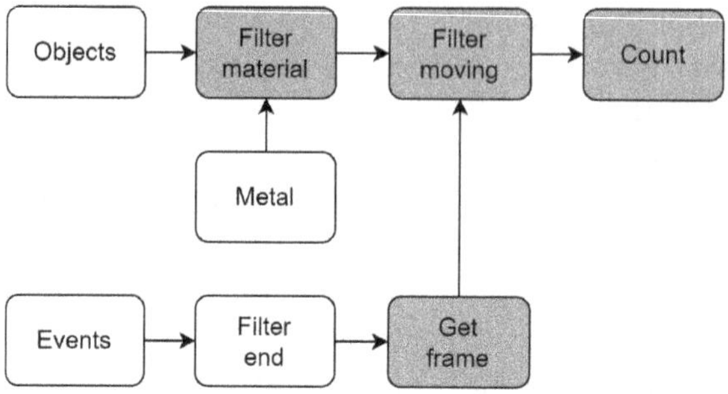

Figure 3: An example of the tokenized representation of a question in the CLEVRER dataset. "Noun/adjective" tokens – features – have white backgrounds, while "verb" tokens – actions – are shaded. Arrows show the dependencies for token processing [5].

3 Methodology

We used function-level profiling to capture statistics such as runtimes, invocation counts, and tensor sizes. We then developed a post-processing tool to partition the profiling results into nine dominant categories of operations.

3.1 Experimental methodology

The characterization of these workloads were performed using the Deep Learning Profiler (DLProf) designed by NVidia for GCN and NLM, and the PyTorch profiler for NSCL, NS-DR, and NLM. DLProf uses NVidia's Nsight Systems to perform performance analysis on deep learning frameworks in PyTorch and TensorFlow. Nsight Systems generates data on CPU and GPU functions: how often a function is called, the processing time, and system resource utilization. DLProf is capable of visualizing Nsight's generated data for more efficient reading. Additionally, DLProf provides Tensor Core usage for kernel operations and make performance recommendations via Expert Systems. Expert Systems is a DLProf beta feature designed to analyze the model and discover potential deficiencies.

Data collection for NLM, NSCL, and NS-DR was performed on a system with two Intel Xeon E5-2698 v3 processors and two NVIDIA Tesla M40s for both the DLProf and PyTorch profilers. Data collection for GCN was performed on a system running an Intel(R) Core(TM) i7-10750H CPU with one NVIDIA GeForce RTX 2060 GPU.

3.2 Workload taxonomy

Since simply reporting per-function profiling results would require us to discuss an overwhelming number of individual functions, we devised a function classification scheme to partition the data into six dominant categories of operations. We drew inspiration for this classification from the Berkeley "seven dwarfs" [29]: a mid-2000s effort to identify what were then the most crucial kernels for parallel computing. In fact, some of our categories are identical to Berkeley "dwarfs," but we identify several others that also appear highly relevant to the workloads discussed in this paper and/or machine learning more generally. Given the wide diversity of models in modern AI research, we do not claim that our classification scheme is universally applicable. However, we do hope to show that workload classification schemes are valuable for exploring the differences between models.

1) *Dense matrix multiplication:* Fast, efficient dense matrix multiplication (GEMM) remains a critical requirement for large DL models. Fully connected layers in neural networks use GEMM as their primary mathematical operation, often with very large input matrices.

Multiplication of large, dense matrices is very computationally intensive: the work to multiply a $m \times k$ matrix with a $k \times m$ matrix $W = O(mnk)$. At the same time, data intensity only grows quadratically with input size: given the two input matrices and $m \times n$ output matrix, the data intensity $Q = O(mk + kn + mn) = O(\max(mk, kn, mn))$. This gives an operational intensity of:

$$I = \frac{W}{Q} = \frac{O(mkn)}{O(\max(mk, kn, mn))} = O(\min(m, k, n))$$

When m, k, and n are all large, operational intensity can be very high. Since there are no internal dependencies in matrix multiplication (the multiply-and-add operations can be performed in any order), the multiplication of large, roughly square matrices is highly parallelizable. The challenge emerges when any one of the dimensions is small relative to the other two; in this case, the operational intensity approaches O(1), requiring highly efficient data movement to avoid becoming memory-bound. Such "tall-and-skinny" matrices are difficult to process efficiently on GPUs [30]. While operational intensity can sometimes be addressed by processing multiple inputs simultaneously via batching, this may not be an option for latency-sensitive inference operations where input must be processed as soon as it is received. An extreme case of tall-and-skinny GEMM is the multiplication of a matrix by a vector, as an n-element vector can be viewed as an $n \times 1$ matrix.

2) *Sparse matrix multiplication:* Sparse matrices are those in which the values of only some elements are specified; all other elements are assumed to be some constant value, typically 0. There are numerous ways to implement sparse matrices; in general,

there is a trade-off between the generality of the sparsity (how much structure is assumed) and how easy it is to implement in hardware [31]. Sparse matrix multiplication requires efficient mechanisms to perform lookups into the tables of nonzero values. Achieving high performance in sparse matrix multiplication involves careful consideration of algorithmic complexity and memory access patterns. This can be especially challenging when dealing with very large sparse matrices. Iterative techniques for solving linear systems rely on efficient sparse matrix multiplication [32]. Hardware implementations of sparse matrix multiplications can significantly accelerate the operation by involving custom hardware or GPUs that are optimized for parallel processing, such as work done on NVIDIA Kepler GPUs [33].

3) *Convolution:* Convolutions are another class of common, computationally intensive operation. Unlike in GEMM, one of the inputs, the filter, is multiplied with a set of submatrices of the other input. In theory, this should lead to excellent operational intensity, since filter weights are reused for each individual multiplication, and weights from the other matrix are reused by overlapping submatrices. In practice, convolutions have proven difficult to parallelize, and are frequently implemented using the im2col algorithm, which reduces the problem to matrix multiplication. Performing im2col requires substantial data movement and duplication, worsening the performance and memory footprint of convolution. Recent efforts have focused on performing convolution "natively", without invoking im2col [34]. The more complex memory access patterns of these approaches have their own considerations for efficient implementation. Whether im2col is used or not, convolution presents a workload distinct from simple GEMM.

4) *Element-wise tensor operations:* Many operations in DL are applied uniformly to all elements in a tensor. These include activation, normalization, tensor addition, Hadamard products, tensor-scalar operations, and relational operations. Activation functions such as sigmoid, hyperbolic tangent (tanh), and rectified linear unit (ReLU) introduce nonlinearities and enable the network to learn complex mappings. Normalization techniques are commonly encountered in element-wise operations to improve the training stability and convergence of deep neural networks, such as batch normalization and layer normalization. While tensor addition involves adding corresponding elements of two tensors, tensorscalar operations involve applying a scalar operation (such as addition or multiplication) to each element of a tensor. These operations are fundamental for combining information or adjusting the scale of the tensor. All of these operations are "embarrassingly parallel" in that they can be applied to every element of a tensor simultaneously, but are challenging in that they have poor operational intensity; each element of the input is only operated on once.

5) *Regional operations:* Some operations act on spatially local regions of tensors. The best-known example of a regional operation is pooling, which reduces the size of a tensor in one or more dimensions by performing some reduction operation (such as

max or average) regionally. This is not the only class of regional operation; other examples include non-maximum suppression and region-of-interest alignment in object detection networks. These operations are distinct from element-wise operations in that they operate on potentially overlapping regions rather than single elements and thus have more complex access patterns; they are distinct from convolution in that they operate on only a single tensor and typically involve less computation.

6) *Embedding lookups:* Embedding are a way of transforming data indices or high-dimensional one-hot vectors into low-dimensional learned vectors. In practice, during inference, embedding function as lookup tables. The irregular memory access patterns of embedding lookups and the large sizes of embedding tables make acceleration a challenge. Traditionally, embedding lookups are latency-sensitive, memory-bound operations for inference [35].

7) *Data movement:* Many types of operations require substantial data movement but little or no computation. This primarily consists of host-device and device-host transfers; we also include operations such as tensor duplication and assignment.

8) *Data transformation:* Data transformation operations are those which reshape or subsample data. This includes matrix transposes, tensor reordering, and masked selection. We also include coalescing in this category, which is a process in which duplicate entries for the same coordinates in a sparse matrix are eliminated by summing their associated values [36].

9) *CUDA:* The final operational category is CUDA. The CUDA Runtime API consists of operations that are used to interface with Nvidia GPUs. The CUDA functions used by these models create the model on the first iteration, as well as a synchronization of memory transfers in the GPU. These functions are not detected by the PyTorch profiler used with NSCL and NSDR, but they are detected by DLProf used with NLM and GCN.

4 Results

In this section, we present and discuss results for the GCN, NLM, NSCL, NS-DR architectures, and individual submodels introduced in Section 3.

4.1 Graph Convolutional Network

The following results are based on data collected from iterations 2–200. The total CPU runtime was approximately 1.985 s, and the total GPU runtime at a significantly lower 0.457 s. As understood by the description, the GCN model requires a considerable amount of matrix multiplication. The results provided by DLProf confirm this profile with the additional context of the type of matrix multiplication being sparse matrix multiplication.

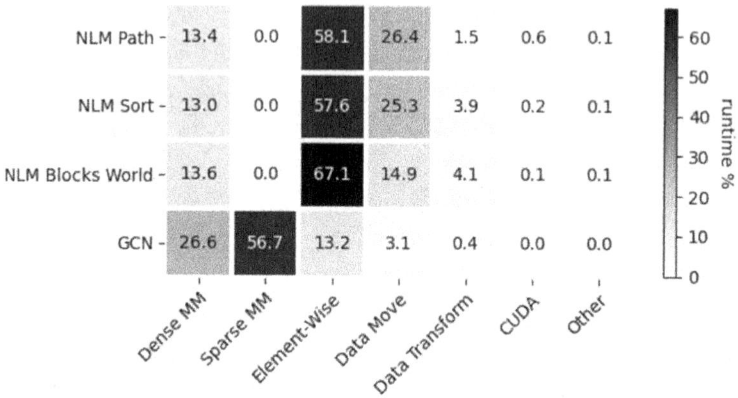

Figure 4: Heatmap showing the proportion of each category of GPU operation of the GCN and NLM using the DLProf profiler.

Table 1: GPU runtimes and runtime breakdowns for GCN and NLM using the DLProf profiler discussed in this paper.

Model	GEMM	Sparse MM	Element-wise	Data movement	Data transform	CUDA	Other	Total
NLM Path	1.37 ms	0 ms	5.95 ms	2.70 ms	0.15 ms	0.06 ms	0.01 ms	10.24 ms
NLM Sort	6.68 ms	0 ms	29.36 ms	12.87 ms	1.97 ms	0.09 ms	0.05 ms	51.02 ms
NLM Blocks World	12.74 ms	0 ms	62.75 ms	13.94 ms	3.83 ms	0.13 ms	0.10 ms	93.49 ms
GCN	121.2 ms	258.6 ms	60.4 ms	13.9 ms	1.78 ms	0 ms	0 ms	457 ms

The CPU profile (Figure 5) is dominated by element-wise operations (52.7%), followed by sparse MM (40.3%), with dense MM (3.1%) a very distant third. Meanwhile, the GPU profile (Figure 4), is noticeably dominated by sparse MM (56.7%), followed by dense MM (26.6%), and element-wise operations (13.2%). The adjacency and the diagonal node degree matrices, used in the GCN model, are largely populated by zeros; this high population leads to the necessary usage of sparse MM in the model. The current implementation of the GCN model has a very low GPU utilization (6%).

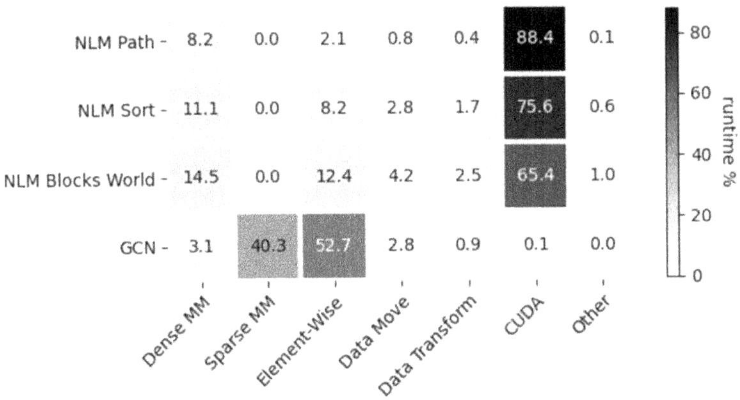

Figure 5: Heatmap showing the proportion of each category of CPU operation of the GCN and NLM using the DLProf profiler.

4.2 Neural Logic Machines

A characterization profile based on the CPU and GPU usage was created for each of the NLM logic tasks: sort, path, and Blocks World using the DLProf profiler shown in Figure 4 for the GPU and Figure 5 for the CPU. Each task provided ranging variations in the produced results. The total GPU runtime of the path, sort, and Blocks World tasks were 10.24 ms, 51.02 ms, and 93.49 ms, respectively, as shown in Table 1. The largest category of GPU operation was the "element-wise operations" category, where a significant amount of overall time was spent. The percentage of time spent in the element-wise operations category can be determined as 57.6% for the sort task, 58.1% for the critical path task, and 67.1% for the Blocks World task. The Blocks World task resulted in the highest percentage of element-wise functions related to the vector multiplication being performed on the numerical task. The high percentage of element-wise functions is also represented in the path task, with approximately a third of the total runtime spent on element-wise tasks.

Data movement represents the second largest GPU category that was characterized during the task executions. The sort task and path task shared the highest percentage of data movement within the model's execution. This is predominantly due to the frequent movement of numerical values in these general tasks. Blocks World yielded the lowest data movement characteristic, which can be perceived by the more symbolic and logical flow of functions and its data.

Following the "data movement" category, the dense matrix multiplication category contributed the third most to the task executions. Dense matrix multiplication plays a vital role in matrix-level multiplications and allows for parallelization.

The CPU results from the DLProf profiler had a significantly longer runtime compared to the GPU operations where the path, sort, and Blocks World tasks runtimes were 2,417.58 ms, 2,842.56 ms, and 3,283.56 ms, respectively. The dominating category of the CPU characterization profiling is the CUDA category. The CUDA runtime totaled 88.4%, 75.6%, and 65.4% of the path, sort, and Blocks World tasks, respectively. This operation consists of the interfacing between the NVIDIA GPUs and the CPU, which includes applications of logic layers, logic interfaces, module lists, and many more operations within the NLM execution. The second most prominent category resulted in the dense matrix multiplication, followed by the third category element-wise operations.

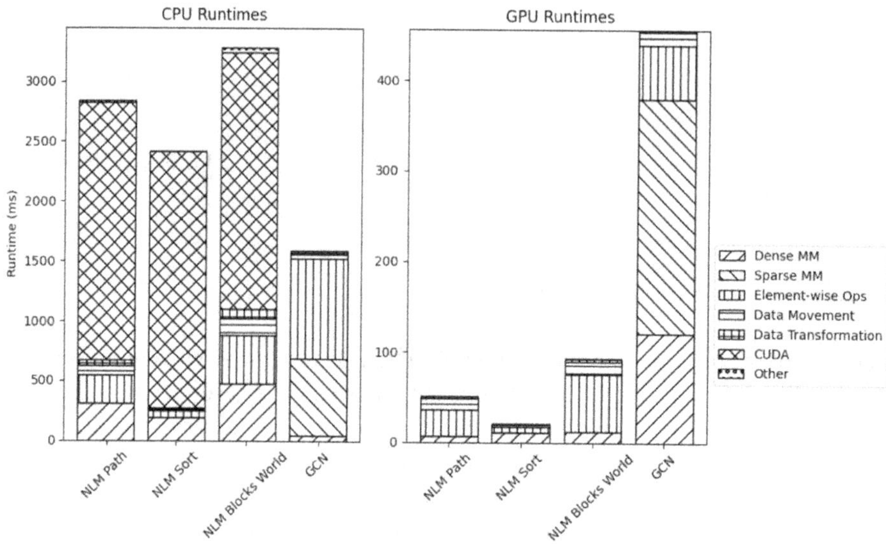

Figure 6: CPU and GPU runtime breakdown of the three distinct tasks for NLM and the GCN model.

Figure 8 shows the characterizing results of the three NLM tasks using the PyTorch profiler. Comparing the PyTorch profiler results to the DLProf profiler results, a slight deviation can be seen. The PyTorch profiler followed a similar trend with the element-wise operations being the highest category of the Blocks World and sort task; however the path task shows the highest percentage in dense matrix multiplication operations at 53.5%. Data movement also presented as another most significant operation category for the three tasks. The runtime results for the three NLM tasks using the PyTorch profiler in Figure 7 show the same trend in longer runtimes for each of the largest categories. The element-wise operations tend to hold the longest runtimes followed by data movement operations.

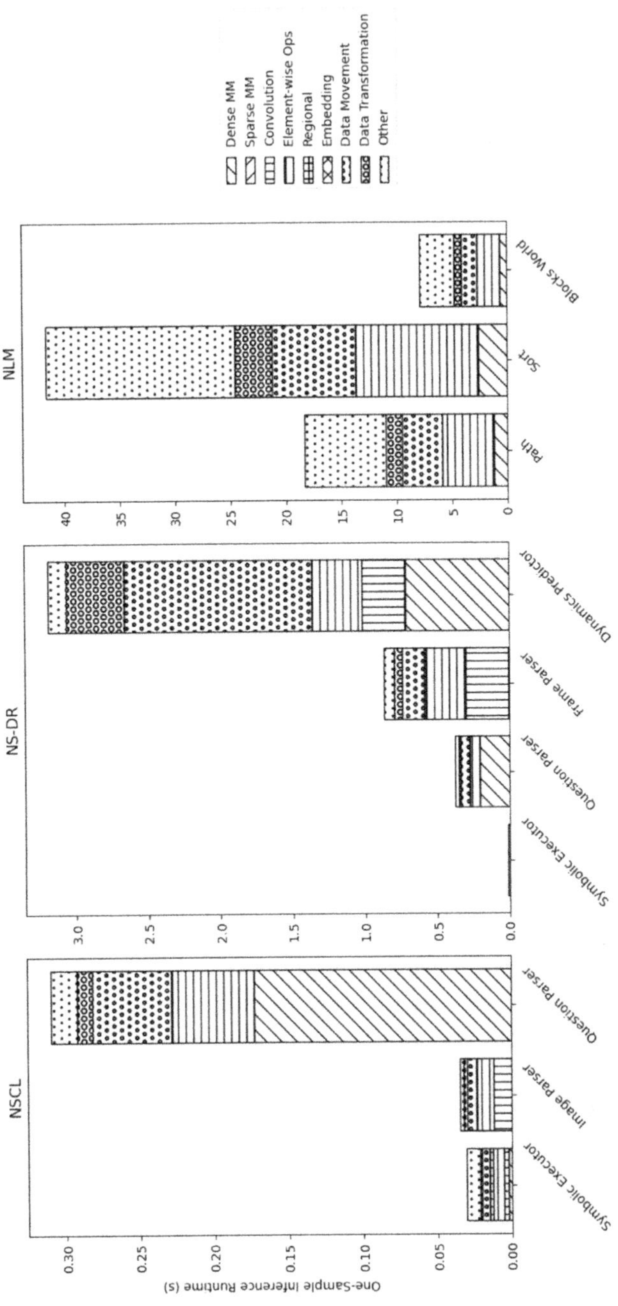

Figure 7: GPU category one-sample runtime results breakdown of NSCL, NS-DR and NLM using the PyTorch profiler.

4.3 Image and video frame parser

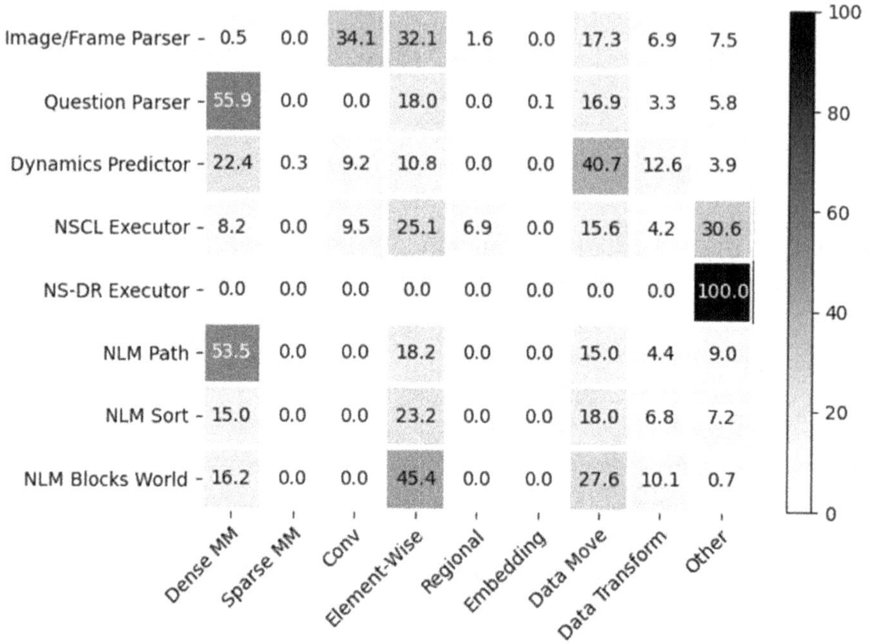

Figure 8: GPU category heatmap results breakdown of NSCL, NS-DR, and NLM using the PyTorch profiler.

Mask R-CNN, the video frame parser, has a well-studied performance profile that is dominated by convolution and activation functions [23]. As mentioned in Section 2, we used Detectron2 due to difficulties bringing up the version of Mask R-CNN used in the NS-DR paper. Mobile-optimized, production-ready implementations of Detectron2 models are provided by the d2go project [37], which we used to collect realistic performance measurements for Mask R-CNN inference latency. The pretrained model we used was *not* trained on the CLEVR or CLEVRER dataset, as there is no pretrained model, model source, or training instructions for CLEVRER, and we wanted to use the same submodel for both the NSCL and NS-DR. However, we do not anticipate that the performance characteristics of the model would be any different if it were trained for a different dataset, particularly since the runtime of this model does not significantly vary with the number of objects in the input image.

Our analysis of Detectron2, as shown in Figure 7, shows that Mask R-CNN spends the most execution time on convolution and element-wise operations (such as activation functions and normalization). This is not particularly surprising, but serves as a simple example of our classification scheme. With an average 34.6 ms inference time on our target machine, a full 25-frame video requires 865 ms of inference time for the

CLEVRER dataset. Thus, the NS-DR video frame parser takes dramatically longer than the NSCL image parser.

4.4 Question parser

NLP models such as Seq2Seq, the original question parser in the NS-DR, are a well-established family, having seen datacenter deployments since at least 2016 [38]. However, transformer-based attention models, such as the OpenNMT model we profiled, typically outperform RNN-based models such as Seq2Seq, and have seen wider adoption since 2017 [39].

The amount of computation required to perform inference for an NMT model is dependent on the input sequence length. Sentences in the CLEVR dataset have an average length of 18.4 words. The CLEVRER dataset is split between open-ended questions with an average length of 10.9 words and multiple-choice questions with an average length of 51.3 words (counting the answer choices), for an overall average of 22.2 words.

Figure 7 shows that the performance of the transformer-based OpenNMT model is dominated by dense matrix multiplication and data movement. The OpenNMT model performed inference in an average of 13.5 ms per input word/token. In practice, the runtime of a transformer model asymptotically grows quadratically with input sequence length [40], but we observed a linear relation on sample input sentences. It is likely that, for the relatively short lengths of the inputs we were observing, linear per-word operations dominated the quadratic attention operations.

4.5 Dynamics predictor

PropNet, the neural dynamics predictor, spends the majority of its runtime on data movement. In fact, while analyzing the behavior of this model using the Nvidia-smi utility, we noticed it was rarely able to exceed 50% utilization of one GPU. The model also spends a substantial amount of time on coalescing, which entails merging duplicate entries in sparse tensors. It is unclear to what extent this coalescing could be avoided through better-optimized code. A substantial portion of the remaining runtime is spent on dense matrix multiplication. This is because the workload involves the multiplication of many very tall and skinny matrices, where m and n are quite large, but k may be as small as 1. As discussed in Section 3, these types of matrices are challenging to multiply efficiently.

Internally, this model consists of several smaller fully connected and convolutional networks. Some of these models must be run multiple times for a single frame in order to compute the propagation of forces. This contributes to this model's large amount of data movement and generally long runtime. The data movement of this model could be reduced by leveraging sparsity and weight reuse; since the same submodel is used for inference many times, it is wasteful to move over the weights for each inference.

Some tensor dimensions are input-dependent, corresponding to the number of objects in the frame. We also observed substantial variation in inference runtimes. Across 400 samples, the fastest inference finished in 1.8 s, and the slowest in 5.3 s, with an average of 3.4 s.

4.6 NSCL symbolic program executor

The symbolic executor for the NSCL spends a large portion of its execution time on element-wise operations, and a smaller but still significant amount of time on data movement. These element-wise operations stem from its manipulation of vectors of probabilities (with entries corresponding to the probabilities of each object in the scene being a correct answer). In addition to simple arithmetic, the model computes numerous Softmax functions over these vectors in order to isolate predictions. Notably, not much time is spent on matrix-matrix operations (GEMM and convolution), which would have better operational intensity and potential for parallelism than the scalar and vector operations we see dominating.

4.7 NS-DR symbolic program executor

The symbolic program executor for the NS-DR has an average runtime of 12.9 ms per sample. While this is not large compared to the other models in the network, all other models must finish before the executor can begin and it therefore lies on the critical path for inference.

Since the NS-DR executor is a CPU-only, scalar model, it would not make sense to categorize it according to the scheme we derived for the other workloads. Therefore, we used Python's built-in cProfile profiler to perform function-level profiling and identified categories of similar functions. Note, however, that the plurality of the runtime still falls into the "Other" category. This is a consequence of there being many miscellaneous functions, which individually contribute little to runtime, but collectively contribute more than any of the major categories.

Much of the executor's runtime is spent on querying the set of extracted features. Examples of these queries include finding the ID of an object with given properties, looking up the properties of an object given its ID, and queries relating multiple objects such as finding an object's closest neighbor. These queries look similar to standard database operations. For a sufficiently large set of symbolic features, it is likely that feature querying could be approached using existing work in parallelizing SQL queries [41]. However, the feature set the NS-DR extracts is relatively small, meaning that the overhead that parallelism incurs would almost certainly overwhelm the speedup for this model.

The next largest category of execution time is spent on scalar arithmetic operations, in particular summing large arrays. In principle these sorts of operations can sometimes benefit from being spread across multiple CPU cores, but in practice, we again expect the small size of the feature set to make this pointless.

The third largest category of operation is in fact JSON parsing. This is an unfortunate artifact of the way data is passed in the NS-DR model: rather than an end-to-end integration, it stores data in JSON files between submodels. This means that the executor has to load in the questions and extracted features from JSONs for each inference sample. This overhead turns out to be substantial for the very short runtime of this model.

5 Analysis

Figure 6 shows the CPU and GPU execution time breakdowns for single input samples for each of the three NLM tasks and GCN model, collected using DLProf. We now discuss each model individually.

5.1 Graph Convolutional Network

The GCN model showed great ability to classify nodes in an undirected graph network. After viewing the data, there is great potential for software and hardware improvements. From the profiler results described in Section 5, most of the processing takes place in the CPU, where element-wise and sparse MM operations take precedent.

GCN spends 25% of the total runtime processing data on the GPU. In this model, all functions use the float32 datatype. Tensor cores are available but require float16 datatype for utilization. It is possible to increase the Tensor Core utilization via Pytorch's Automatic Mixed Precision (AMP). GCN operates using the float32 datatype; however the GPU's Tensor cores require the float16 (half precision) datatype. AMP enables usage of mixed precision. Some functions are much faster using half precision, while other functions require the range provided by full precision. Pytorch's AMP is designed to appropriately assign each function to the necessary datatype. Using AMP could increase efficiency within the heavily utilized matrix multiplication calls. Dense MM uses an internal matrix multiply-and-add function (torch.addmm). However, torch.addmm is not natively supported for half precision. There is a potential fix by way of the cuSPARSE library, which does support addmm in half precision. Sparse MM does not have the same parallelism potential as dense MM, but the techniques described in [42, 43] are capable of further parallelizing Sparse MM.

A possibility to aid in hardware acceleration is aimed at the high level of element-wise operations. Ghose et al. [44] provides insight into implementations of processing-in-memory (PIM) architectures. The concept of PIM has existed since the 1960s [44]; however with the advent of 3D-stacked memory PIM has more recently become viable. AMD research further explores the application of PIM in their approach for collaborative machine learning acceleration (Co-ML) [45]. A design for near-data processing (NDP) enables the NDP units to process data intensive functions that exhibit low compute-to-byte ratios. Through their implementation, they observe speedups between 1% and 25% over normal GPU architectures [45].

5.2 Neural Logic Machines

The NLM showcased the ability for the architecture to train a model on small-scale tasks and generalize to solve large-scale tasks with lifted rules and added premises, which shows the expressive power of the network. The ability to scale the rule set from a small-sized rule set to a large-sized rule set has proven to be difficult for ILP systems. The combination of using both symbols and probabilities helps solve these issues.

The DLProf profiler shows that element-wise operations comprise the majority of the kernel executions on the GPU for the NLM. While element-wise operations apply uniformly to all elements in a tensor, further improvement in these operations can have a significant impact on the NLM's performance. It is worth discovering which specific element-wise tensor operations are used to build a strategy for an efficient performance enhancement. Many of these tensor operations can possibly be improved by utilizing PyTorch's automatic mixed precision to enable tensor core usage. Data movement also serves as the second highest contributing category in the GPU performance. Improving the data movement operations involving memory-intensive interacts could enhance data movement efficiency. Most of the CPU runtime spent for the NLM is on CUDA support operations, which ranges between 65% and 88% for the different NLM tasks. Determining how impactful these CUDA operations are on the CPU could have a significant impact on the overall operations for the NLM.

Characterization of the three NLM tasks using the PyTorch profiler reveals a similar trend in the breakdown of each prominent operation category. Element-wise operations represent the majority of operations for the Blocks World and sort tasks, which have potential for improvement. The path task shows dense matrix multiplications comprise over 53% of operations evaluated using the PyTorch profiler. These dense matrix multiplications use computationally intensive operations, which can be improved by exploring further parallelization options. Operations that do not fall into the established categories are placed into the “Other” category. The “Other” category operations show the longest runtimes for all three tasks when compared to the other operations during the tasks' execution.

Discovering these other operations and optimizing them may prove to have a significant impact in the runtime performance of the system. The PyTorch profiler overhead during profiling the NLM tasks may fall into this category.

6 Conclusion

While the models investigated in this research look topologically distinct from traditional deep learning models, our analysis suggests that their performance characteristics can be largely viewed as a combination of existing workloads. The analyses of the models show mixed results for opportunities for acceleration of computation. NLM has low operational intensities in GPU, consisting of vector and/or scalar operations, and exhibit complex control flow. These factors combined greatly limit the potential for parallelism. NLM models require numerous element-wise operations for inference, with the computational demand increasing with the complexity of the task-specific logical rule set. However, GCN uses a significant level of dense and sparse MM in both GPU and CPU. Dense MM may be further parallelized and sped up by using the tensor cores available in GPUs. While the analysis of the NSCL and NS-DR show that there are relatively few opportunities for acceleration of symbolic computation, the symbolic workloads of these two models have low operational intensities, consisting of vector and/or scalar operations, and exhibit complex control flow. These factors combined greatly limit the potential for parallelism. However, the symbolic components do not make up large portions of the execution times of either workload, and are therefore unlikely to pose a bottleneck.

Appendix

Repository links for the models referenced in this paper:

NSCL	https://github.com/vacancy/NSCL-PyTorch-Release
NS-DR	https://github.com/chuangg/CLEVRER
Detectron2 (d2go)	https://github.com/facebookresearch/d2go
OpenNMT	https://github.com/OpenNMT/OpenNMT-py
Propnet	https://github.com/YunzhuLi/PropNet
NLM	https://github.com/google/neural-logic-machines
GCN	https://github.com/tkipf/pygcn

References

[1] M. Garnelo and M. Shanahan, "(Introduction Support for Nsai) Reconciling Deep Learning with Symbolic Artificial Intelligence: Representing Objects and Relations," *Current Opinion in Behavioral Sciences*, vol. 29, pp. 17–23, 2019, https://www.sciencedirect.com/science/article/pii/S2352154618301943

[2] J. Lighthill, "Artificial Intelligence: A General Survey," in *Artificial Intelligence: a paper symposium*, 1973.

[3] J. Hendler, "Avoiding Another Ai Winter," *IEEE Annals of the History of Computing*, vol. 23, no. 2, pp. 2–4, 2008.

[4] J. Mao, C. Gan, P. Kohli, J. B. Tenenbaum, and J. Wu, "The Neurosymbolic Concept Learner: Interpreting Scenes, Words, and Sentences from Natural Supervision," 2019.

[5] K. Yi, C. Gan, Y. Li, P. Kohli, J. Wu, A. Torralba, and J. B. Tenenbaum, "Clevrer: Collision Events for Video Representation and Reasoning," 2020.

[6] H. Dong, J. Mao, T. Lin, C. Wang, L. Li, and D. Zhou, "Neural Logic Machines," in *International Conference on Learning Representations*, 2019. [Online]. Available: https://openreview.net/forum?id=B1xY-hRctX

[7] N. J. Nilsson, "Principles of Artificial Intelligence," in *Principles of Artificial Intelligence*. Berlin Heidelberg: Springer-Verlag, pp. 1–476, 1982, https://www.springer.com/gp/book/9783540113409

[8] Y. Zhu, A. Fathi, and L. Fei-Fei, *Reasoning about Object Affordances in a Knowledge Base Representation*. Springer International Publishing, pp. 408–424, 2014.

[9] M. Seeger, "Learning with Labeled and Unlabeled Data," 2001.

[10] X. Zhu, Z. Ghahramani, and J. Lafferty, "Semi-supervised Learning Using Gaussian Fields and Harmonic Functions," *ICML*, pp. 912–919, 2003.

[11] P. Sen, G. M. Namata, M. Bilgic, L. Getoor, B. Gallagher, and T. Eliassi-Rad, "Collective Classification in Network Data," *AI Magazine*, vol. 29, no. 3, pp. 93–106, 2008, http://www.cs.iit.edu/ml/pdfs/sen-aimag08.pdf

[12] J. Weston, F. Ratle, and R. Collobert, "Deep Learning via Semisupervised Embedding," in *Proceedings of the 25th International Conference on Machine Learning*, ser. ICML '08. New York, NY, USA: Association for Computing Machinery, pp. 1168–1175, 2008, [Online]. Available: https://doi.org/10.1145/1390156.1390303

[13] K. Xu, W. Hu, J. Leskovec, and S. Jegelka, "How Powerful are Graph Neural Networks?" in *7th International Conference on Learning Representations, ICLR 2019, New Orleans, LA, USA, May 6–9, 2019*. OpenReview.net, 2019. [Online]. Available: https://openreview.net/forum?id=ryGs6iA5Km

[14] T. N. Kipf and M. Welling, "Semi-Supervised Classification with Graph Convolutional Networks," in *Proceedings of the 5th International Conference on Learning Representations*, ser. ICLR '17, 2017. [Online]. Available: https://openreview.net/forum?id=SJU4ayYgl

[15] N. Rauschmayr, S. Kama, M. Kim, M. Choi, and K. Kenthapadi, "Profiling Deep Learning Workloads at Scale Using Amazon Sagemaker," in *KDD 2022*, 2022. [Online]. Available: https://www.amazon.science/publications/profiling-deep-learningworkloads-at-scale-using-amazon-sagemaker

[16] J. Gu, H. Liu, Y. Zhou, and X. Wang, "Deepprof: Performance Analysis for Deep Learning Applications via Mining Gpu Execution Patterns," 2017. [Online]. Available: https://arxiv.org/abs/1707.03750

[17] M. Gori, G. Monfardini, and F. Scarselli, "A New Model for Learning in Graph Domains," in *Proceedings 2005 IEEE International Joint Conference on Neural Networks*, vol. 2, pp. 729–734, 2005.

[18] F. Scarselli, M. Gori, A. C. Tsoi, M. Hagenbuchner, and G. Monfardini, "The Graph Neural Network Model," *IEEE Transactions on Neural Networks*, vol. 20, no. 1, pp. 61–80, 2009.

[19] Y. Li, D. Tarlow, M. Brockschmidt, and R. Zemel, "Gated Graph Sequence Neural Networks," 2015. [Online]. Available: https://arxiv.org/abs/1511.05493

[20] D. K. Duvenaud, D. Maclaurin, J. Iparraguirre, R. Bombarell, T. Hirzel, A. Aspuru-Guzik, and R. P. Adams, "Convolutional Networks on Graphs for Learning Molecular Fingerprints," in *Advances in Neural Information Processing Systems*, C. Cortes, N. Lawrence, D. Lee, M. Sugiyama, and R. Garnett, Eds., vol. 28. Curran Associates, Inc., 2015, https://proceedings.neurips.cc/paper/2015/file/f9be311e65d81a9ad8150a60844bb94c-Paper.pdf
[21] S. Muggleton, "Inductive Logic Programming," *New Generation Computing*, vol. 8, no. 02, pp. 295–318, 1991.
[22] J. Johnson, B. Hariharan, L. Van der Maaten, L. Fei-Fei, C. L. Zitnick, and R. B. Girshick, "CLEVR: A Diagnostic Dataset for Compositional Language and Elementary Visual Reasoning," *CoRR*, vol. abs/1612.06890, 2016, http://arxiv.org/abs/1612.06890
[23] K. He, G. Gkioxari, P. Dollar, and R. Girshick, "Mask R-cnn," 2018.
[24] Y. Wu, A. Kirillov, F. Massa, W.-Y. Lo, and R. Girshick, "Detectron2," https://github.com/facebookresearch/detectron2, 2019.
[25] K. Cho, B. Van Merrienboer, C. Gulcehre, D. Bahdanau, F. Boügares, H. Schwenk, and Y. Bengio, "Learning Phrase Representations Using RNN Encoder–decoder for Statistical Machine Translation," in *Proceedings of the 2014 Conference on Empirical Methods in Natural Language Processing (EMNLP)*. Doha, Qatar: Association for Computational Linguistics, pp. 1724–1734, Oct. 2014. [Online]. Available: https://aclanthology.org/D14-1179
[26] G. Klein, Y. Kim, Y. Deng, J. Senellart, and A. M. Rush, "Opennmt: Open-source Toolkit for Neural Machine Translation," 2017.
[27] D. Bahdanau, K. Cho, and Y. Bengio, "Neural Machine Translation by Jointly Learning to Align and Translate," 2016.
[28] Y. Li, J. Wu, J.-Y. Zhu, J. B. Tenenbaum, A. Torralba, and R. Tedrake, "Propagation Networks for Model-based Control under Partial Observation," *ICRA*, 2019.
[29] K. Asanovic, R. Bodik, B. Catanzaro, J. Gebis, P. Husbands, K. Keutzer, D. Patterson, W. Plishker, J. Shalf, S. Williams, and K. Yelick, "The Landscape of Parallel Computing Research: A View from Berkeley," 2006.
[30] C. Rivera, J. Chen, N. Xiong, J. Zhang, S. L. Song, and D. Tao, "Tsm2x: High-performance Tall-and-skinny Matrix–matrix Multiplication on Gpus," *Journal of Parallel and Distributed Computing*, vol. 151, pp. 70–85, May 2021, doi: http://dx.doi.org/10.1016/j.jpdc.2021.02.013
[31] A. Zhou, Y. Ma, J. Zhu, J. Liu, Z. Zhang, K. Yuan, W. Sun, and H. Li, "Learning N: M Fine-grained Structured Sparse Neural Networks from Scratch," *CoRR*, vol. abs/2102.04010, 2021, https://arxiv.org/abs/2102.04010
[32] Y. Saad, *Iterative Methods for Sparse Linear Systems*, 2nd ed. Society for Industrial and Applied Mathematics, 2003. [Online]. Available: https://epubs.siam.org/doi/abs/10.1137/1.9780898718003
[33] Z. Zhang, H. Wang, S. Han, and W. J. Dally, "Sparch: Efficient Architecture for Sparse Matrix Multiplication," 2020.
[34] A. Vasudevan, A. Anderson, and D. Gregg, "Parallel Multi Channel Convolution Using General Matrix Multiplication," in *2017 IEEE 28th International Conference on Application-specific Systems, Architectures and Processors (ASAP)*, pp. 19–24, 2017.
[35] L. Ke, U. Gupta, B. Y. Cho, D. Brooks, V. Chandra, U. Diril, A. Firoozshahian, K. Hazelwood, B. Jia, H.-H. S. Lee, M. Li, B. Maher, D. Mudigere, M. Naumov, M. Schatz, M. Smelyanskiy, X. Wang, B. Reagen, C.-J. Wu, M. Hempstead, and X. Zhang, "Recnmp: Accelerating Personalized Recommendation with Near-memory Processing," in *2020 ACM/IEEE 47th Annual International Symposium on Computer Architecture (ISCA)*, pp. 790–803, 2020.
[36] "Torch.sparse." [Online]. Available: https://pytorch.org/docs/stable/sparse.html
[37] "D2go Brings Detectron2 to Mobile," https://ai.facebook.com/blog/d2gobrings-detectron2-to-mobile/, accessed: 2021-05-09.
[38] I. Sutskever, O. Vinyals, and Q. V. Le, "Sequence to Sequence Learning with Neural Networks," 2014.

[39] A. Vaswani, N. Shazeer, N. Parmar, J. Uszkoreit, L. Jones, A. N. Gomez, L. Kaiser, and I. Polosukhin, "Attention Is All You Need," 2017.

[40] A. Katharopoulos, A. Vyas, N. Pappas, and F. Fleuret, "Transformers are RNNs: Fast Autoregressive Transformers with Linear Attention," *CoRR*, vol. abs/2006.16236, 2020, https://arxiv.org/abs/2006.16236

[41] T. Cruanes, B. Dageville, and B. Ghosh, "Parallel Sql Execution in Oracle 10g," in *Proceedings of the 2004 ACM SIGMOD International Conference on Management of Data*, ser. SIGMOD '04. New York, NY, USA: Association for Computing Machinery, pp. 850–854, 2004. [Online]. Available: https://doi.org/10.1145/1007568.1007666

[42] A. Buluc¸ and J. R. Gilbert, "Parallel Sparse Matrix-matrix Multiplication and Indexing: Implementation and Experiments," *SIAM Journal on Scientific Computing*, vol. 34, no. 4, pp. C170–C191, 2012, doi: https://doi.org/10.1137/110848244.

[43] A. Azad, G. Ballard, A. Buluc¸, J. Demmel, L. Grigori, O. Schwartz, S. Toledo, and S. Williams, "Exploiting Multiple Levels of Parallelism in Sparse Matrix-matrix Multiplication," *SIAM Journal on Scientific Computing*, vol. 38, no. 6, pp. C624–C651, Jan. 2016, doi: https://doi.org/10.1137%2F15m104253x.

[44] S. Ghose, A. Boroumand, J. S. Kim, J. Gomez-Luna, and O. Mutlu, "Processing-in-memory: A Workload-driven Perspective," *IBM Journal of Research and Development*, vol. 63, no. 6, pp. 3:1–3:19, 2019.

[45] S. Aga, N. Jayasena, and M. Ignatowski, "Co-ml: A Case for Collaborative Ml Acceleration Using Near-data Processing," in *Proceedings of the International Symposium on Memory Systems*, ser. MEMSYS '19. New York, NY, USA: Association for Computing Machinery, pp. 506–517, 2019. [Online]. Available: https://doi-org.libweb.lib.utsa.edu/10.1145/3357526.3357532

Nikhila Vintha and Devinder Kaur

Multivariant time series prediction using variants of LSTM deep neural networks

Abstract: Deep Learning models have emerged as powerful tools for managing complex temporal data. In this chapter, comparative analysis of three deep learning architectures based on Long Short-term Memory Networks (LSTM), viz., Vanilla LSTM, Stacked LSTM, and Bidirectional LSTM (Bi-LSTM) is conducted. LSTM networks can learn long-term dependency. They are used for the analysis of sequential data such as time series, speech, and text data. The chapter begins by explaining the unique characteristics of multivariate time series data, emphasizing the importance of choosing appropriate modeling techniques. It then delves into the detailed description of LSTM, Stacked LSTM, and Bi-LSTM architectures, and the description of the internal functionalities. The three models were analyzed using three data sets viz., FedEx, Ford, and Meta. Accuracy of prediction is measured in terms of performance metrics of root mean square error (RMSE) and R^2. It was found the best performance was given by the Bi-LSTM model across the three data sets.

Keywords: multivariate time series, deep learning models, long short-term memory, Stacked LSTM, Bi-LSTM

1 Introduction

Multivariate time series data is a high-dimensional data where all the input parameters change with time to predict a specific output. It exhibits itself in many diverse domains such as finance, healthcare, climate monitoring, and industrial processes. Analyzing and predicting multivariate time series data poses unique challenges due to its complex temporal dependencies.

The goal of this research is to develop Long Short-Term Memory (LSTM)-based deep learning models to predict the outcome of multivariate time series data. Time series data used is the stock prediction data retrieved from the Yahoo website (https://finance.yahoo.com/). The dataset consists of date, high, low, volume, open, and close of the stock price. We are predicting the closing stock price of the day by taking into consideration three stock prices, viz, opening, high, and low for a given day. We have

Nikhila Vintha, Electrical and Computer Science Engineering, University of Toledo, Toledo, Ohio, USA, e-mail: nvintha@rockets.utoledo.edu
Devinder Kaur, Electrical and Computer Science Engineering, University of Toledo, Toledo, Ohio, USA, e-mail: devinder.kaur@utoledo.edu

https://doi.org/10.1515/9783111344126-010

taken three datasets Ford, FedEx, and Meta. The datasets have the data from the year 2010 to Feb 2023. The data was parsed as 70% for training and 30% for testing.

There are many other statistical approaches in machine learning like Auto regressor, ARIMA, SARIMAX, and many other techniques, but deep learning models can deal with high-dimensional data and the ability of gaining nonlinear relationship tend to outperform the statistical methods. LSTM (Long Short-Term Memory) is a type of RNN model that consists of many layers. There are three types of LSTM, viz, vanilla LSTM, Stacked LSTM, and Bidirectional LSTM. The performance metrics are RMSE and R^2. RMSE is also used for model comparison, where lower RMSE values indicate better-performing models and R^2 value should be between 0 and 1, with higher values indicating better model performance.

The chapter is organized into 6 sections. Section 2 describes data preprocessing, Section 3 Recurrent neural networks, Section 4 methodology and implementation, Section 5 optimizers and modal accuracy, and Section 6 conclusion and future work.

2 Data preprocessing techniques

This section describes various data processing techniques that were used before developing the model. The features of various time series data are described. The description of other procedural steps like data cleaning, choosing the sliding window, and using the Python libraries is given.

2.1 Time series data

Time series data is also known as time-stamped data; examples are stock values, changes in temperature, aircraft flights trajectories, etc. Time series data is of two types – one is univariate and the other is multivariate. The univariate time series consists of one input variable and one output variable with respect to time. Multivariate time series data consists of two or more input variables and one output variable with respect to time.

Time series data is used in machine learning applications for various tasks such as classification, clustering, regression, and prediction. In this chapter, we used multivariate time series data, which is cyclic as it holds more than one year.

2.2 Data collection

Stock market data is retrieved from Yahoo website. Three datasets are taken to predict the closing stock price of the data. The sample data sets for Ford are shown in Figure 1, FedEx is shown in Figure 2 and Meta is shown in Figure 3.

	Date	Open	High	Low	Close	Adj Close	Volume
0	2010-01-04	10.17	10.28	10.05	10.28	6.316511	60855800
1	2010-01-05	10.45	11.24	10.40	10.96	6.734339	215620200
2	2010-01-06	11.21	11.46	11.13	11.37	6.986262	200070600
3	2010-01-07	11.46	11.69	11.32	11.66	7.164451	130201700
4	2010-01-08	11.67	11.74	11.46	11.69	7.182884	130463000
...	...	...	...	...	...	...	...
3314	2023-03-06	13.08	13.18	12.93	12.98	12.980000	50571200
3315	2023-03-07	12.94	12.98	12.78	12.83	12.830000	46540400
3316	2023-03-08	12.78	13.01	12.71	12.98	12.980000	46421000
3317	2023-03-09	13.02	13.08	12.44	12.45	12.450000	65423500
3318	2023-03-10	12.43	12.51	12.00	12.10	12.100000	73798900

Figure 1: Sample data of Ford dataset.

	Date	Open	High	Low	Close
0	2010-01-04	84.209999	85.000000	83.110001	83.449997
1	2010-01-05	83.440002	84.589996	82.639999	84.540001
2	2010-01-06	84.330002	85.120003	83.610001	83.839996
3	2010-01-07	83.849998	83.849998	82.400002	82.930000
4	2010-01-08	84.080002	85.260002	82.949997	84.989998
...	...	...	...	...	...
3314	2023-03-06	209.600006	211.889999	209.350006	210.130005
3315	2023-03-07	209.720001	209.720001	206.160004	206.320007
3316	2023-03-08	205.899994	209.970001	205.389999	209.539993
3317	2023-03-09	211.080002	213.309998	207.300003	207.830002
3318	2023-03-10	207.190002	207.190002	198.500000	201.190002

Figure 2: Sample data of FedEx dataset.

	Date	Open	High	Low	Close
0	2012-05-18	42.049999	45.000000	38.000000	38.230000
1	2012-05-21	36.529999	36.660000	33.000000	34.029999
2	2012-05-22	32.610001	33.590000	30.940001	31.000000
3	2012-05-23	31.370001	32.500000	31.360001	32.000000
4	2012-05-24	32.950001	33.209999	31.770000	33.029999
...	...	...	...	...	...
2715	2023-03-06	188.000000	189.660004	184.639999	184.899994
2716	2023-03-07	189.000000	190.360001	184.009995	184.509995
2717	2023-03-08	182.869995	185.259995	181.339996	184.970001
2718	2023-03-09	186.350006	188.929993	180.300003	181.690002
2719	2023-03-10	181.009995	184.770004	178.800003	179.509995

Figure 3: Sample data of Meta dataset.

Stock market data consists of stock value of close, volume, open, high, and low, in addition to date. As a multivariate time series is used in this research, our target value is the closing stock price. So, the closing stock value is predicted using open, high, low, and volume of the stocks traded. Figure 4 shows the closing stocks of Ford Dataset. Figure 5 shows the closing stocks of FedEx dataset. Figure 6 shows the closing stocks of the Meta dataset.

Figure 4: Closing stock price of Ford Dataset.

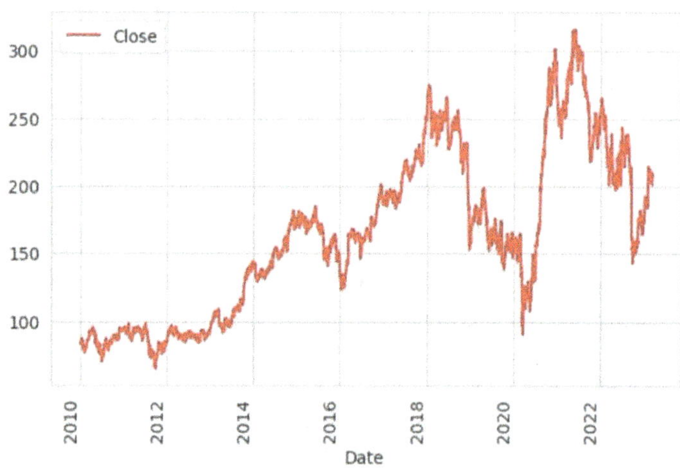

Figure 5: Closing stock price of FedEx Dataset.

Figure 6: Closing stock price of Meta Dataset.

2.3 Python for machine learning

Python is a high-level programming language that is widely used for machine learning problems, and it is also a general purpose programming language. Python is proposed for machine learning because it is an independent platform. Python has many built-in machine learning libraries such as Pandas, NumPy, Matplotlib, Kera's, Tenser flow, etc., which are used to develop these models.

2.3.1 Pandas

It is primarily used for data analysis and manipulation, and is widely used for working with tabular data, such as data from CSV files and Excel spreadsheets. Pandas are better for performance and can handle large datasets with ease. Here, the data frame used is the pandas. The dataset is downloaded from the Yahoo website in csv format, and it is imported to Panda's data frame using Panda's library. Figure 7 shows the structure of Panda's data frame. The data frame consists of rows and columns, where rows indicate the number of records and columns indicate the number of variables in the data frame, and Id indicates the column that is used to identify each row.

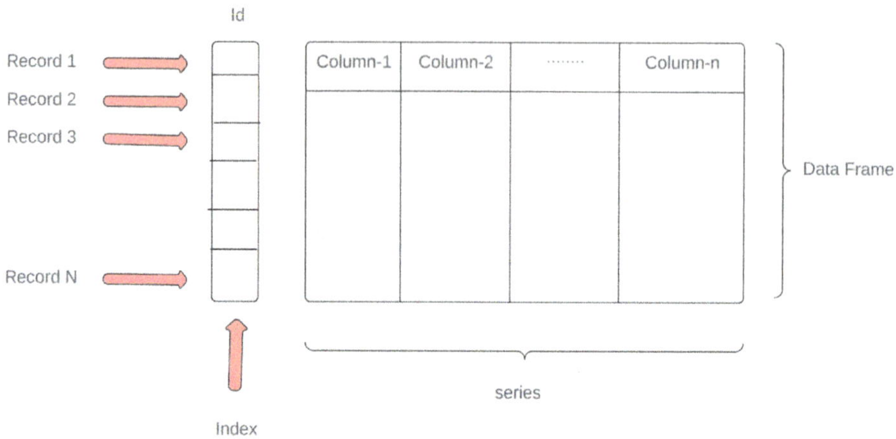

Figure 7: Pandas Data frame [13].

2.3.2 NumPy

NumPy has tools for analyzing numerical data and it is widely used for transforming large datasets. It is also used for creating and transforming arrays, matrices, and vectors. NumPy is fast, powerful, and easy to use, making it an essential part of any machine learning project.

2.3.3 Matplotlib

Matplotlib is used to create data visualizations and allows users to quickly create several types of charts and graphs. Matplotlib is also used to generate histograms, scatter plots, bar charts, and more. It supports a wide range of data formats, including CSV, Excel, JSON, and HTML.

2.3.4 Kera's

Kera's is an open-source library in python used for deep learning. It can be used to build and train models in a few lines of code. Kera's makes it easy to build complex deep learning models on top of powerful deep learning libraries such as TensorFlow and Theano.

2.3.5 Tensor flow

TensorFlow is an open-source software library designed by Google for machine learning applications such as neural networks. It is used by researchers and developers to build models and algorithms to solve a wide variety of tasks such as image recognition, natural language processing, and predictive analytics.

2.4 Data preprocessing

Data preprocessing techniques include data cleaning, data normalization, feature selection, feature engineering, and data transformation. Data cleaning involves detecting and removing outliers, missing values, and duplicate records. Data normalization is the process of scaling the data to a specific range. Feature engineering involves creating new features from existing ones.

2.4.1 Moving average

Moving average is a common data preprocessing technique used in time series analysis to smooth out noisy data and identify underlying trends or patterns. It involves calculating the average of a fixed window of data points, where the window "moves" through the time series data.

Here, the simple moving average (SMA) is used. Moving average or rolling mean is applied using Panda's statistical function, rolling mean. The window size of the moving average is taken as '7'. This window size is passed as the parameter in the function. Below is the syntax used to calculate the rolling mean of the Panda's data frame.

– data['column_name']. rolling(rolling_window).mean()

2.4.2 Train and test sets

Data is parsed into training and test sets. The training set is used to train the model with known data, while the test set is used to evaluate the performance of the model on unseen data. The training set should be larger than the test set and should be representative of the data that the model will be expected to work with. The test set should have data unseen by the model and should be designed to simulate the real-world data the model will be expected to work with. Here, 70% of the data is used for training and 20% of the data is used for testing. Figure 8 shows the code that splits into training and test sets.

```
# Split the data into training and testing sets
train_size = math.ceil(len(output_data) * 0.7)
train_input = input_data[:train_size, :]
train_output = output_data[:train_size, :]
test_input = input_data[train_size:, :]
test_output = output_data[train_size:, :]
```

Figure 8: Python code used for splitting data into training and test sets.

2.4.3 Scaling data

Scaling is a major step in preprocessing. It helps to ensure that all features are treated equally, and that the data is in the proper range for the specific algorithm. Scaling can also help to improve the accuracy of some algorithms and can help to reduce the amount of time needed to train and test the model.

Here, data is scaled using the min-max scalar. The min-max scalar is a type of scaling used to transform features or samples by scaling them to a range. In the min-max scalar, the data is scaled to fit within the range of 0 to 1. This range can be adjusted to fit the specific dataset, but the goal is to have all features scaled to the same range. Figure 9 shows the scaling of the dataset.

```
# Scale the input and target columns between 0 and 1
input_scaler = MinMaxScaler(feature_range=(0, 1))
output_scaler = MinMaxScaler(feature_range=(0, 1))
input_data = input_scaler.fit_transform(input_data)
output_data = output_scaler.fit_transform(output_data)
```

Figure 9: Scaling of data using Minmax scaler.

3 Recurrent neural networks

Recurrent neural network is a multilayer neural network where the output of the previous step is fed back to the input of the next step, so it is called recurrent neural network. A recurrent neural network (RNN) is an artificial neural network used for sequence processing. It is designed to recognize patterns in sequential data, such as text, audio, video, and time series data. The RNN's architecture includes a memory part, which is in hidden state, which allows it to remember past inputs, and a feedback loop, which enables the network to pass information from one layer to the previous one. This allows the RNN to learn from its past inputs and make decisions based on the patterns it recognizes.

3.1 Working of RNN

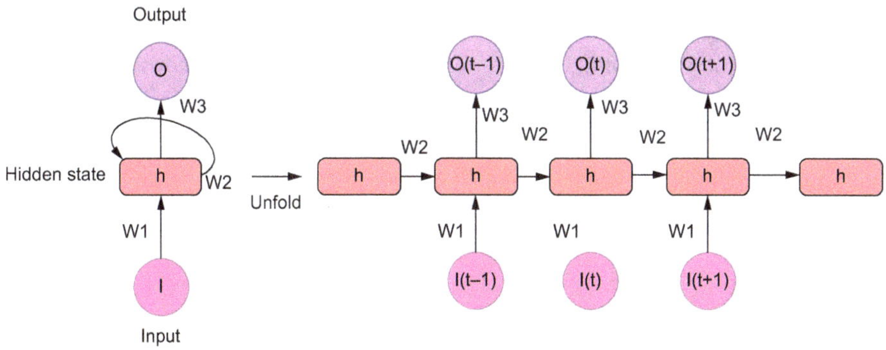

Figure 10: Recurrent Neural Network.

Figure 10 illustrates the architecture of RNN. The basic architecture of an RNN consists of input layers with weight 'W1', hidden layers with weights 'W2', and output layers with weights 'W3'. Shallow neural network has the weight, and biases are independent to one another while in Recurrent Neural Networks (RRN), what happens is that the network converts independent weights and biases into dependent weights and biases. So, all the layers consist of the same weights and biases than the previous output of the hidden layer, which gives input to the next hidden layer. Thus, it helps RNN to memorize, reducing the cost and complexity. Finally, all layers are joined together, forming a single recurrent layer.

3.2 Drawbacks of RNN

Like other neural networks, standard Recurrent Neural networks also have exploding gradients and vanishing gradient problems. Training RNN is also difficult as it cannot process long sequences. So, we can use Deep Recurrent neural networks such as LSTM, Bidirectional LSTM and GRU.

3.2.1 Exploding gradient problem

Exploding gradient is when the Recurrent Neural Network (RNN) becomes so large that they produce numerical overflow, and the weight of the network suddenly increases so that it can no longer be updated. This can occur when the network is trained with many time steps.

3.2.2 Vanishing gradient problem

Vanishing gradient problem is a well-known issue in training recurrent neural networks (RNNs). This issue occurs when the gradient of the loss function with respect to the weights becomes too small to be effective in training the network. So, the network's weights cannot be updated, and it does not learn from the data. As a result, many RNNs are unable to learn from long-term dependencies in the data. A variant of RNN, called the LSTM (Long Short-Term Memory), can overcome this problem.

4 Methodology and implementation

This section describes the components of LSTM cell structure like memory cell, forget gate, input gate, output gate, and the implementation of Vanilla LSTM, Stacked LSTM, and Bi-directional LSTM.

4.1 Long short-term memory

LSTMs can learn long-term dependencies that are difficult for traditional RNNs. They have a memory cell that can preserve information for extended periods of time, and gates that control the flow of information into and out of this memory cell. This allows the network to focus on the most essential information and ignore extraneous data. This makes LSTMs ideal for tasks such as predicting stock prices, predicting the next word in a sentence, and other complicated tasks that require long-term memory.

The cell state of LSTM is shown in Figure 11. The parameters represented in the diagram are:
- New cell state, represented as C_t
- Previous cell state, represented as C_{t-1}
- New cell input, represented as X_t
- New cell output, represented as h_t
- Previous cell output, represented as h_{t-1}

The detailed working of forget gate, input gate, and output gate of memory cell in LSTM is explained.

4.1.1 Memory cell

A memory cell in Long short-term memory (LSTM) is a computational unit used to store information for prolonged periods of time and to store its state across time

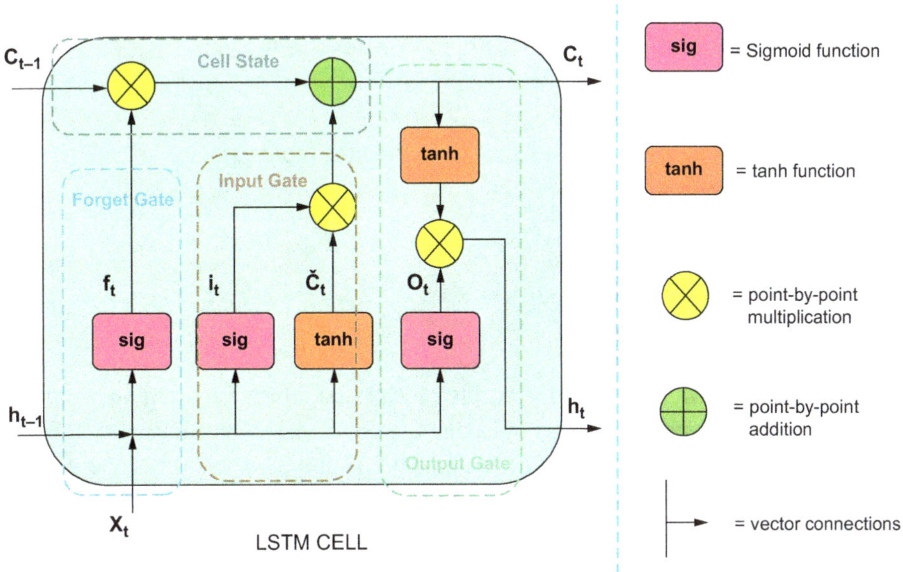

Figure 11: LSTM Cell State [12].

steps. The memory cell is composed of three gates: the first one is the forget gate. It helps to de-emphasize irrelevant information and store the relevant information. The second one is the input gate – used to store the current value of information, and the third one is the output gate, which is used to send the new value of the information. Figure 8 shows the architecture of memory cells.

4.1.2 Forget gate

The forget gate in Recurrent Neural Networks, shown in Figure 11, controls the flow of information in the network's memory cell. The forget gate helps the network decide which information should be forgotten or which should be kept in the memory cell. It is implemented by using sigmoid activation function, which outputs values between 0 and 1. If the sigmoid function generates the value '1', it shows that the value should be passed from the previous hidden state (h_{t-1}) to the current hidden state (h_t). If the sigmoid function generates the value '0', it shows that the information should be forgotten.

Equation (1) shows the sigmoid activation function for the forget gate [1]. Sigmoid activation function:

$$f_t = \sigma\left(W_f \times [h_{t-1}, x_t] + b_f\right) \tag{1}$$

The terms in the eq. (1) are represented as [3]

t = timestamp

f_t = forget gate at t

h_{t-1} = Previous hidden state

W_f = Weight matrix between the forget gate and the input gate

b_t = connection bias at t

X_t = Current input

4.1.3 Input gate

The input gate, shown in Figure 11, controls which latest information should be added to the current cell state. It does this by taking the previous hidden state (h_{t-1}), and the current input (x_t) as inputs to the second sigmoid function (i_t). Then, the sigmoid function generates binary a mask between '0' and '1'. A value of '1' shows that the information should be passed on to the hidden state, while a value of '0' shows that the information should be discarded. The same inputs, previous hidden state and current input, are also passed to the hyperbolic tangent function (tanh), which produces output ($C\sim_t$). Equation (2) shows the sigmoid activation function of input gate and eq. (3) shows the tanh function of the input gate.

Sigmoid activation function of the input gate is:

$$i_t = \sigma(W_i \times [h_{t-1,} x_t] + b_i) \tag{2}$$

$$Ct = tanh\left((w_c \times h_{t-1} + w_c \times x_t) + b_c\right) \tag{3}$$

The terms in the eqs. (2) and (3) are represented as [3]

t = timestamp

i_t = input gate at t

h_{t-1} = previous hidden state

x_t = current input

W_i = weight matrix of the sigmoid function between the input gate and the output gate

b_t = bias vector at t

$C\sim_t$ = value generated by tanh

W_c = weight matrix of tanh operator between cell state information and network output

b_c = bias vector at t with respect to W_c

4.1.4 Cell state

The cell state, shown in Figure 11, performs element-wise multiplication using the input from the previous timestep (C_{t-1}) and the output from the forget gate (f_t). This multiplication is done to decide which information should be ignored. Next element-wise addition is done by using the output from the element-wise multiplication and the output from the sigmoid (i_t) and tanh function ($C\sim_t$). Finally, after element-wise multiplication and element-wise addition, the cell state information is updated, giving the new cell state (C_t) to the LSTM network. After the LSTM is trained, then the network understands which patterns should be remembered and which should be forgotten. Equation (4) shows the cell state function.

Formula for operating cell state:

$$C_t = f_t \times C_{t-1} + i_i \times C\sim_t \tag{4}$$

The terms in the eq. (4) are represented as [3]

t = timestep
C_t = cell state information
f_t = forget gate at t
i_t = input gate at t
C_{t-1} = previous timestep
$C\sim_t$ = value generated by tanh

4.1.5 Output gate

The output gate, shown in Figure 11, decides what information should be passed to the next step. This is done by taking inputs from the previous hidden state (h_{t-1}) and the current state (x_t). These are passed to the sigmoid activation function (O_t). Next, the updated cell state is passed over to hyperbolic tangent function (tanh). Then, the element-wise multiplication is done by using output from the sigmoid and tanh activation functions. Finally, the output from the element-wise multiplication is used to decide the hidden state value. This hidden state value decides the information that should be carried. The new cell state and hidden state value are used to produce the final LSTM output (h_t). Equation (5) shows the sigmoid activation function for the output gate and eq. (6) shows the tanh activation function for output gate.

Sigmoid activation function of the output gate is:

$$O_t = \sigma(W_o[h_{t-1,} x_t] + b_o) \tag{5}$$

$$h_t = O_t \times tanh(C_t) \tag{6}$$

The terms in the eqs. (5) and (6) are represented as [3]

t = timestep
O_t = output gate at t
W_o = weight matrix of output gate
b_o = bias vector with respect to Wo
h_t = LSTM output

4.2 Implementation of Vanilla LSTM

Figure 12 shows the architecture of the Vanilla LSTM. The inputs are sent into the LSTM layer as it holds 50-time stamps. Then, the output of the LSTM layer passes it to the dense layer. The output from dense layer is passed to the final layer. The final layer predicts the final output.

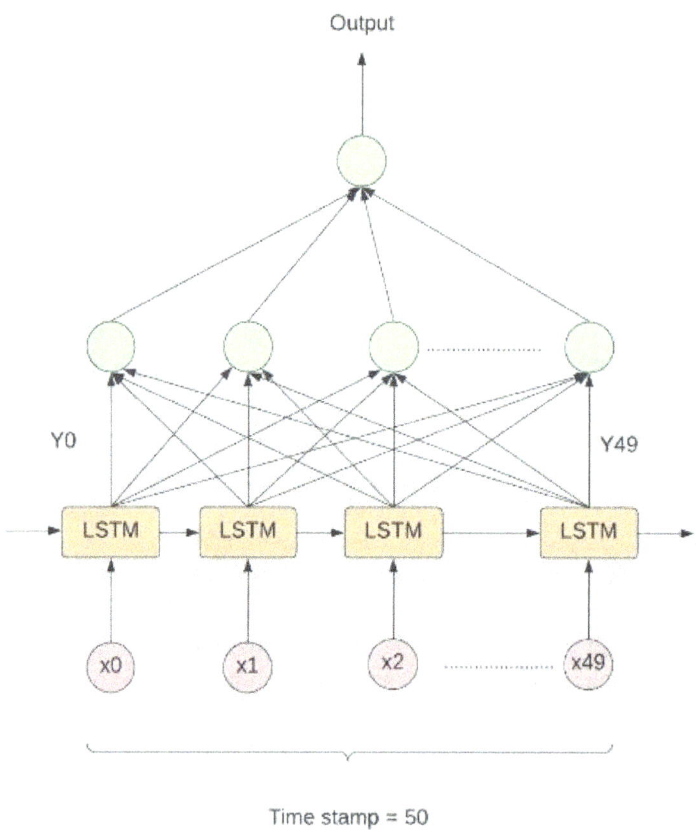

Figure 12: Proposed Vanilla LSTM Architecture.

4.2.1 Model training of Vanilla LSTM

The model is trained by using the proposed Vanilla LSTM model. It is defined using the Sequential API. We initialize the model variable with the Sequential () method.

```
Model: "sequential"
_________________________________________________________________
 Layer (type)                Output Shape              Param #
=================================================================
 lstm (LSTM)                 (None, 100)               40800

 dropout (Dropout)           (None, 100)               0

 dense (Dense)               (None, 1)                 101

=================================================================
Total params: 40,901
Trainable params: 40,901
Non-trainable params: 0
_________________________________________________________________
```

Figure 13: Vanilla LSTM Model.

Figure 13 shows the Vanilla LSTM model. The first layer is an LSTM layer, with the recurrent segment holding 100 neurons, followed by a dropout layer and the fully connected dense layer. The Linear function is used to predict the output of the model. The model is tested on three datasets of stock, which are FedEx, Ford, and Meta.

The trainable parameters in the LSTM layer are calculated according to the formula as shown in eq. (7).

These trainable parameters are the weights of the connections between the units in the network.

$$4((\text{n nodes} + \text{input feature}) \times \text{n nodes} + \text{n nodes}) \tag{7}$$

In this Vanilla LSTM model, we have 100 nodes and one input feature.

$$4((100 + 1) \times 100 + 100) = 40800 \tag{8}$$

The next layer is the dropout layer. The dropout layer is used in neural networks to avoid over fitting. It sets the input units to '0' during training.

The other layer is the dense layer, with trainable parameters as shown in (4.9)

$$\text{inputs} \times \text{outputs} + \text{biases} \tag{9}$$

Here, the input is 100, 1 is output, and bias is 1.

$$100 \times 1 + 1 = 101$$

The total parameters are: 40,800 + 101 = 40,901.

Next, we compile and train the model. For model compilation, we use Compile () method. In this method, we use optimizer and loss to specify the loss function. And for model training, we use the fit () method. In this method, we specify the batch size and the number of epochs. Predict () method is used to make predictions. Figure 14 shows the actual and predicted graph for the Ford stock. Figure 15 shows the actual and predicted graph for FedEx stock, Figure 16 shows the actual and predicted graph for Meta stock.

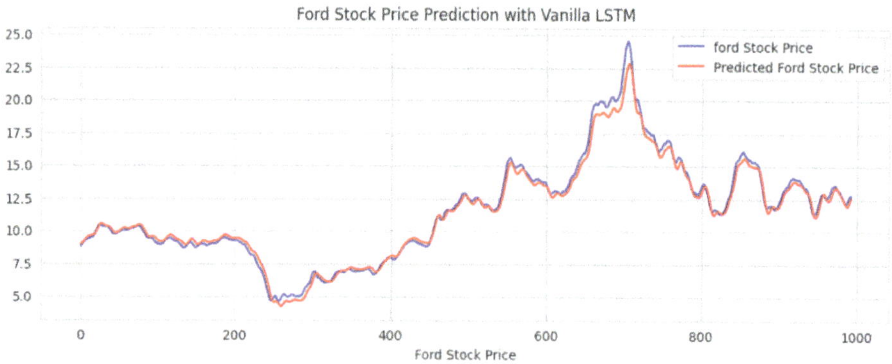

Figure 14: Ford Stock Actual and Predicted graph for Vanilla LSTM Model.

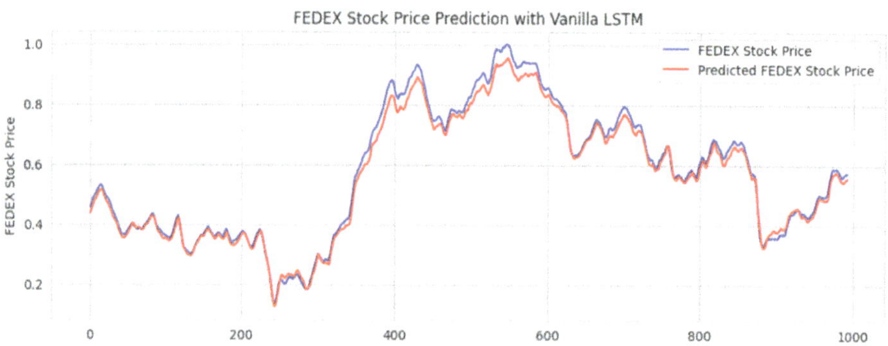

Figure 15: FedEx Stock Actual and Predicted graph for Vanilla LSTM Model.

4.3 Stacked Long Short-Term Memory (LSTM)

The stacked LSTM consists of an input layer, LSTM layer, and an output layer. In a stacked LSTM model, the first layer can learn basic features from the input sequence and the next layers can learn more complex features, based on the output sequence from the previous layer. The last layer in the stack generates the final output sequence. During training, the model learns to adjust the weights of the gates to decide

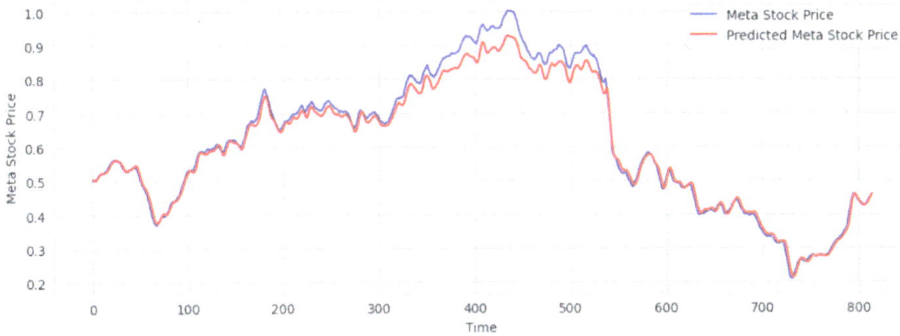

Figure 16: Meta Stock Actual and Predicted graph for Vanilla LSTM Model.

which information to keep or forget and which latest information to add to the memory cells. The gates in each LSTM layer control the flow of information through the network.

4.3.1 Implementation of Stacked LSTM

Figure 17 shows the architecture of Stacked LSTM. Stacked LSTM consists of multiple hidden layers and each hidden layer consists of multiple memory cells. In stacked LSTM, we have multiple layers where the output of one layer is input to another layer in three-dimensional format. After every LSTM layer, we have a dropout layer. And we have a dense layer, where the output from the dense layer is passed to the final layer. The final layer predicts the final output.

4.3.2 Model training of Stacked LSTM

Figure 18 shows the Stacked LSTM model. It consists of three LSTM layers, two dropout layers, and one dense layer. Here, we use (return _sequence = True) for the first two LSTM layers, as the output of one layer should provide input to another layer in three-dimensional format. All the models are the same as Vanilla LSTM model except for the number of layers. The first LSTM layer consists of 150 neurons, the second LSTM layer consists of 100 neurons, and the third layer consists of 50 neurons, followed by a fully connected dense layer.

Figure 19 shows the actual and predicted graph for the Ford stock, Figure 20 shows the actual and predicted graph for the FedEx stock, Figure 21 shows the actual and predicted graph for the Meta stock.

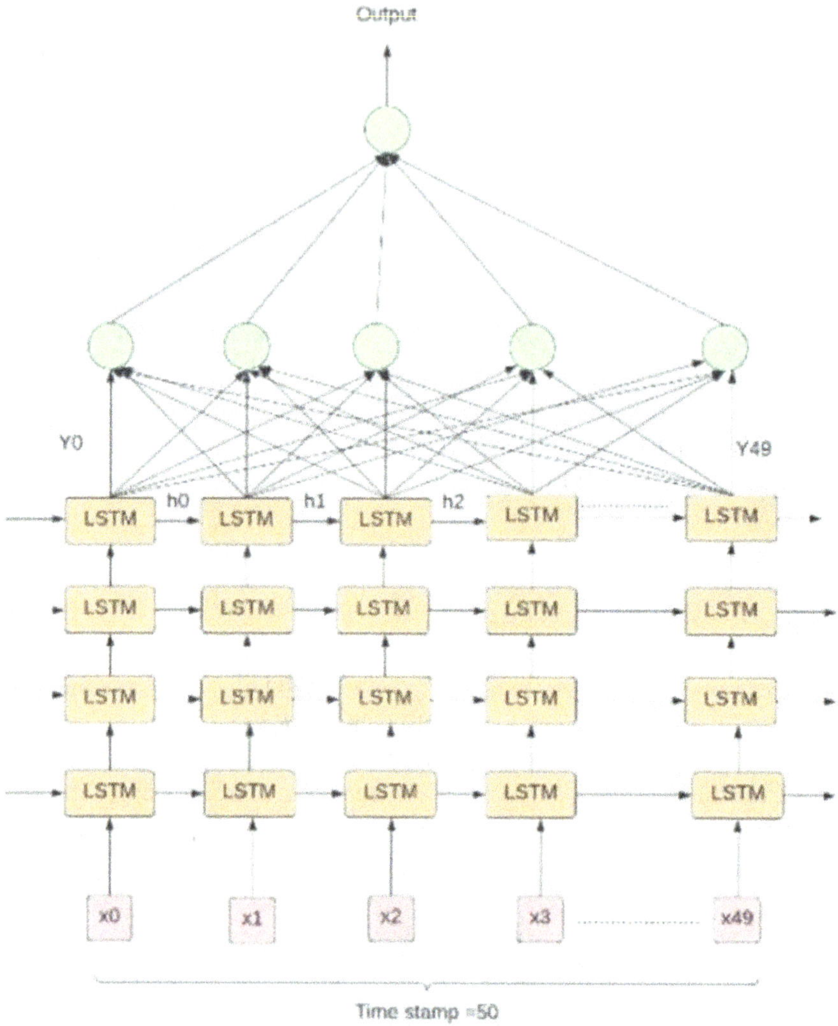

Figure 17: Proposed Stacked LSTM Architecture.

```
Model: "sequential_1"
_________________________________________________________________
 Layer (type)                Output Shape              Param #
=================================================================
 lstm_1 (LSTM)               (None, 60, 100)           40800

 dropout_1 (Dropout)         (None, 60, 100)           0

 lstm_2 (LSTM)               (None, 60, 64)            42240

 dropout_2 (Dropout)         (None, 60, 64)            0

 lstm_3 (LSTM)               (None, 32)                12416

 dense_1 (Dense)             (None, 1)                 33

=================================================================
Total params: 95,489
Trainable params: 95,489
Non-trainable params: 0
_________________________________________________________________
```

Figure 18: Stacked LSTM Model.

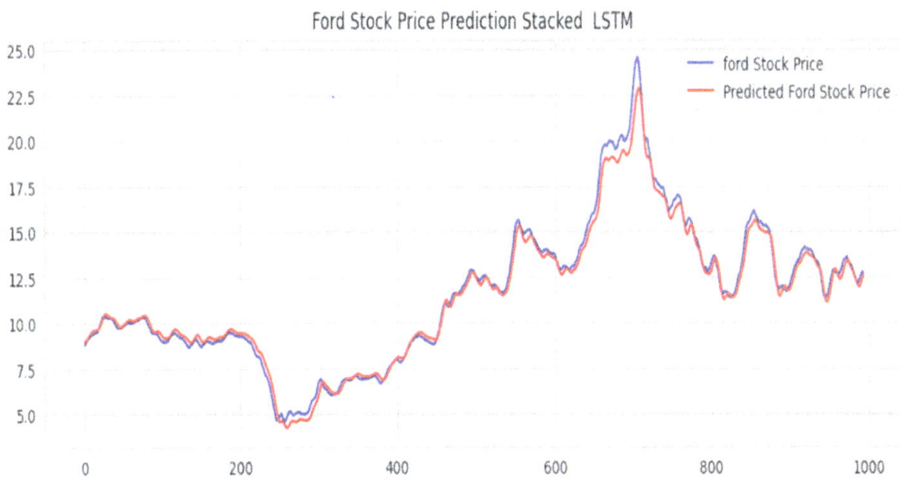

Figure 19: Ford Stock Actual and Predicted graph for Stacked LSTM Model.

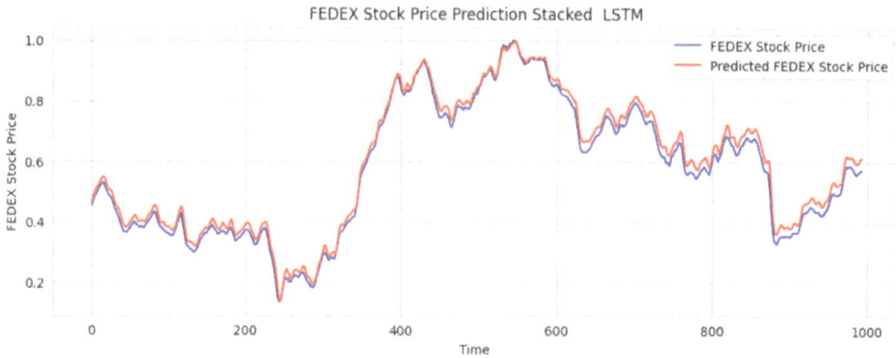

Figure 20: FedEx Stock Actual and Predicted graph for Stacked LSTM Model.

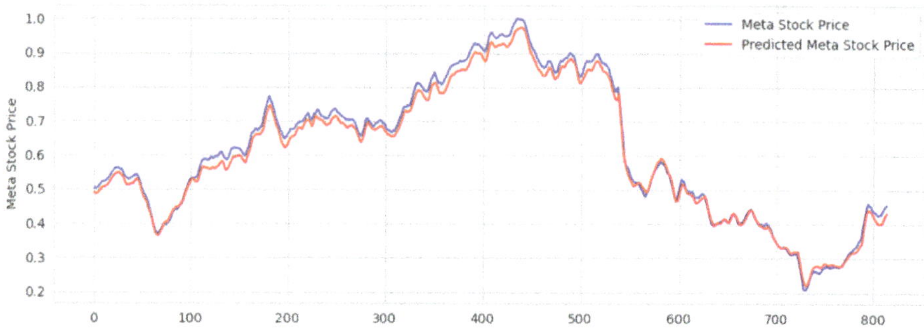

Figure 21: Meta Stock Actual and Predicted graph for Stacked LSTM Model.

4.4 Bidirectional Long Short-Term Memory (Bi-LSTM)

The Bidirectional LSTM consists of an Input layer, Forward layer, Backward layer, Activation layer, and Output layer.

Input Layer: Input layer takes the input in the form of sequence and sends it to the forward layer.

Forward Layer: The forward layer takes the current input, previously hidden state, and cell states, and produces a new hidden state and new cell state as output. The hidden state is used to stand for the current memory, and it makes prediction about the next item in the sequence.

Backward Layer: The backward layer of a Bi-LSTM (Bidirectional Long Short-Term Memory) processes the input sequence in the opposite direction, starting from the last item in the sequence and going ahead to the first. At each time step, the backward layer takes in the current input item and the previously hidden and cell states, and produces a new hidden and cell state as output. The backward layer hidden state is the future context of the input sequence, and it is used to make predictions for the previous item.

Activation Layer: The activation layer uses sigmoid or tanh function and produces the final output by taking the input from final hidden states of both the forward and backward layers.

Output Layer: The output layer takes the output from the activation function and gives us the final output.

4.4.1 Implementation of Bidirectional LSTM

Figures 5–17 shows the Bi-LSTM model. Bidirectional LSTM has input flows from both the directions, forward and backward. Bidirectional LSTM can learn from both directions, past and future. It can hold long-term dependencies from both left and right contexts.

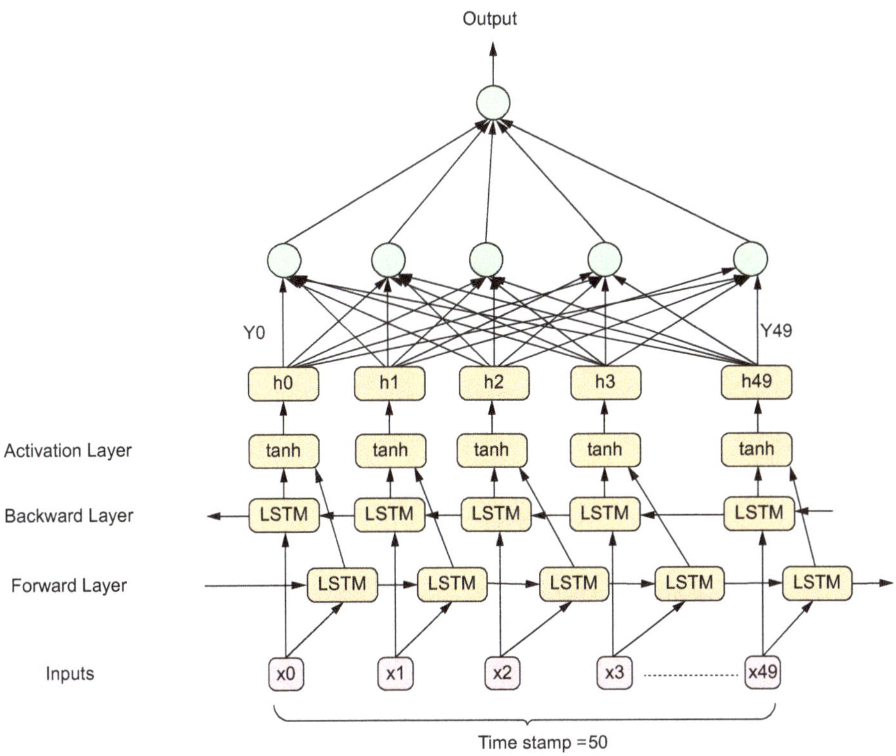

Figure 22: Proposed Bidirectional LSTM Architecture.

4.4.2 Model training of Bi-LSTM

Figure 22 and 23 shows the Bidirectional LSTM model. It consists of one Bidirectional layer, one dropout layer, and one dense layer. We use tanh activation function. The Bidirectional LSTM layer consists of 50 neurons and is followed by a fully connected dense layer.

Figure 24 shows the actual and prediction graph for the Ford stock, Figure 25 shows the actual and predicted graph for the FedEx stock, Figure 26 shows the actual and predicted graph for the Meta stock.

```
Model: "sequential_4"
_________________________________________________________________
 Layer (type)                Output Shape              Param #
=================================================================
 bidirectional_2 (Bidirectio  (None, 100)              20800
 nal)

 dropout_5 (Dropout)         (None, 100)               0

 dense_4 (Dense)             (None, 1)                 101

=================================================================
Total params: 20,901
Trainable params: 20,901
Non-trainable params: 0
_________________________________________________________________
```

Figure 23: Bidirectional LSTM Model.

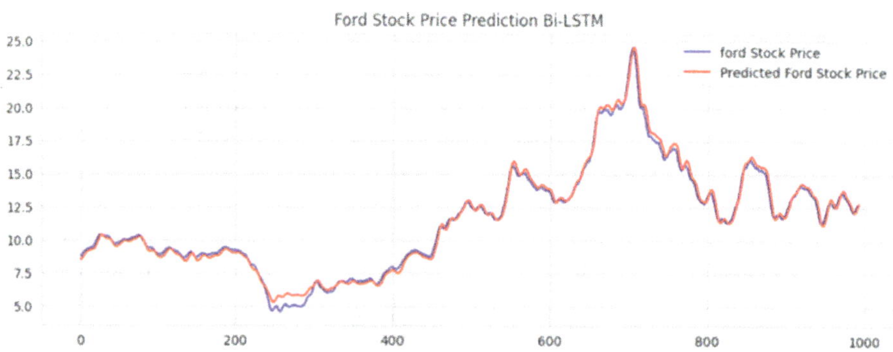

Figure 24: Ford Stock's actual and prediction graph for Bidirectional LSTM Model.

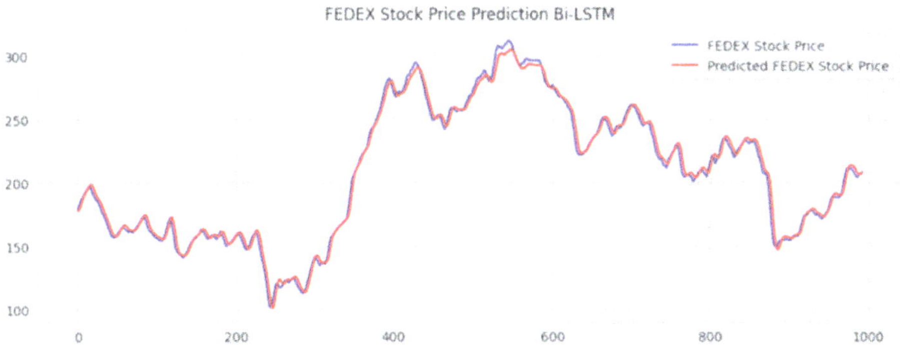

Figure 25: FedEx Stock's actual and prediction graph for Bidirectional LSTM Model.

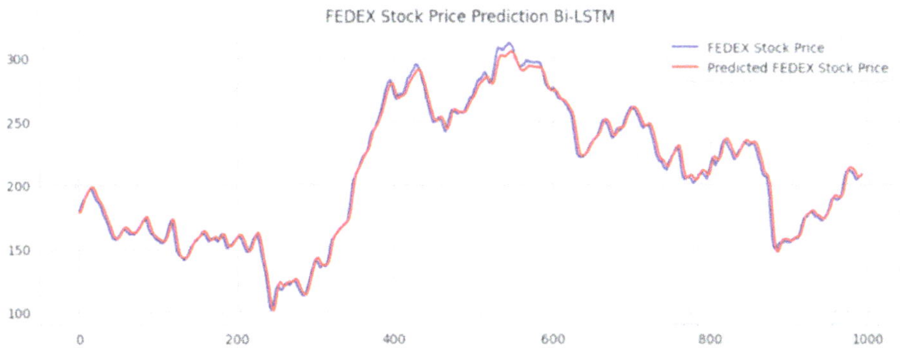

Figure 26: Meta Stock's actual and prediction graph for Bidirectional LSTM Model.

5 Optimizers and model accuracy

In this section, the performance of three models, vanilla LSTM, Stacked LSTM, and Bidirectional LSTM is compared on the performance metric of accuracy prediction, pertaining to the closing stock price of three different datasets. The accuracy is calculated using Root Mean Square Error (RMSE), which is the most common metric used for regression tasks. According to the rule of thumb, an RMSE value between 0.2 and 0.5 has the best accuracy. These models are implemented using python using Kera's and Tenser flow. The loss function used in these models is Root Mean Square Error (RMSE). The optimizer used to fit the model is ADAM (Adaptive Movement Estimation).

5.1 Optimizers in deep learning

Optimizer in deep learning is an algorithm or function that is used to change the neural network attributes. There are many optimizers in deep learning, such as Gradient Descent, Stochastic Gradient Descent, Mini- batch Gradient Descent, Adagrad, RMS Prop, AdaDelta, and ADAM. In this section, we used Adam optimizer, as it trains the neural network with less time. and more efficiently.

5.1.1 ADAM (Adaptive Movement Estimation) optimizer

The Adam (Adaptive Movement Estimation) is an optimization algorithm used to train deep learning models. It is a variant of stochastic gradient descent (SGD) that combines the advantages of two other optimization algorithms, namely, Adaptive Gradient Algorithm (AdaGrad) and Root Mean Square Propagation (RMSProp). In this optimizer, the learning rate is updated for each parameter, based on its past gradients and their magnitudes. Specifically, the algorithm calculates the exponential moving averages of the gradient and the squared gradient, and uses these estimates to update the parameters. The algorithm also includes a bias correction mechanism to account for the fact that the moving averages are initialized to zero.

5.2 Model accuracy

In conclusion, we find that Bi-directional LSTM has the best performance accuracy of all the three datasets. Here, the accuracy is concluded using the RMSE and R_2 values as shown in figures 27–29.

RMSE: RMSE is defined as the square root of the mean of the squared differences between the predicted and the actual values. It is calculated using the following formula:

$$RMSE = \sqrt{\bar{1}} \div N \sum_{i=0}^{N} (Y - Y^{\sim})^2 \tag{10}$$

R-square (R^2): R-square is a regression error metric used to measure the performance of the model. It is used to measure how best the data fits the regression line. R-square value should be close to '1'. If it not close to '1' (0), it shows that the regression line does not fit the data.

R^2 = 1-sum squared regression (SSR) × total sum of squares (SST) [27]

$$R_2 = 1 - \Sigma(Y_i - Y^{\sim})_i^2 (Y_i - Y^{-})^2 \tag{11}$$

5.2.1 Comparison of error metrics

Ford

	RMSE	MSE	R2
Model			
Vanilla LSTM	0.399293	0.159435	0.964402
Stacked LSTM	0.250504	0.062752	0.983018
Bi-LSTM	0.065258	0.004259	0.992739

Figure 27: Error metrics of Ford Stock.

FedEx

	RMSE	MSE	R2
Model			
Vanilla LSTM	0.002384	0.000006	0.998105
Stacked LSTM	0.002532	0.000006	0.989712
Bi-LSTM	0.001299	0.000002	0.995191

Figure 28: Error metrics for FedEx Stock.

Meta

	RMSE	MSE	R2
Model			
Vanilla LSTM	0.399293	0.159435	0.964402
Stacked LSTM	0.250504	0.062752	0.983018
Bi-LSTM	0.142143	0.020205	0.988767

Figure 29: Error metrics of Meta Stock.

5.3 Forecasting

Bi-directional LSTM is used to forecast the closing stock price of three datasets. The forecasting is done for the next 30 days (about 4-and-half weeks), from March 11, 2023, to April 11, 2023. Figure 30 shows the actual vs forecast graph for the Ford stock. Figure 31 shows the actual vs forecast graph for the FedEx stock. Figure 32 shows the actual vs forecast graph for the Meta stock.

Ford

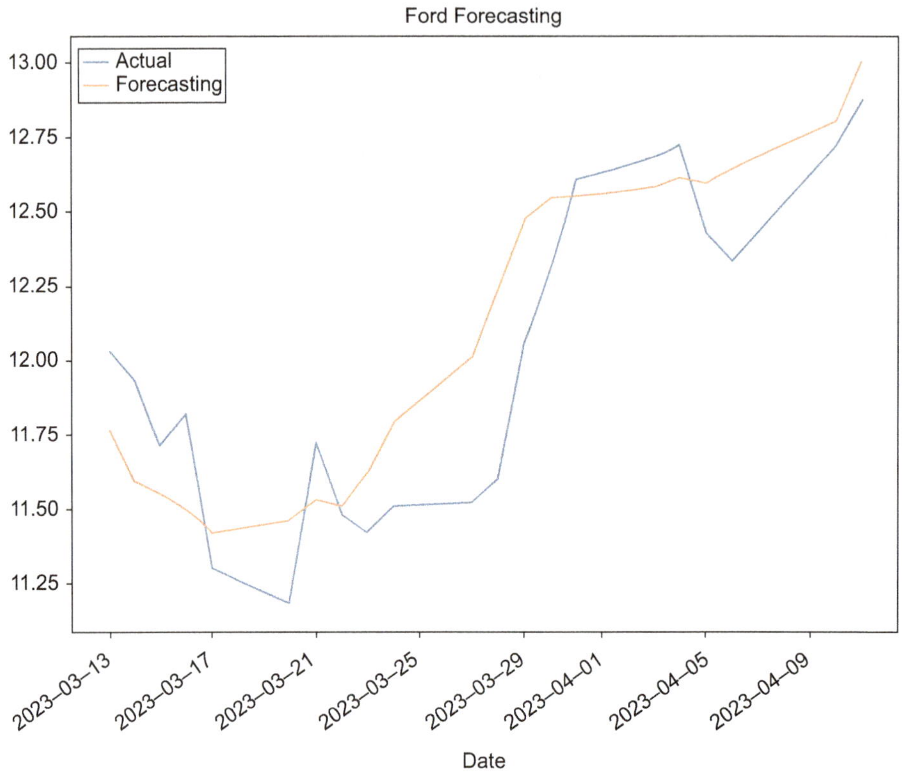

Figure 30: Actual vs Forecast graph for Ford Stock.

FedEx

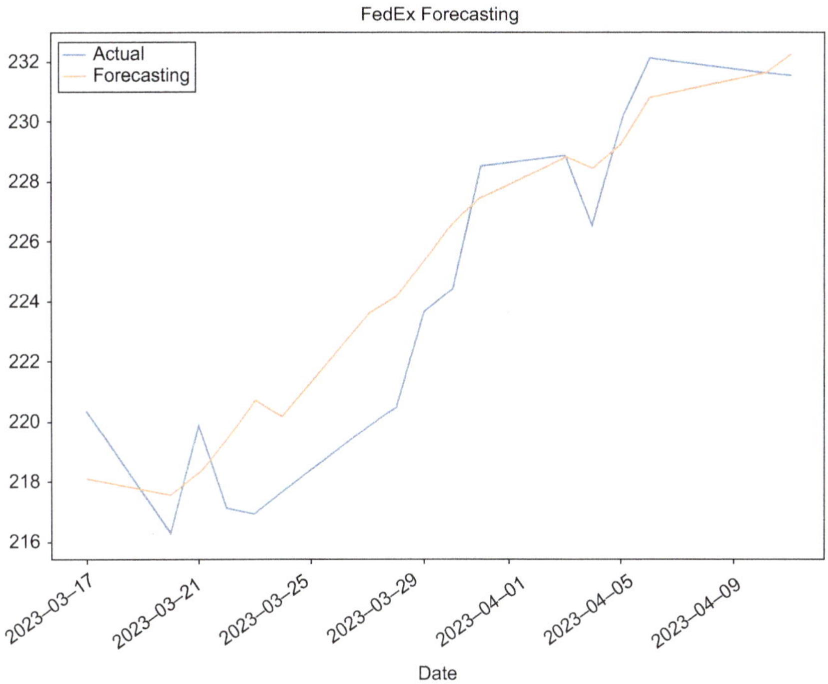

Figure 31: Actual vs Forecast graph for FedEx Stock.

Meta

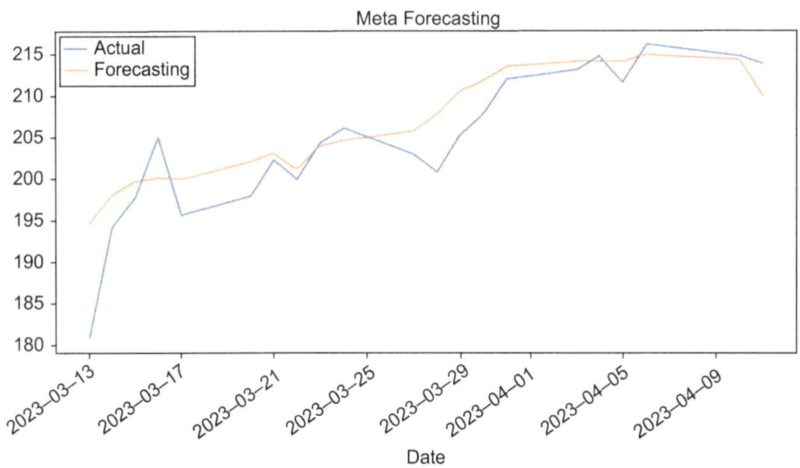

Figure 32: Actual vs Forecast values for Meta Stock.

6 Conclusion and future work

6.1 Optimizers in deep learning

Stock market prediction modeling requires us to learn long-term time dependencies in the data. In this research, we have used deep learning techniques that can learn long-term dependencies to solve complex problems. In this paper, three LSTM-based models, viz. Vanilla LSTM, Stacked LSTM, and Bidirectional LSTM are used to forecast the closing price of the stock market for a given day. Three datasets, viz., Ford, FedEx, and Meta were used to train and test the models. These models are implemented in Python using Kera's Library, and ADAM optimizer is used to build the models. The models were compared using the Root Mean Square Error (RMSE) and R-square (R^2) values. Bi-directional LSTM has the best performance accuracy.

6.2 Future work

In future, the work can be expanded by exploring various upcoming deep learning models such as deep fuzzy neural networks and Transformers.

References

[1] J. Herrmann, C. Garrett, J. Sheetz and M. Shekaramiz, "Stock Market Prediction Using Machine Learning Algorithms: The Case of Ford Motor Company," July 2020.

[2] Introduction to LSTM Units in RNN, https://www.pluralsight.com/guides.

[3] https://www.w3resource.com/python-exercises/pandas/index.php.

Anthony C. Brunson, Ryan D. Clendening, Richard Dill,
Brett J. Borghetti, Brett Smolenski, Darren Haddad,
and Douglas D. Hodson

Cellphone-based sUAS range estimation: a deep-learning classification and regression approach

Abstract: Small Unmanned Aircraft Systems (sUAS) are accessible platforms that pose security threat. These threats warrant affordable and accurate methods for tracking sUAS. We apply a novel approach to estimate the sUAS range using neural network-based solutions by processing cell phone acoustic recordings, without requiring statistical methods like TDoA or DoA. The data comes from twenty-eight cellphones recording of three different sUAS that fly over the devices. We conduct three experiments as a part of this research. In the first two experiments from [1], the audio data is converted into 0.5s Mel-spectrograms frames and 0.5s raw audio frames, to separate predictions into four range classes. We sequester the data into an 80/20 training test split. The 2DCNN architecture outperforms the other architectures (1DCNN and 2DCRNN). The 2DCNN is then retrained to generalize the sUAS range across various sUAS types to achieve an average Macro-F1 score of 0.7492. In the third experiment, the audio is transformed into 0.1s Mel Frequency Cepstral Coefficient (MFCC) frames to predict the actual distance in meters that the sUAS is from the audio source. A 2DCNN architecture is created that is tested with regression to predict the actual distance in meters that the sUAS is from the audio source. In all scenarios, truth values are calculated from the Euclidean distance between the sUAS and a cell phone. The results show that deep-learning-based sUAS-ranging with cellphones is an effective and low-cost method for accurately tracking sUAS.

Keywords: drone detection, machine learning, acoustic, range detection

Anthony C. Brunson, Department of Electrical and Computer Engineering, Air Force Institute of Technology, Dayton, OH, USA, e-mail: anthony.brunson@afit.edu
Ryan D. Clendening, Department of Electrical and Computer Engineering, Air Force Institute of Technology, Dayton, OH, USA, e-mail: ryan.clendening@afit.edu
Richard Dill, Department of Electrical and Computer Engineering, Air Force Institute of Technology, Dayton, OH, USA, e-mail: Richard.Dill@afit.edu
Brett J. Borghetti, Department of Electrical and Computer Engineering, Air Force Institute of Technology, Dayton, OH, USA, e-mail: Brett.Borghetti@afit.edu
Brett Smolenski, North Point Defense, Rome, NY, USA, e-mail: bsmolenski@northpointdefense.com
Darren Haddad, Info. Exploit. Branch Air Force Research Labs, Air Force Institute of Technology, Rome, NY, USA, e-mail: darren.haddad@us.af.mil
Douglas D. Hodson, Department of Electrical and Computer Engineering, Air Force Institute of Technology, Dayton, OH, USA, e-mail: Douglas.Hodson@afit.edu

https://doi.org/10.1515/9783111344126-011

1 Introduction

The accessibility of small Unmanned Aircraft Systems (sUAS) presents significant security risks to the public and military operations. In 2017, a Canadian passenger jet collided with a hobbyist drone, causing damage to the wing and risking passenger lives [2]. In the Russian-Ukraine Conflict, sUAS played a significant role in reconnaissance collection and artillery attacks [3]. The low profile and highly accessible nature of sUAS demands sUAS defense strategies that are affordable, scalable, and accurate. Although other more expensive tracking methods exist, this effort uses affordable and scalable sensing technology, with the overall goal of protecting, defending, and tracking sUAS throughout a restricted airspace.

This research investigates how cellphones can provide sUAS tracking capabilities by estimating the sUAS range from a mobile device's microphone. sUAS emit sound from the rotating motors and propellers that produce lift and velocity. Sound is generated, resulting in a fundamental frequency (or frequencies) between 0 and 2 kHz range and its harmonics [4]. Additionally, the physical vibration of sUAS produces additional acoustic noise, which tends to be at high frequencies (3 KHz–4 KHz) [5]. In the first two experiments, we use four datasets of recorded sUAS flights, named after the recorded sUAS, with corresponding range truth data: IF [6], Matrice [7, 8], Phantom, and Combined. Combined is the superset of the single sUAS datasets. We begin range predictions by separating each category into 20 meter range buckets. The acoustics data is converted into 0.5s Mel-spectrogram format for the 2-dimensional models and 0.5s raw audio for the 1D model. In this experiment, the 2D convolutional neural network (2DCNN), 1D convolutional neural network (1DCNN), and the 2D convolutional recurrent neural network (2DCRNN) are each trained from the combined data set. The best performing architecture is then retrained using all four datasets and evaluated on three sequestered test sets. We choose 40 meters as the demarcation point because a cellphone needs to recognize when an sUAS is close to a sensor and in a high-threat airspace. In the third experiment, we use a subset of the Matrice dataset, split into 75/25 train and test sets. We use this data to create a separate 2DCNN, whereby the output is a single predicted distance in meters that the sUAS is away from the audio source. Experiment 3 is different from experiments 1 and 2 due to each prediction in the classification problem either falling in the range or out of the range, whereas, in the regression problem, we can estimate the precise distance an sUAS is from an audio source.

This research contributes a method to predict the Euclidean distance an sUAS is from low-fidelity acoustic sensors within cellphones, which is scalable and accurate. The paper is organized as follows: Section 2 presents related sUAS tracking research. Section 3 offers the research methodology and model architectures. Lastly, Section 4 provides the results.

2 Related works

Researchers commonly employ two acoustics source localization methods: Direction of arrival (DoA) and time difference of arrival (TDoA) [9]. DoA is calculated using multiple signal classification (MUSIC). TDoA is calculated using generalized cross-correlation (GCC) and can produce highly accurate localization results for systems with multiple nodes [10]. Researchers have proposed using deep learning to supplement TDoA calculation; however, these techniques require fixed sensor locations [11]. A summary of sUAS localization efforts follows. Sedunov et al. developed an sUAS detection and localization system using a collection of acoustic arrays [4]. Each array consisted of 15 custom-built microphone arrays, spaced at an 80–120 m distance. The researchers applied the Steered-Response Phase Transform (SRP-PHAT) to produce the direction-of-arrival (DoA). Sedunov et al. achieved an average 4.7 degree DOA precision and 200 m range.

Kyritsis et al. developed an sUAS localization technique using DoA estimation from a four-element acoustic array [12]. The researchers determined that the maximum detectable range of the sUAS in a rural environment was 77 m and that they could achieve accurate DoA estimation.

Additionally, previous contributions have investigated sUAS ranging with a high-fidelity microphone. These efforts used high-quality recording equipment with relatively large bit depths and sample rates to achieve accurate sUAS range estimation [13–15].

Although limited, previous efforts on sUAS localization rely on sophisticated acoustic sensors and arrays to produce impressive results. In contrast, this research is the first effort to estimate the sUAS range from ordinary cellphones, devoid of sophisticated microphones. Furthermore, we introduce a machine learning method to estimate sUAS location without explicitly calculating TDoA or DoA.

3 Methodology

This section describes a novel method to estimate the sUAS range without requiring statistical methods like TDoA or DoA. These methods provide a deep learning network, capable of estimating the sUAS range from a single acoustic device (i.e., a cellphone). This contribution enables constellations of cellphones to provide persistent sUAS awareness without being limited to fixed, high-fidelity acoustic sensor configurations. We first state our research assumptions and then examine the dataset used. We then explain the features extracted as the inputs to the networks and the deep learning models used in the experiments. Finally, we present an overview of the experiment design and objectives, demonstrating that cellphone-based sUAS range estimation is achievable with a deep learning-based approach.

This research makes the following assumptions.

- sUAS targets have different acoustic signatures
- sUAS acoustic signatures change in predictable ways, depending on the range that their acoustic emissions propagate
- Accuracy decreases with an increase in environmental noise
- 80 m is the maximum detection range for an sUAS

Three sUAS flight scenarios source the datasets. There are ten hover passes, 36 short passes, and 19 long passes. The sUAS fly at 30.48 m AGL (above ground level) for each flight, move between 10 and 20 kn, and fly directly over the sensor constellation during every pass. The sensor constellation contains three clusters of cellphones that span a range of approximately 300 m. In hover flights, an sUAS hovers at 33 m. In short flights, a single sUAS flies in a straight line for 410 m. Lastly, in long flights, an sUAS flies 1.5 km across the constellation of cell phones. The entire data collection is conducted at an active airfield; thus, environmental noise (i.e., propeller noise from airplanes) is present throughout the data. Twenty-eight cell phones are positioned across the sUAS flight path. The cellphone positioning ensures that the sUAS range is generalized for varying Doppler effects, internal microphones, and device orientations. The phones capture acoustic data using RedVox, a multimodal data collection tool [16]. RedVox records acoustic data with microsecond granularity. All acoustic data is sampled at 8 KHz and converted to audio (i.e., .wav) files. Table 1 displays the cellphone and app configurations. The cellphones collect acoustic data for three different sUAS: Inspired Flight 1200 [6], DJI Matrice 600 [7], and DJI Phantom 4 Pro [8]. These sUAS differ in shape, weight, and acoustic signatures. The DJI Phantom 4 flies short and long flights, the Inspired Flight flies hover and short flights, and the Matrice flies hover, short, and long flights. Therefore, the range distributions across each sUAS vary slightly.

Table 1: Sensor configuration.

Device Count	13 Apple iPhones
	15 Samsung Androids
Apple App version	V4.0.2.4
Android App Version	V3.3.1.2
File Type	Red Vox API 1000
Audio Format	PCM Floating Point

This research uses four datasets. Each dataset contains 0.5s raw audio samples and a range class for the given frame. Although frame length is fixed for this effort, we expect the sUAS range truth data to become more ambiguous as frame length per sample increases. Therefore, an increase in frame length would likely cause a decrease in classification performance. The first three datasets consist of flights flown by each sUAS model, the Inspired Flight 1200, DJI Matrice 600, and the DJI Phantom 4 Pro. Each of the three individual sUAS datasets has a 20% test set split, sequestered before training. To maintain evaluation integrity, the training data of the combined dataset exclusively contains the training data from the three other datasets (and vice versa for test data). The truth data is separated into four classes by distance in meters: $y \leq 40$ (Class 0), $40 < y \leq 60$ (Class 1), $60 < y \leq 80$ (Class 2), or $y > 80$ (Class 3). These classes are chosen to represent valuable proximity dividers for an sUAS in flight. If an sUAS is over 80 m away, the cellphone receives little to no acoustic signal from the sUAS. However, if an sUAS is directly above a cellphone, the distance is less than 40 m away (all flights are conducted at an altitude of 33 m). Truth data within two meters of the class separations are removed to account for sUAS movement within the 0.5s frames. Table 2 provides class breakdown and dataset sizes. Figure 1 introduces the workflow of the three experiments.

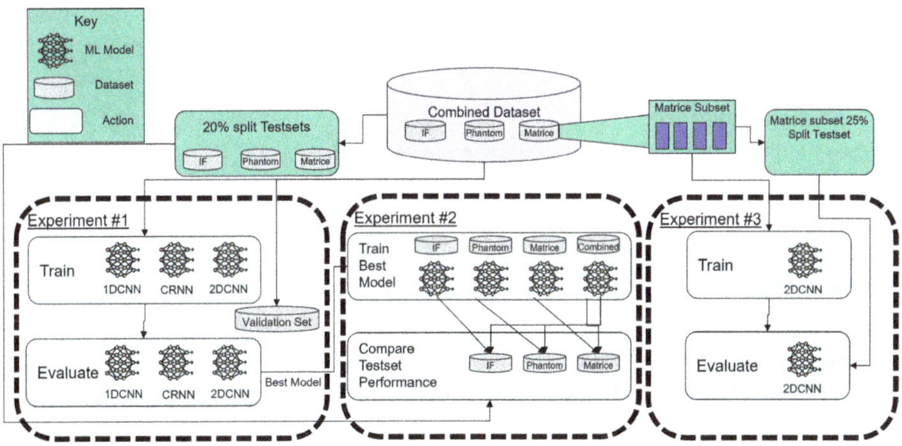

Figure 1: Methodology flow diagram.

Table 2: Dataset sizes and class distributions.

Dataset	<40 m	40–60 m	60–80 m	>80 m	Total time
IF	590.5 s (40.5%)	250.5 s (17.2%)	187.5 s (12.9%)	429.5 s (29.4%)	1,458 s
Matrice	1,929.5 s (37.6%)	657 s (12.8%)	1,048.5 s (20.4%)	1,493 s (29.0%)	5,128 s
Phantom	1,045 s (25.7%)	850 s (21.0%)	676 s (16.7%)	1,485.5 s (36.6%)	4,056.5 s
Combined	3,565 s (33.5%)	1,757.5 s (16.5%)	1,912 s (18.0%)	3,408 s (32.0%)	10,642.5 s

Experiment 1

The first experiment determines which model architectures best classify the sUAS range from the Combined dataset. This experiment uses the Combined dataset as the baseline to train and evaluate the 2DCNN, 2DCRN, and 1DCNN.

We report each architecture's score in terms of Macro F1-score, the arithmetic mean of the class 1-scores, which is a combination of recall and precision $\left(\frac{TP}{TP+1/2(FP+FN)}\right)$. We use three different model architectures throughout the experimentation process. The first is a 2DCNN, a specific type of neural network that has shown particular strength in sUAS acoustics tasks [17, 18]. The second architecture is a 2DCRNN, which has shown promise in various sound localization tasks, and combines the strengths of a CNN with the temporal memory of recurrent neural networks [19, 20]. The last architecture is a 1D convolutional neural network (1DCNN). A 1DCNN exhibits similar performance as RNNs in various time series prediction tasks; however, a 1DCNN trains in a fraction of the time. 1DCNN is chosen for this research effort as it is much more computationally efficient to train than other RNN architectures, while achieving positive results in raw audio classification problems.

The model architectures have different input formats. The 2DCNN and 2DCRNN receive data input in Mel-Spectrogram format, whereas the 1DCNN has raw formatted input. The Mel-spectrogram represents raw acoustics data in the frequency domain while preserving the time domain. In line with previous research efforts [19, 21], Librosa [22] converts the 0.5s raw acoustic frames 8x128, which contains 128 Mel-frequency bins with an FFT length of 2048 and a hop length of 512. However, the 1DCNN input is raw 0.5s audio clips (4000x1 array); thus, the data is not transformed into the Mel-Spectrogram form before inputting into the network. A 20% validation set split is used to determine architecture. All architectures' validation results are compared to a naive alternative in which the model predicts the majority class in the training dataset. The architecture with the highest validation set performance is selected for Experiment Two. Tables 3–5 display the three model architectures; each layer of the respective networks is ordered sequentially, from top to bottom. The convolution layers in the model architecture tables are represented by (number of filters) @(receptive field). Additionally, BN stands for batch normalization, which is applied prior to the activation function. In the dense layers, the number in the size column signifies the number of perceptrons within the layer.

Experiment 2

Experiment Two determines how accurately the best performing model generalizes across the three different sUAS types and evaluates if the model meets the hypothesized criteria established in Section 1. The best model is trained on each of the four datasets, and then each of the individual sUAS-trained models is compared to the

Combined trained model. Three tests evaluate the test set performance of each sUAS-trained model to the model trained using Combined. F1 and balanced accuracy (the arithmetic mean of the recall scores of the four range classes) are used to compare the performance of the deep learning models. The Individual sUAS dataset test sets are withheld from Combined and preserved for model evaluation. These tests determine if the model trained using multiple sUAS models can extract model-agnostic sUAS range features that enhance the network's ability to generalize across the different sUAS models. Additionally, this experiment assesses the merits of deep learning-based sUAS ranging with cellphones.

Table 3: 2DCNN architecture.

2DCNN architecture	Size	Activation	Strides
Input	128x8x1	–	–
Batch Norm	–	–	–
2DConv	8 @ 4x4	BN/Relu	–
Max Pool	2x2	–	–
2DConv	16 @ 2x2	BN/Relu	2x2
Dropout	0.3	–	–
2DConv	32@2x2	BN/Relu	2x2
Dropout	0.3	–	–
Flatten	–	–	–
Dense	512	Relu	–
Dense	256	Relu	–
Dropout	0.5	–	–
Output	4	Softmax	–

Table 4: 2DCRNN architecture.

2DCRNN architecture	Size	Activation	Strides
Input	128x8.1	–	–
Batch Norm	–	–	–
2DConv	8 @ 2x2	BN/Relu	1x1
2DConv	16@4x4	BN/Relu	3x1
Dropout	0.3	–	–
2DConv	32@4x4	BN/Relu	4x1
Max Pool	12x1	–	–
Dropout	0.3	–	–
Reshape	–	–	–
GRU	32	–	–
Dense	64	Relu	–
Dropout	0.5	–	–
Output	4	Softmax	–

Table 5: 1DCNN architecture.

1DCNN architecture	Size	Activation	Strides
Input	4000x1	–	–
1DConv	64@80	BN/Relu	4
Max Pool	4	–	–
1DConv	64@3	BN/Relu	2
Max Pool	4	–	–
Dropout	0.3	–	–
1DConv	128@3	BN/Relu	–
Max Pool	4	–	–
1DConv	256@3	BN/Relu	–
Max Pool	2	–	–
Reshape	–	–	–
GAP	1	–	–
Output	4	Softmax	–

Experiment 3

This experiment aims to predict the range of a drone from the cell phone audio source through a 2D Convolutional Neural Network (2DCNN). The input is four audio files that are preprocessed and converted into MFCCs, with each observation within each file consisting of 31.9 ms time steps and 20 coefficients. We use the model mentioned above with this input, which includes one output at the end, so we can use regression to predict an actual number that represents the distance in meters that the drone is from the audio source. Three loss functions, including Mean Squared Logarithmic Error (MSLE), Mean Squared Error (MSE), and Mean Absolute Error (MAE), are compiled with the model, and the varying losses are compared and contrasted. While the first two experiments established 80 meters as the maximum predicted range, this experiment truncates the audio files to include portions of the flyovers to include ranges up to 180 meters.

The data is split into training and validation sets using a 75/25 split. To avoid causality violations, we use four audio files that represent four different passes of the sUAS. The first three files are used for training, and the last file is used for testing, resulting in a 75/25 split. The separation ensures no overlap between the partitions and that no neighboring observations influence the predictions. We use batch normalization to reduce overfitting and improve the generalization ability of the model.

The 2DCNN architecture is composed of two Conv2D layers with relu activation functions. The layers contain a filter size of 32 and a kernel size of (2,2). Each layer uses Batch Normalization, Max Pooling, and a Dense function, with values of 160 and 480, respectively, followed by a Dropout layer of 0.5. Finally, the data is flattened and sent to two more dense layers, with the output set at 256 for the first and the final dense layer's output set at 1, the single predicted value needed for regression. The first dense layer uses a relu activation function, and the final uses linear, which is standard for regression problems, yielding one output. The basic architecture is selected due to its ability to solve a range of regression problems accurately. The summary of this architecture is displayed in Table 6.

Table 6: 2DCNN architecture for regression.

2DCNN architecture	Size	Activation	Strides
Input	1x20x1	–	–
2DConv	32 @ 2x2	BN/Relu	2x2
Max Pool	2x2	–	–
Dense	160		
Dropout	0.5		
2DConv	32 @ 2x2	BN/Relu	2x2
Max Pool	2x2	–	–
Dense	480	–	–
Dropout	0.5	–	–
Flatten	–	–	–
Dense	256	Relu	–
Dense	1	linear	–
Output	1		

Table 7 shows the lowest values of each loss function saved in the models after 200 epochs. MSLE, MSE, and MAE are used to evaluate model performance. In reviewing the errors displayed in Table 7, it is apparent that the best performing loss function to use with this model is MSLE. The calculated loss is displayed in Figure 2.

Table 7: Loss table.

				Result		
	Loss function			MSLE	MSE	MAE
Conv2D	MSLE	0.041	→	0.005	5.708	63.58
	MSE	3800.791	→	0.241	38.719	1972.661
	MAE	27.21	→	0.007	6.198	76.76

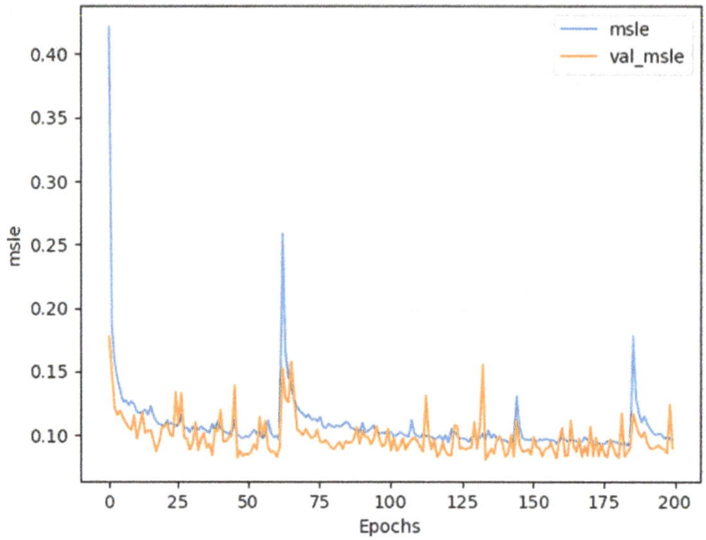

Figure 2: Conv2D – MSLE Loss.

4 Results

Experiments 1 and 2

Experiments 1 and 2 evaluate the research objective of determining if acoustics-based sUAS range estimation is achievable using deep learning. Additionally, these experiments demonstrate if deep learning-based sUAS range prediction is generalizable across multiple sUAS targets. The first experiment reports that the 2DCNN deep learning architecture best achieves sUAS range prediction. The second experiment aims to determine if the 2DCNN model generalizes the sUAS range across various sUAS types and evaluates how accurately the model predicts the range. Experiment 2 also assesses the guiding research hypothesis that a deep learning model can achieve over 70% balanced accuracy and a Macro F1 score within 40 meters above 80%. Table 8 shows the validation set performance of the three model architectures. The 2DCRNN and the 1DCNN have similar validation set performances, while the 2DCNN outperforms the other architectures. These results confirm the hypothesis that the 2DCNN is best equipped for sUAS ranging. Thus, the 2DCNN is used in Experiment Two. The results of the 2DCRNN also confirm that the dimensionality of the input data needs to effectively allow the 2DCRNN to learn time-based dependencies in the Mel-spectrogram. Other researchers that have developed 2DCRNNs for acoustics deep learning have input shapes that are much wider in the time domain (128x128) [19, 20]; thus, there is more information available in the time-domain axis of the Mel-spectrogram. Unfortunately, the acoustics data used in our experiment has an 8 KHz sampling rate and does not have the time-domain resolution, available to researchers with high-fidelity audio. Additionally, the dimensionality of a 4000 × 1 audio frame inherently presents challenges that make machine learning much more challenging than a concise representation of audio data. From Dai's paper that developed the 1DCNN for raw audio, the researchers concluded that, at best, the 1DCNN performed similarly to a 2DCNN on sound event classification tasks [23].

Experiment 2 evaluates how effectively the 2DCNN generalizes the sUAS range across various sUAS models through three different comparisons. The first result of Experiment Two compares the 2DCNN trained with the IF dataset and the 2DCNN trained with Combined on the IF test set. Table 9 presents the Balanced Accuracy and F1 scores of the two models' performances. The 2DCNN trained on Combined improves IF range accuracy across all four classes. Additionally, massive mispredictions with 2DCNN (Combined dataset) are reduced (e.g., predicting 80 m +, but the sUAS is within 40 m). It achieves an F1 score within 40 m of 0.90 and a balanced accuracy of 0.824, beyond the hypothesized success thresholds.

The second result compares the 2DCNN trained with the Matrice dataset and the 2DCNN trained with the combined dataset on the Matrice test set. Table 10 presents the Balanced Accuracy and F1 scores of the two models' performances. The 2DCNN trained with Combined outperforms the classification capability of the model trained with the Matrice dataset. It achieves an F1 score within 40 m of 0.91 and a Balanced Accuracy of 0.793, which exceeds the performance thresholds. The third result compares the 2DCNN trained with the Phantom dataset and the 2DCNN.

Table 8: Model architecture validation set performance.

Architecture	**Input type**	**Macro F1-score**
2DCNN	Mel-Spect	0.72
2DCRNN	Mel-Spect	0.63
1DCNN	Raw Audio	0.58
Naïve	–	0.13

Table 9: IF test set performance.

2DCNN trained w/	**IF**	**Combined**
<40 m F1	0.89	0.90
40–60 m F1	0.69	0.79
60–80 m F1	0.65	0.73
80 m + F1	0.82	0.84
Balanced Accuracy	0.769	0.824

trained with the combined data set on the Phantom test set. The Balanced Accuracy and F1 scores of the two models' performances are in Table 11. The model does not achieve the performance goals with an F1 score of less than 0.80 within 40 m and a Balanced Accuracy score of less than 0.70. This performance degradation is likely caused by the design of the DJI Phantom. The Phantom is a small sUAS with low rotor power that yields a smaller acoustic footprint than the other sUAS have. Although the 2DCNN trained using Combined only confirms the hypothesis on two of three sUAS models, there are important takeaways from the test set performances regarding the generalizability of sUAS range estimation and the usefulness of deep learning-based sUAS tracking. The first is that sUAS ranging is generalizable across different sUAS types. The performance increase from training the model with Combined versus a single sUAS implies that the 2DCNN learns sUAS type-agnostic features in the convolution layers that improve ranging performance across all sUAS types. This concept challenges how humans perceive sound and further demonstrates the power of using deep learning to recognize patterns not easily recognizable by human perception. These results demonstrate that deep learning is an effective method to localize

Table 10: Matrice Test set performance.

2DCNN Trained w/	Matrice	Combined
<40 m F1	0.90	0.91
40–60 m F1	0.60	0.64
60–80 m F1	0.73	0.75
80 m + F1	0.81	0.84
Balanced Accuracy	0.768	0.793

Table 11: Phantom test set performance.

2DCNN Trained w/	Phantom	Combined
<40 m F1	0.73	0.72
40–60 m F1	0.57	0.60
60–80 m F1	0.46	0.52
80 m + F1	0.75	0.75
Balanced Accuracy	0.626	0.647

sUAS with cellphones, when presented with low-fidelity data and a suboptimal data collection environment. These results imply that if given a large constellation of cellphones, an sUAS range estimation model distributed across all devices effectively distinguishes which devices are close (within 40 m), moderately close (40–60 m), moderately far (60–80 m), and far (farther than 80 m) of the sUAS. Combining the results, the sUAS can be effectively tracked within the constellation of cell phones. These methods provide an effective sUAS defense strategy that is de-burdened from relying exclusively on expensive sensing methods.

Experiment 3

The results of experiment 3 are displayed in Figures 3 through 5. The figures show the predicted vs. actual range of the sUAS in meters. Darker blue regions are multiple predictions stacked on the same point. As displayed, MSLE performs superior to MAE and MSE. Another view of the test dataset in Figures 6 through 8 shows the truncated audio file with time in seconds as the x-axis and range in meters as the y-axis. The solid line represents the truth data, and the scatter plots represent the predicted points. As the sUAS is closer to the recording device, accuracy increases, ultimately predicting ranges within 5 meters of the actual distance. This accuracy is consistent until the sUAS is 45 meters away, at which point the accuracy gradually begins to decline. Once beyond 180 meters, the sUAS is nearly inaudible to the recording device and cannot make accurate predictions.

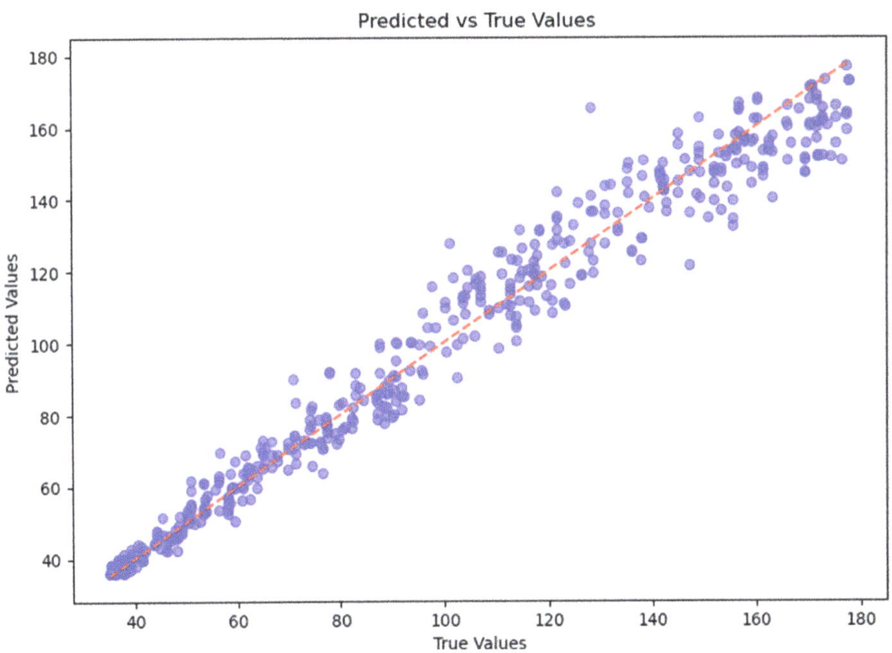

Figure 3: Conv2D – Predicted vs. True 2/MSLE Loss.

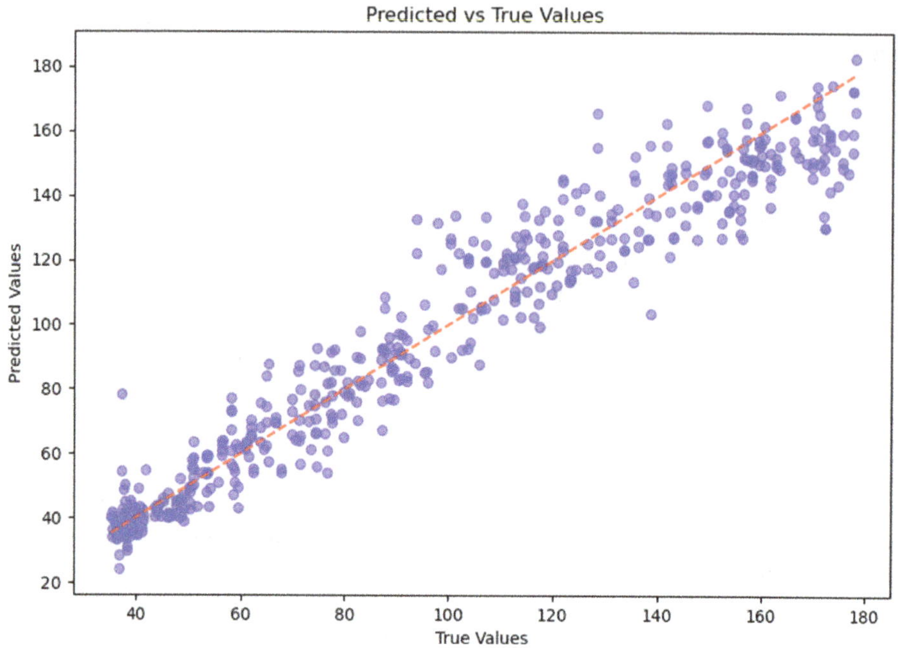

Figure 4: Conv2D – Predicted vs. True 2/MSE Loss.

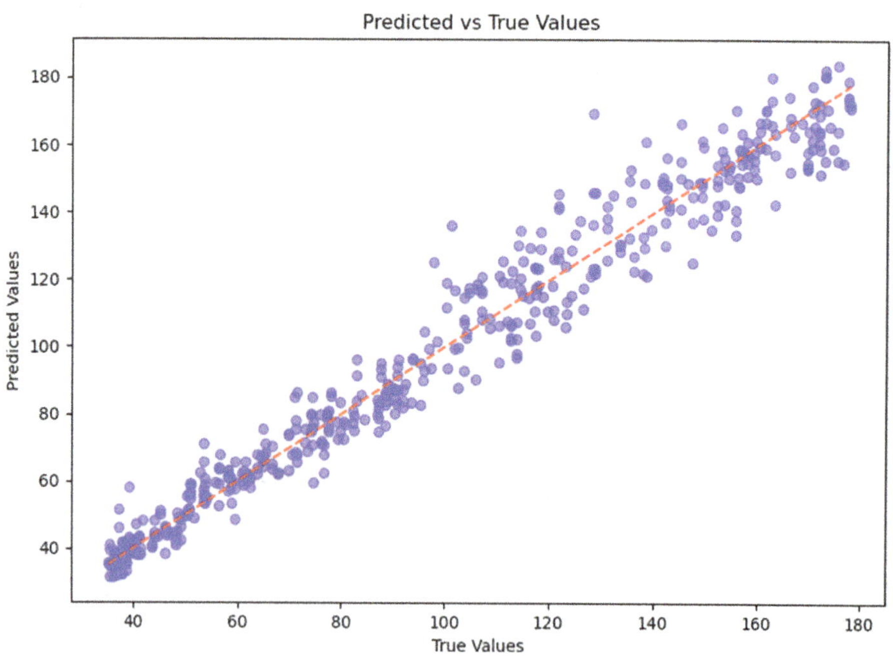

Figure 5: Conv2D – Predicted vs. True 2/MAE Loss.

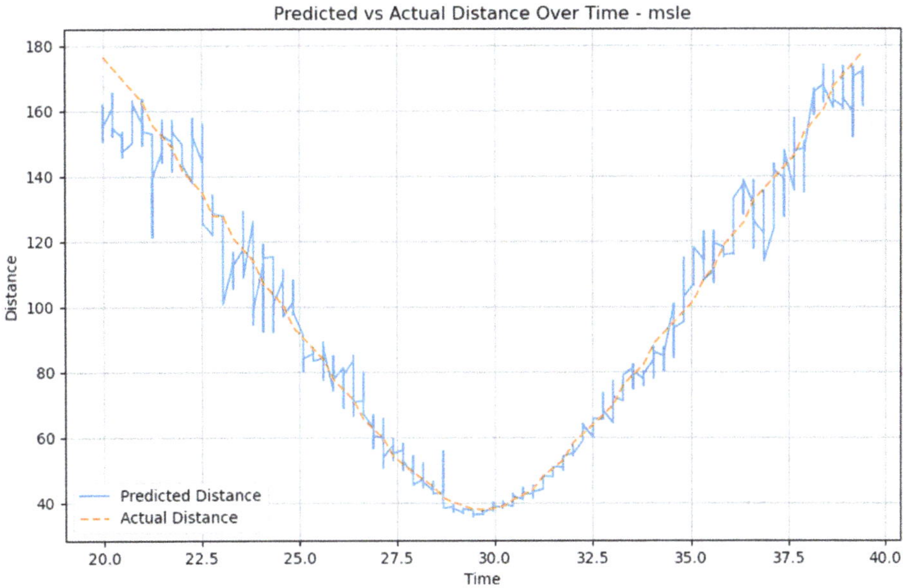

Figure 6: Conv2D – Predicted vs. True Over Time w/MSLE Loss.

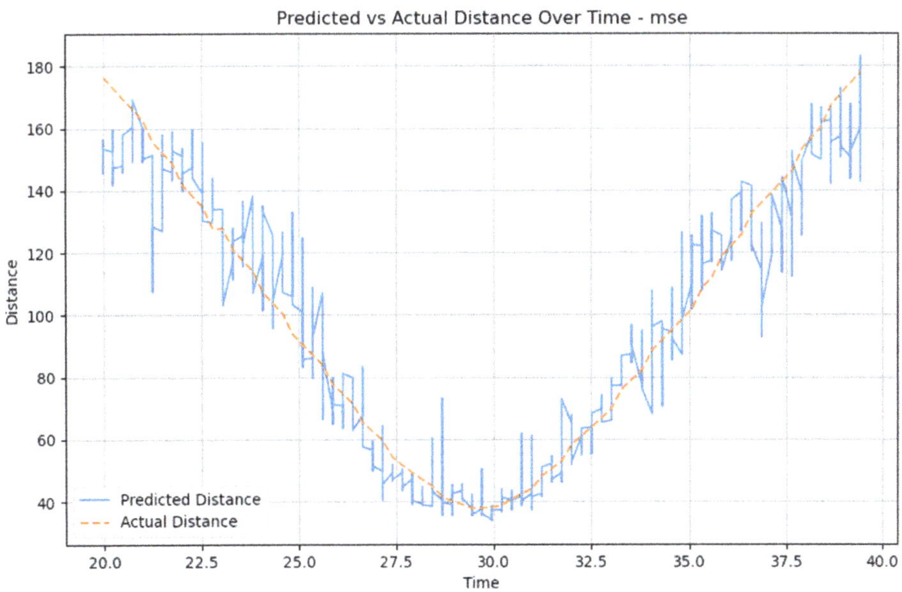

Figure 7: Conv2D – Predicted vs. True Over Time w/MSE Loss.

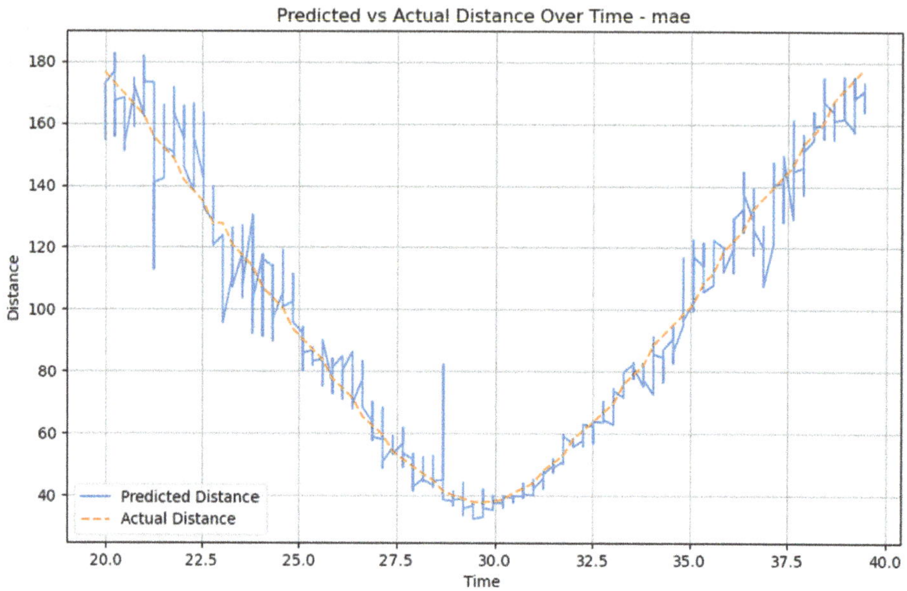

Figure 8: Conv2D – Predicted vs. True Over Time w/MAE Loss.

5 Conclusion

The threat of sUAS by state and non-state actors demands sUAS countermeasures with equally accessible defense resources. As a result, the research effort develops a deep learning-based sUAS ranging method that can be used for sUAS tracking within a constellation of low-cost sensing devices. First, three different model architectures are trained, and the best performing model, the 2DCNN, is selected for further testing. Test sets from each of the three individual sUAS datasets evaluate the performance of the 2DCNN trained using the Combined dataset against the 2DCNN trained on individual sUAS datasets. The results show that additional data from various sUAS help the model better achieve sUAS ranging on each sUAS model. The Balanced Accuracies and F1 scores within 40 m for both the IF and Matrice sUAS models are above the threshold of 0.70 and 0.80, respectively; however, the Balanced Accuracy and Macro F1 score for the Phantom sUAS are not above either threshold. In the regression task, the range in meters is predicted, which results in an error rate of less than 5 meters when the sUAS is less than 45 meters away. As the sUAS increases in distance from the cell phone, the error rate increases as expected due to the microphone's lost ability to pick up sound, but out to 180 meters, Figure 3 shows that predictions are still highly accurate.

The next step in this research is to build a model to predict the GPS coordinates. To accomplish this, directionality must be included in the truth dataset. Additionally, including altitude will provide the ability to obtain a more localized result because, with this paper, since the measure is Euclidean distance, a horizontal range of 100 meters has the same results as a vertical range of 100 meters. Ultimately, this research is a decisive step forward in accomplishing the ultimate goal of using audio to predict the actual location of a drone.

References

[1] R. Clendening, *Cellphone-Acoustics Based sUAS Detection and Tracking.* Dayton OH: Air Force Institute of Technology, 2023.

[2] T. M. Andrews, *A Commercial Airplane Collided with A Drone in Canada, A First in North America.* Washington DC: Washington Post, 2017.

[3] C. Vallance, "Ukraine Sent Dozens of 'Dronations' to Build Army of Drones," *BBC*, 8 July 2022.

[4] A. Sedunov, D. Haddad, H. Salloum, A. Sutin and N. Sedunov, "Stevens Drone Detection Acoustic System and Experiments in Acoustics UAV Tracking," *2019 IEEE International Symposium on Technologies for Homeland Security (HST)*, 2019.

[5] H. Kolamunna, T. Dahanayaka, J. Li, S. Seneviratne, K. Thilakaratne, A. Zomaya and A. Seneviratne, "DronePrint: Acoustic Signatures for Open-set Drone Detection and Identification with Online Data," *Proceedings of the ACM on Interactive, Mobile, Wearable and Ubiquitous Technologies*, vol. 5, no. 1, pp. 1–31, 2021.

[6] "Inspired Flight," 2023, [Online]. Available: https://www.inspiredflight.com/if1200a.php. [Accessed 17 10 2023].

[7] DJI, "Matrice 600," [Online]. Available: https://www.dji.com/matrice600. [Accessed 17 10 2023].

[8] DJI, "Phantom 4 PRO," [Online]. Available: https://www.dji.com/phantom-4-pro-v2. [Accessed 17 10 2023].

[9] X. Shi, G. Mao, B. Anderson, Z. Yang and J. Chen, "Robust Localization Using Range Measurements with Unknown and Bounded Errors," *IEEE Transactions on Wireless Communications*, vol. 16, no. 6, pp. 4065–4078, 2017.

[10] X. Chang, C. Yang, J. Wu, X. Shi and Z. Shi, "A Surveillance System for Drone Localization and Tracking Using Acoustic Arrays," *2018 IEEE 10th Sensor Array and Multichannel Signal Processing Workshop*, pp. 573–577, 2018.

[11] Z. Wang, D. Hu, Y. Zhao, Z. Hu and Z. Liu, "Real-Time Passive Localization of TDOA via Neural Networks," *IEEE Communications Letters*, vol. 25, no. 10, pp. 3320–3324, 2021.

[12] A. Kyritsis, R. Makri and N. Uzunoglu, "Small UAS Online Audio DOA Estimation and Real-Time Identification Using Machine Learning," *Machine Learning and Signal Processing Based Acoustic Sensors*, 2022.

[13] K. Gopalan, B. Smolenski and D. Haddad, "Acoustic Detection of Drone Range and Type Using Nonuniform Band Energy Features," *Journal of the Acoustic Society of America*, vol. 152, no. 4, p. 152, 2022.

[14] K. Gopalan, B. Smolenski and D. Haddad, "Detection and Classification of Drones Using Fourier-Bessel Series Representation of Acoustic Emissions," *Journal of the Acoustic Society of America*, vol. 152, no. 4, p. 152, 2022.

[15] M. Tan, B. Smolenski and D. Haddad, "Real-time Acoustic Detection and Identification of Drones in Operational Conditions," *Journal of the Acoustic Society of America*, vol. 152, no. 4, p. 152, 2022.

[16] M. Isla, A. Christe and T. Yoshiyama, "Redvox-python-sdk," [Online].

[17] Y. Seo, B. Jang and S. Im, "Drone Detection Using Convolutional Neural Networks with Acoustic STFT Features," in *15th IEEE International Conference on Advanced Video and Signal Based Surveillance (AVSS)*. Auckland, 2018.
[18] S. Al-Emadi, A. Al-Ali and A. Al-Ali, "Audio-Based Drone Detection and Identification Using Deep Learning Techniques with Dataset Enhancement through Generative Adversarial Networks," *UAV Detection, Classification, and Tracking)*, no. Special, 2021.
[19] M. Yiwere and E. Joo Rhee, "Sound Source Distance Estimation Using Deep Learning: An Image Classification Approach," *Speech, Acoustics, Audio Signal Processing and Applications in Sensors*, no. Special, 2020.
[20] S. Adavanne, A. Politis, J. Nikunen and T. Virtanen, "Sound Event Localization and Detection of Overlapping Sources Using Convolutional Recurrent Neural Networks," *Journal of Selected Topics in Signal Processing*, 2018.
[21] Y. Pandeya, D. Kim and J. Lee, "Domestic Cat Sound Classification Using Learned Features from Deep Neural Nets," *Applied Science*, vol. 8, 2018.
[22] B. McFee, C. Raffel, D. Liang, D. Ellis, M. McVicar, E. Battenberg and O. Nieto, "Librosa: Audio and Music Signal Analysis in Python," *14th Python in Science Conferences*, 2015.
[23] W. Dai, C. Dai, S. Qu, J. Li and S. Das, "Very Deep Convolutional Neural Networks for Raw Waveforms," *International Conference on Acoustics, Speech and Signal Processing*, 2017.

B. Chandra, Kushal Pal Singh, Prem Kalra, and Rajiv Narang

Automatic diagnosis of 12-lead ECG using DINOv2

Abstract: The electrocardiogram (ECG), though a cheap, reliable, and fast diagnostic test for detecting several heart anomalies, needs accurate interpretation by a skilled professional. The proposed method in this paper uses a visual transformer, DINOv2, for automatic ECG diagnosis. The DINOv2, pre-trained on many real-world object images, has been fine-tuned with ECG images. This model is used for extracting relevant features required by the classification method for anomaly detection. The main aim of this paper is to classify the widely known CODE-15 dataset with increased accuracy. This dataset is 15% of data from a larger CODE dataset. This 12-lead ECG dataset has samples for six heart conditions: 1dAVb, RBBB, LBBB, SB, AF, and ST. The model has achieved a test accuracy of 96.30%. This accuracy has been achieved with a short duration (2 s) of ECG data only. Ribeiro et al. [1] have tried to classify the entire CODE data and have been able to achieve an average precision of 92.37%. The analysis of cases where predictions do not match with the ground truth shows that 28% of such cases are indeed classified correctly by the model. The efficacy of the method has been tested on the China Physiological Signal Challenge (CPSC) 2018 dataset also. This dataset contains ECG data for 8 anomalies, viz., AF, 1dAVB, LBBB, RBBB, PAC, PVC, STD, and STE, along with normal sinus rhythm. We achieved an f1-score of 0.92 as against 0.76 by Yang et al. [11], for nine classes.

Keywords: ECG classification, vision transformer, DINOv2, fine-tuning, machine learning

1 Introduction

Heart diseases, also known as cardiovascular diseases (CVDs), are the most prevalent noncommunicable, chronic diseases and are the leading cause of death globally. If the disease is detected in early stages, the progression of the disease can be checked and most of these deaths can be avoided. This requires a fast, reliable, and affordable di-

B. Chandra, ANSK School of Information Technology, Indian Institute of Technology Delhi, New Delhi, India, e-mail: bchandra@maths.iitd.ac.in
Kushal Pal Singh, ANSK School of Information Technology, Indian Institute of Technology Delhi, New Delhi, India
Prem Kalra, Department of Computer Science and Engineering, Indian Institute of Technology Delhi, New Delhi, India
Rajiv Narang, Department of Cardiology, All India Institute of Medical Sciences, New Delhi, India

https://doi.org/10.1515/9783111344126-012

agnostic method for timely detection and intervention. For it to be available to the general population, it needs to be affordable.

An electrocardiogram (ECG) is one such method. It is a record of electrical activity in the heart and captures the propagation of signal. The heart can be considered as an electrical device that contracts and expands by propagation of electrical signals. By analyzing the pattern of propagation, such as time taken from traversing one point to another, voltage at different points, etc., an expert cardiologist can assess the health of the heart and can identify several anomalies. It is a noninvasive, fast, and affordable method to reveal many heart conditions. However, it requires an expert cardiologist to interpret the ECGs; otherwise, the subtle information, needed for accurate diagnosis, might be missed [1]. Automatic diagnosis can solve this issue of accurate and timely interpretation of ECG, even at remote centers, where expert cardiologists might not be available.

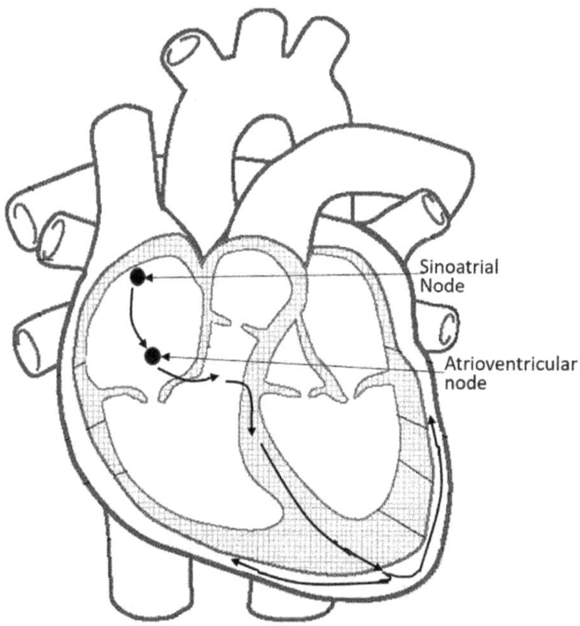

Figure 1: Electrical signal propagation in the heart.

For auto diagnosis, the system needs to understand the heartbeat pattern and extract key features for diagnosis. As discussed above, ECG is the record of electrical signal propagation in the heart. The signal is recorded by placing multiple temporary electrodes on the heart and limbs. The signal originates from the sinoatrial node. The propagation of the signal is depicted in Figure 1 [20].

Different sections in a single cycle indicate electrical activity in different parts of the heart. The recorded signal is plotted in the form of a graph. Figure 2 shows impor-

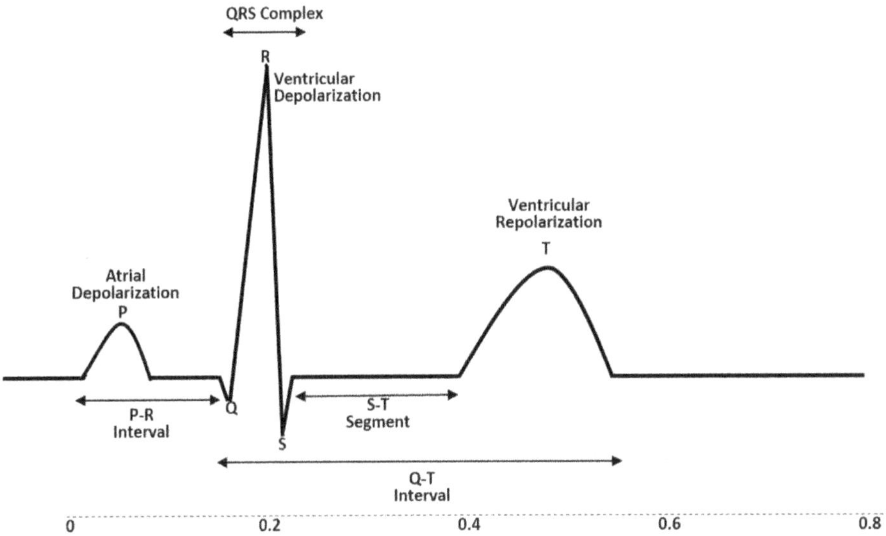

Figure 2: ECG tracing.

tant waves and intervals in a single-cycle waveform of the ECG signal [4]. An ECG captures information about the rhythm of the heart, blood flow to heart muscles, electrical conduction, etc. Any irregularity in the conduction of electrical signals manifests in the ECG in terms of waveform distortions of different waves, e.g., P, Q, R, etc., heart beats/minute or elongation/shortening of different time intervals like duration of P wave, the time interval between P and R waves, etc. These irregularities are caused by different heart conditions, which can be diagnosed by analysis of the ECG signals. The diagnostic system needs to extract these irregularities to identify heart anomalies.

This chapter is an extended version of the paper "Amazing power of DINOv2 for Automatic diagnosis of 12-Lead ECG" published in CSCI 2023. In this chapter, a brief survey of different methods for feature extraction and automatic diagnosis is discussed in Section 2. In Section 3, the background of the state-of-the-art computer vision model for image encoding and feature extraction is discussed. Section 4 discussed the experiments performed. In Section 5, the results are discussed, while the conclusion and future scope are discussed in Section 6.

2 Related work

Traditionally, ECG diagnosis has been attempted by extracting features from ECG signals using signal processing and applying fixed rules or statistical methods for classifying them into different classes [2, 3]. Several attempts have also been made to classify such extracted features using classical Machine Learning (ML) methods like logistic regression, SVM, and decision trees [5]. An approach based on Gradient Boosting Tree (GBT) and Extreme Gradient Boosting Tree (EGBT) classification methods for predicting four classes, viz., atrial fibrillation (AFIB), general supraventricular tachycardia (GSVT), sinus bradycardia (SB), and sinus rhythm (SR) was proposed in [6].

As these methods require manual feature extraction and selection, their performance depends on the accuracy of the human-level accuracy during annotation. This is a resource and skill-intensive task and hence is prone to errors [8, 14]. So, end-to-end feature extraction and selection has been attempted using deep learning-based methods like Convolutional Neural Networks (CNNs) and Recurrent Neural Networks (RNNs). These methods [17, 18] mostly consider the ECG signal as a multichannel time series data. They apply 1-dimensional CNN, RNN, and even transformer models for feature extraction and a DNN-based classifier for final classification. In [7], the authors used a capsule network [9] to detect if an ECG signal has a disease or not, i.e., a binary classification. They designed a 1D-CADCapsNet, which consists of Convolutional layers, followed by Primary and ECG layers of Capsule Networks. The final layer was a binary classification layer. In [10], the authors use a model with 5 1D-CNN hidden layers for classifying ECG signals into 3 classes, viz., "normal", "atrial premature beat", and "premature ventricular contraction." A 1D Batch Normalization follows each hidden layer. In [11], the authors developed a 1D convolution-based Deep Neural Network model for predicting nine classes of heart anomalies: normal sinus rhythm, atrial fibrillation, first-degree atrioventricular block (IVAB), left bundle branch block (LBBB), right bundle branch block (RBBB), premature atrial contraction (PAC), premature ventricular contraction (PVC), ST-segment depression (STD), and ST-segment elevation (STE). The network is based on ResNet with 4 residual blocks. In [1], the authors used the CODE dataset to detect 1st-degree AV block (1dAVb), RBBB, LBBB, SB, AF, and ST. They used the 1D ResNet-based model for classification. They achieved an f1-score of 92.55% and an average precision of 92.37%. They trained the model on the complete CODE [1] dataset, which contains more than 2.3 million records.

There have been some attempts to treat the ECG plots as images and interpret these just like humans do, i.e., by analyzing the visual features of the plot for extracting different time intervals and waveform shape anomalies. These methods apply 2D convolution for feature extraction and DNN layers for classification. In [13], the authors used VGG16 for automatic feature extraction from ECG images, instead of signal, and a classifier to predict the nine anomaly classes. In [19], deep domain adaptation has been applied for ECG diagnosis. This method proposes a subdomain adaptive deep network (SADN) for learning the feature extraction from one dataset and applying it for feature extraction on a different dataset. This paper processes the ECG as an image by plotting the data as a graph. The image is then fed into a CNN network, followed by Residual and SE-Residual blocks [11], and finally to the classifier [19].

In this chapter, we propose a method termed "ECG-DinoV2" that treats ECG plots as images for the diagnosis, using DINOv2 [16] as the backbone. We fine-tuned the DINOv2 for ECG plots for extracting relevant features. We trained the classifier and achieved substantial performance gain, over 3%, as compared to the state-of-the-art model, including the previous attempt with ViT. The results have been compared with relevant state-of-the-art solutions on relevant datasets.

3 Vision Transformer (ViT) and DINOv2

3.1 Vision Transformer (ViT)

Self-supervised learning has revolutionized the field of Natural Language Processing (NLP). It removes the need for labeled data for model training. The labels are generated from the training data itself by way of masking part of the text, and training by predicting the missing/next text. It enables the creation of enormously large datasets. The need for self-supervised learning was felt in the area of Computer Vision also. The Vision Transformer (ViT) was the first transformer for the computer vision domain. The ViT is trained in a similar way as a transformer, as shown in Figure 3. The input image is divided into multiple patches. Each patch is projected into a linear space. These projections of patches are treated like word embeddings. Some of the patches are masked during training and the objective is to predict the masked patch. ViTs have been very successful in extracting relevant features for various computer vision tasks like object detection and recognition.

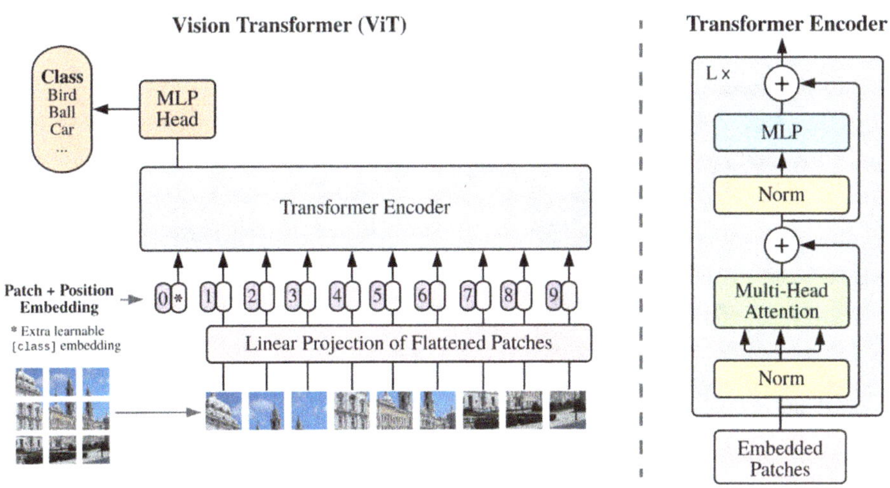

Figure 3: Vision transformer architecture [14].

3.2 DINOv2

Meta developed DINO (Deeper Into Neural Networks) [15], a self-supervised learning-based model for Computer Vision tasks. As against the ViT(Visual Transformer, the DINO model is trained by way of self-distillation. As shown in Figure 4, the teacher and student start with the same architecture but with different weights. They are presented with different views, v_1 and v_2, of the same image v. The view presented to the teacher, i.e., v_1 is larger as compared to that of the student, i.e., v_2. Thus, the teacher can extract the global context while the student learns the local context. The objective is that the student is also able to learn in the same context that the teacher has learned. So, the loss is calculated by computing cross entropy between the embeddings of student and teacher. The loss is propagated back to the student only. The weights of the teacher are updated using the exponential moving average (ema) of the weights of the student:

$$\theta_t \leftarrow \lambda\theta_t + (1-\lambda)\theta_s \tag{1}$$

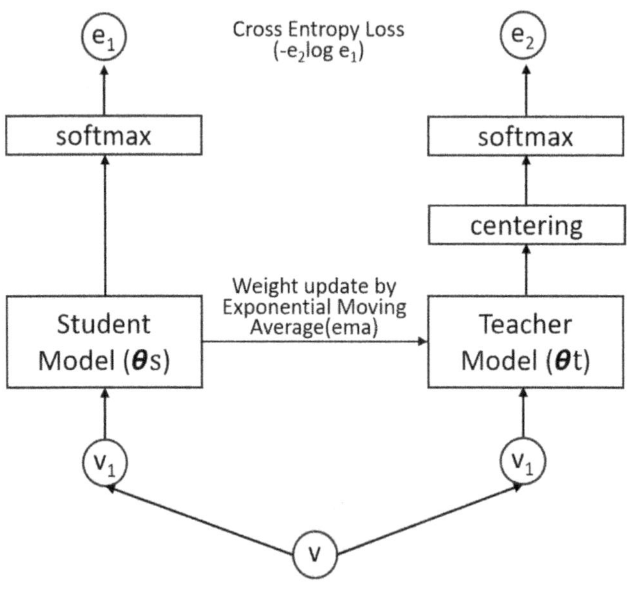

Figure 4: Training of DINO with self-distillation [15].

After training, the teacher model is used for inferencing. DINOv2 is an update over DINO, in terms of efficient implementation and use of curated datasets. This model has been trained on a large and diverse dataset containing 142 million images. The model has shown very good performance on out-of-distribution images, without fine-tuning, for most domains. This resulted in two times faster speed and 1/3rd less memory consumption.

4 Methodology

4.1 Dataset

For fine-tuning the DINOv2 model, we used the CODE15 [12] dataset. This dataset is 15% of a publicly available larger CODE dataset [1]. The distribution of samples for each class is shown in Figure 5. The dataset has 345,779 samples of 12-lead ECG records.

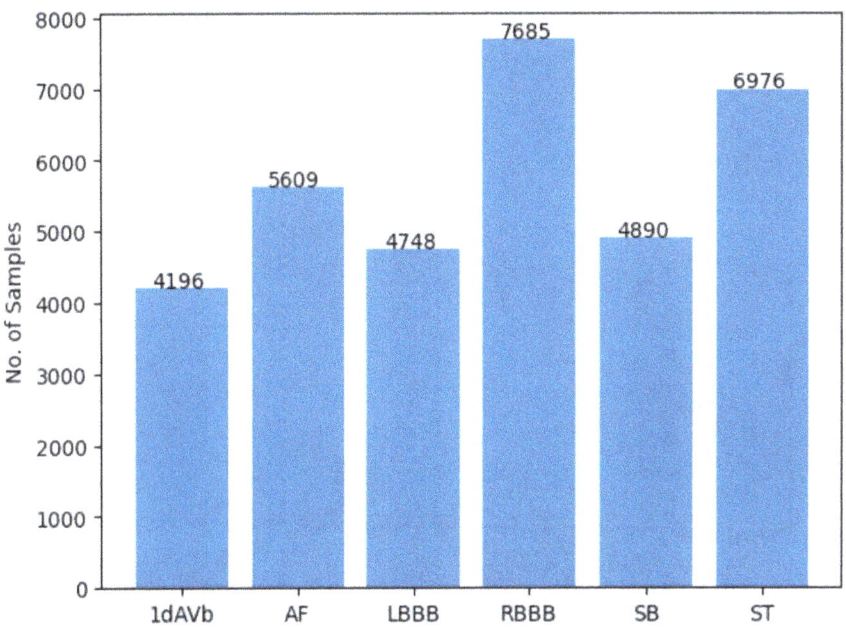

Figure 5: Class distribution in CODE-15% dataset.

The ECG data was sampled at a 400 Hz sample rate. Each record has been made of 10 s duration, after padding, if required. The dataset has samples for seven classes, namely, 1dAVB(1st-degree Atrioventricular Block), AF(Atrial Fibrillation), LBBB(Left Branch Bundle Block), RBBB(Right Branch Bundle Block), SB(Sinus Bradycardia), ST(Sinus Tachycardia), and Normal. Some of the samples have no labels. The number of normal samples (approximately 269 K) is far greater than the rest of the classes (about 4 K to 7 K samples per class). To have a balanced dataset, we selected only the first six classes from this dataset. We split the dataset into train, validation, and test splits in an 80:10:10 ratio.

We also fine-tuned and tested the method with the China Physiological Signal Challenge (CPSC) 2018 dataset [21]. This dataset contains samples with nine heart anomalies, viz., AF, 1dAVB, LBBB, RBBB, PAC (Premature Atrial Contractions), PVC (Premature Ventricular Contraction), STD (ST-segment Depression), STE (ST-segment Elevated), and the normal sinus rhythm (SNR). The distribution of samples in training data is shown in

Figure 6. The training data was split into a 90:10 ratio in train and validation sets. For testing, we used the validation set provided in the challenge. As the number of samples in this dataset is very small for fine-tuning the DINOv2 model, we applied augmentation to bring the number of samples of each class to about 4,000. The augmentation was done by a. selecting multiple 2 s segments from the 10 s ECG signal and b. by adding random noise to the sampled segment. The random noise was generated from a uniform distribution in the range of −4 to +4 mV, for each sampled segment separately. Thus, each segment is different from the others, concerning the starting position of the signal and the different noise added to each segment, with a probability of 0.5.

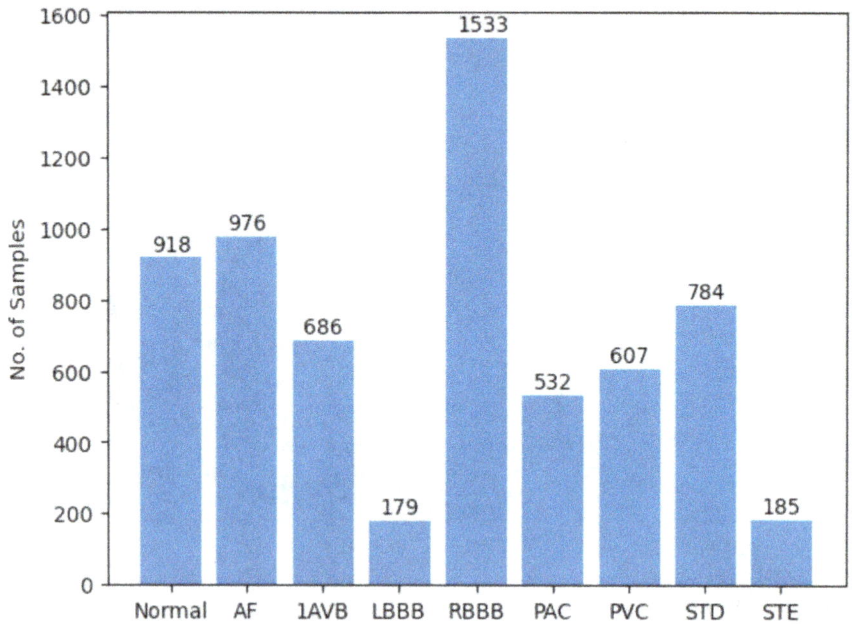

Figure 6: Class distribution in CPSC2018 dataset.

4.2 Preprocessing

Raw ECG scans have noise due to multiple sources. So, for standardization, removing this noise in the preprocessing step is essential. A sample preprocessed image is shown in Figure 7. We applied bandpass filters on the raw signal to filter out the high- and low-frequency noise components like baseline wander, muscle artifact, interference due to power lines, etc. Also, as all the samples should have the same sampling rate, a resampling step has been added to the preprocessing pipeline, for sampling at 400 samples per second. The DINOv2 model takes as input an image of size 518 × 518. So, the image is resized to this size. But if the full 10 s data is plotted and resized to

518 × 518, the image becomes distorted, and many features are lost. So, we limited the signal length to 2 s only to ensure that the features are preserved on resize. We used 2 s data, after experimenting with time durations of 2 s, 3 s, 4 s, and 5 s. The 2 s duration was found sufficient to give the best accuracy. As we are using only 2 s-long ECG data, we discard the first 2 s data as some records might been padded with 3 s data on both ends. The samples in the CPSC 2018 dataset are 10 s in duration, so segments were selected randomly multiple times for augmentation, as discussed in the previous section. After that, the signal was plotted on a graph to get the ECG image. We removed the grid pattern to improve learning before passing it to the model.

4.3 Model & training

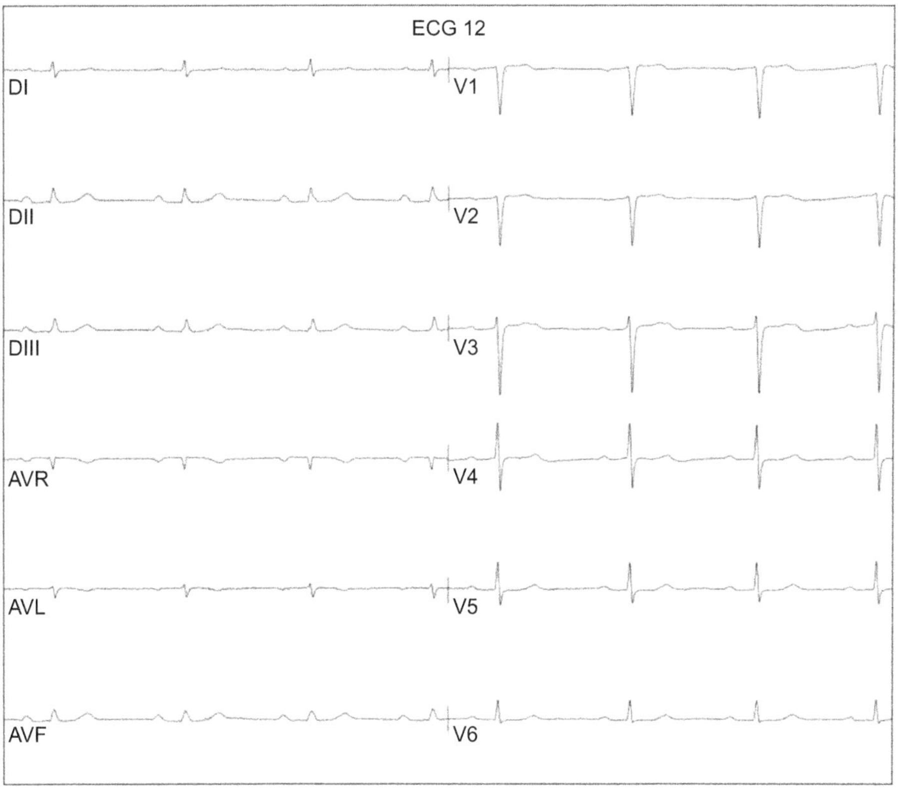

Figure 7: Preprocessed and resized ECG image.

In our model pipeline, we used DINOv2 as the backbone. The complete training pipeline for training is depicted in Figure 8. We selected the small model DINOv2_ViT-S (dinov2_vits14) from "hugging face" for pretrained weights. The small architecture (embeddings of size 384) has sufficient capacity to learn from the ECG images, as it is sufficient for ECG feature extraction. After preprocessing, the ECG image is passed to the DINOv2 model. The DINOv2 model returns embeddings of size 384. These embeddings are then passed to a classifier, which returns probabilities of the six classes. The classifier is DNN-based, consisting of two dense layers. The first dense layer takes the 384-sized embedding vector from the backbone as input and outputs a 256-size vector. This is passed to the ReLu activation layer. The final layer is also a dense layer that outputs the probabilities of the target class, after applying SoftMax to the values. The loss is computed by calculating the cross entropy between the predicted and the target values. The mode was fine-tuned for 14 epochs using the Adam optimizer, with a learning rate of 1e-6, and beta values of 0.90 and 0.999 respectively. The training and validation loss and accuracy at the training are presented in Figure 9.

5 Results and analysis

After training, the checkpoint with the highest validation accuracy (0.9627) was selected. For model evaluation, performance has been evaluated using precision,

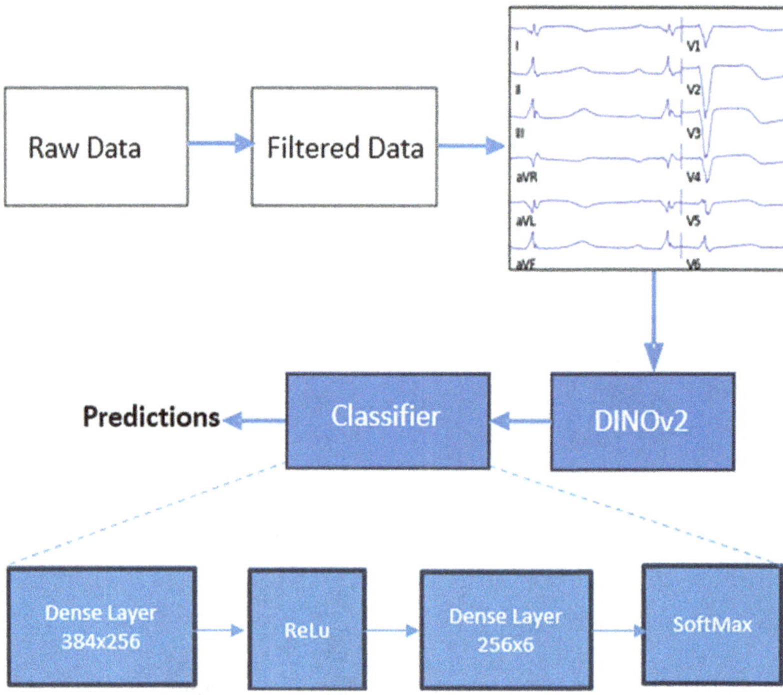

Figure 8: Model training pipeline.

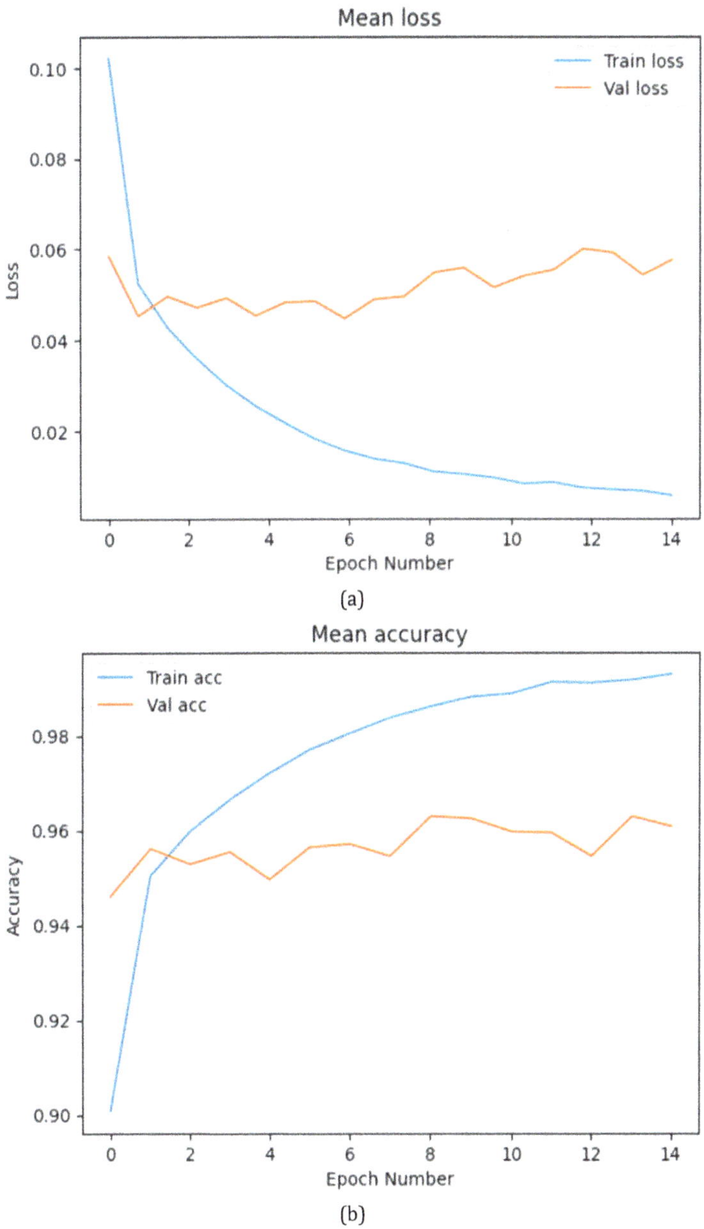

Figure 9: Training (a) loss and (b) accuracy per epoch.

sensitivity(recall), specificity (true negative rate), F1 score, and accuracy metrics. These metrics are defined in the following subsections.

5.1 Precision

The precision measures the proportion of correctly predicted samples for a class out of the total number of samples predicted to belong to that class. In other words, it is the ratio of total true positives to the total samples predicted as positive for the class, including the false positives. Mathematically, it is defined as follows:

$$Precision = \frac{No.\ of\ True\ Positives}{No.\ of\ True\ Positives + No.\ of\ False\ Positives}$$

5.2 Sensitivity

The metric of Sensitivity is also called Recall. It measure the proportion of positive samples of a class that we are able to predict correctly, i.e., the ratio of correctly predicted samples of a class to the total positive samples of that class in the dataset. Mathematically, it can be represented as below:

$$Sensitivity = \frac{No.\ of\ True\ Positives}{No.\ of\ True\ Positives + No.\ of\ False\ Negatives}$$

5.3 Specificity

This metric measures how accurate the method is in identifying the True Negatives. In other words, it is the proportion of samples predicted not belonging to a class accurately to the total number of samples predicted as such, i.e., the negative samples. Specificity is also called True Negative Rate. Mathematically, it is defined as follows:

$$Specificity = \frac{No.\ of\ True\ Negatives}{No.\ of\ True\ Negatives + No.\ of\ False\ Positives}$$

5.4 F1-score

This metric combines Sensitivity and Precision to assess the performance of a method. It is the harmonic mean of Precision and Sensitivity. Higher f1-score indicates better precision and recall. Mathematically, it is defined as follows:

$$F1 - Score = \frac{2 * Precision * Sensitivity}{Precision + Recall}$$

5.5 Accuracy

The accuracy metric measure how accurate the predictions are, i.e., out of the total predictions made, what is the proportion of correct predictions. Mathematically, it is represented as follows:

$$Accuracy = \frac{No.\ of\ True\ Positives + No.\ of\ True\ Negatives}{No.\ of\ Total\ Predictions}$$

5.6 Performance on CODE 15% dataset

On these parameters, class-wise performance results on the CODE15% dataset are summarized in Table 1. We achieved an average test accuracy of 95.85%.

The recall is a minimum of 92%(SB) and a maximum of 99%(RBB). The specificity is a minimum of 99% and 100% for SB and ST. The confusion matrix and the ROC curves are shown in Figures 10 and 11, respectively.

We have achieved better results in terms of both recall(sensitivity) and accuracy, resulting in a better f1-score, as compared to results that are published for a comparable number of classes. We compared the results with [1, 19].

The comparison has been summarized in Table 1. In terms of precision, we achieved a minimum of 0.92. Our method has outperformed the SOTA for all anomalies except for AF and LBBB. Even ViT has outperformed SOTA for these anomalies. In terms of sensitivity, we have achieved better results for 1dAVB and AF; while have equaled SOTA results for ST. ECG-DINOv2 has performed better than ViT for all the anomalies except for SB. In terms of specificity, we are either similar to SOTA or very close to it for all the categories while outperforming the ViT. In terms of the f1-score, we have outperformed SOTA by a minimum of 2%, except for LBBB. Also, the ECG-DINOV2 has outperformed the ViT on all the parameters and for all the classes.

To understand the reasons for predictions not matching the ground truth, we performed a detailed analysis of such cases. There were 141 ECGs from the dataset for which the predictions did not match the ground truth labels. On further review of these ECGs by an experienced cardiologist, it was found that 28% percent of these predictions were correct, i.e., the predictions matched the labels provided by the expert. Another 41% of the samples did not fall into any of the six anomalies, hence these ECGs were not considered for performance calculation. After taking this into account, we calculated the performance again and achieved an average accuracy of 98% and an average f1-score of 0.98. Specificity is 1 for LBBB, RBBB, SB, and ST and it is 0.99 for

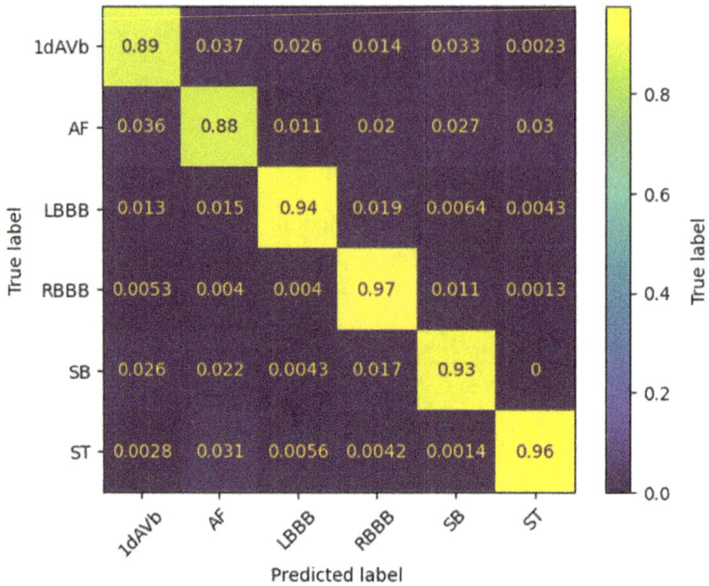

(a)

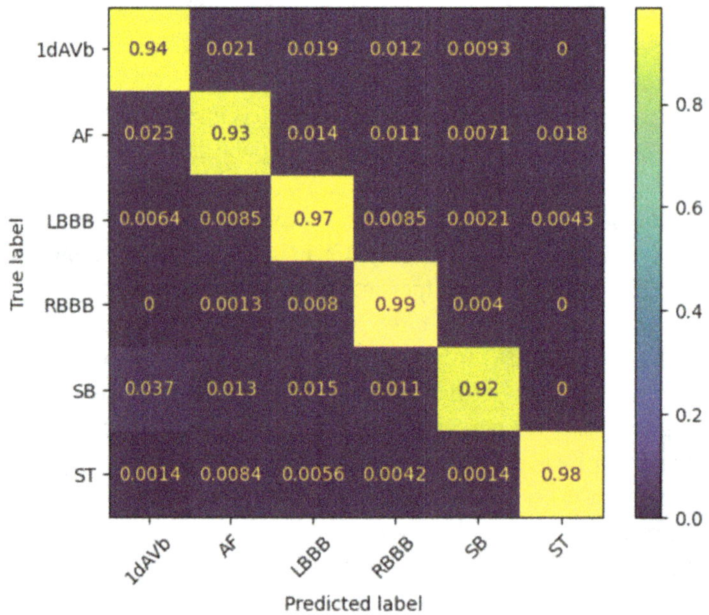

(b)

Figure 10: Confusion matrix for (a) ViT and (b) DINOv2 with CODE15% dataset.

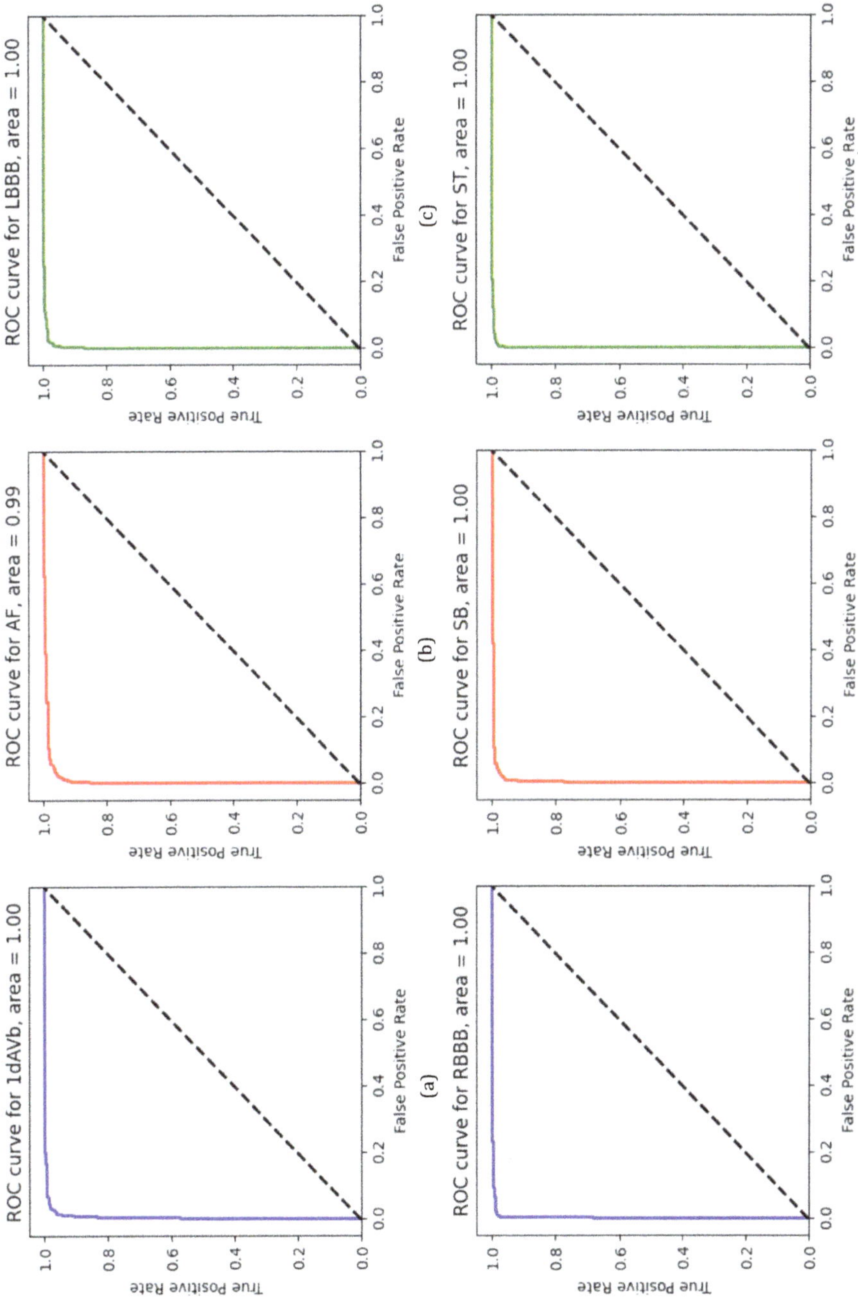

Figure 11: ROC Curves for (a) 1dAVb, (b) AF, (c) LBBB, (d) RBBB, (e) SB and, (f) ST.

Table 1: Performance comparison (Reibero and Jin – CODE, ViT, and ECG-DinoV2 – CODE 15% dataset).

Metric	Method	1dAVb	AF	LBBB	RBBB	SB	ST
Precision	Reibero	0.87	1.0	1.0	0.90	0.83	0.95
	Jin	–	–	–	–	–	–
	ViT	0.90	0.89	0.94	0.95	0.91	0.97
	ECG-DinoV2	0.92(0.96)	0.95(0.97)	0.93(0.99)	0.97(0.99)	0.97(0.99)	0.98(0.99)
Sensitivity	Reibero	0.93	0.77	1.00	1.00	0.94	0.98
	Jin	–	–	–	–	–	–
	Vit	0.89	0.88	0.94	0.97	0.93	0.96
	ECG-DinoV2	0.94(0.97)	0.93(0.98)	0.97(0.99)	0.99(1.00)	0.92(0.96)	0.98(0.99)
Specificity	Reibero	1.00	1.00	1.00	1.00	1.00	1.00
	Jin	–	–	–	–	–	–
	ViT	0.99	0.99	0.98	0.99	0.99	0.99
	ECG-DinoV2	0.99(0.99)	0.99(0.99)	0.99(1.00)	0.99(1.00)	1.00(1.00)	1.00(1.00)
F1-Score	Reibero	0.90	0.87	1.00	0.94	0.88	0.96
	Jin	0.68	0.90	0.95	0.95	–	–
	ViT	0.89	0.89	0.94	0.96	0.92	0.96
	ECG-DinoV2	0.93(0.96)	0.94(0.97)	0.95(0.99)	0.98(1.00)	0.95(0.97)	0.98(0.99)

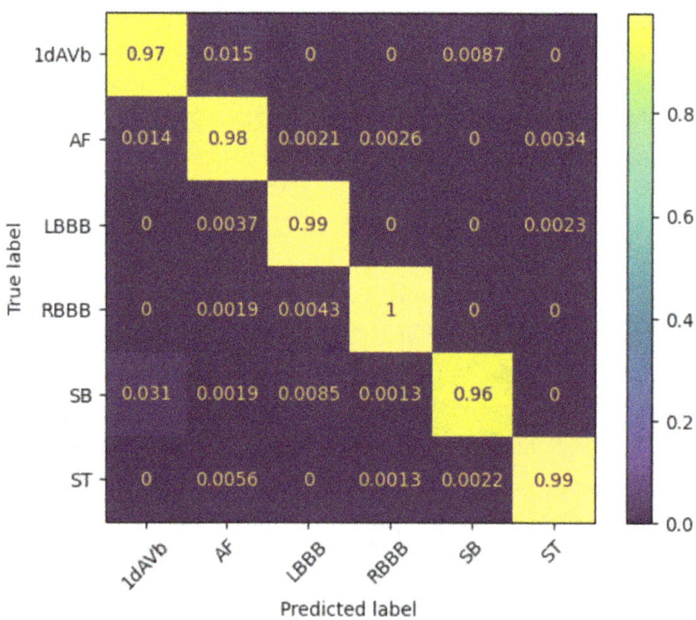

Figure 12: Confusion matrix DINOv2 as per expert labels.

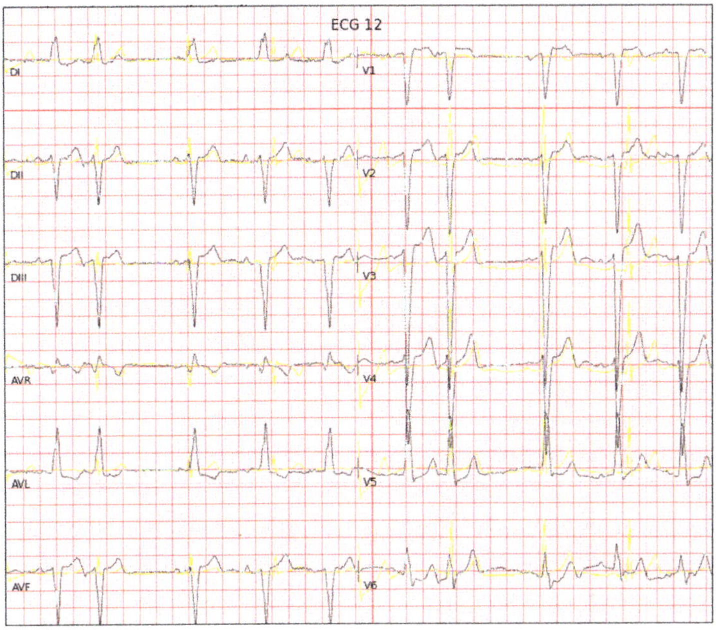

(a)

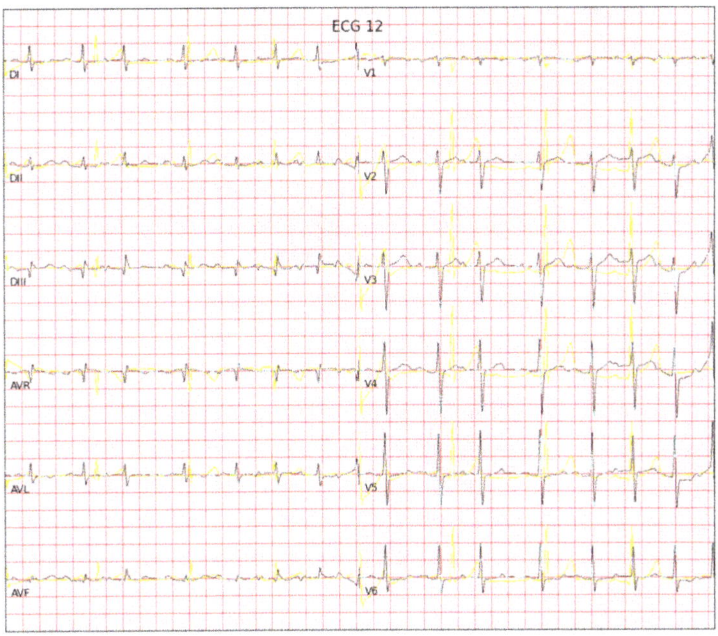

(b)

Figure 13: Examples of ECGs where model predictions did not match the dataset ground truth. (a) dataset label: AF, expert and predicted label: LBBB, (b) dataset label: ST, expert, and predicted label: 1dAVb, (c) dataset label: LBBB, expert, and predicted label: AF, and (d) dataset label: 1dAVb, expert, and predicted label: SB.

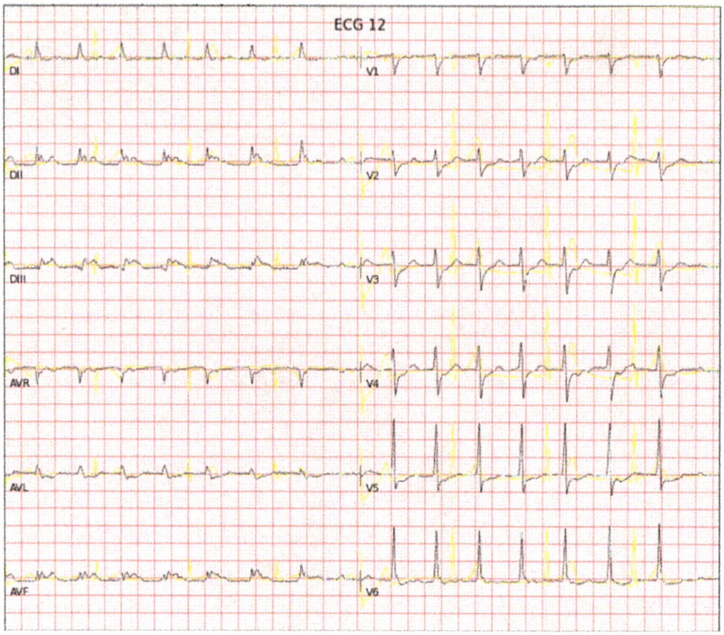

(c)

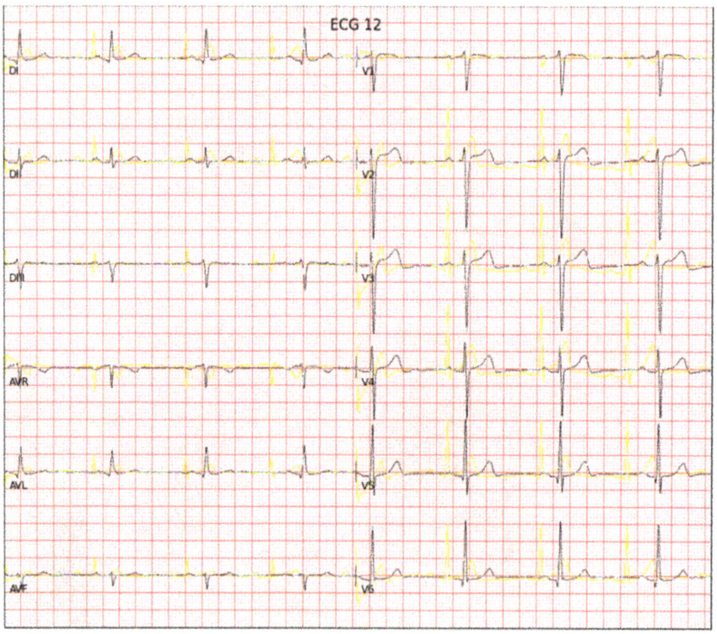

(d)

Figure 13 (continued)

1dAVb and AF. The detailed results are recorded in Table 1, next to the original results for ECG-DinoV2. The confusion matrix, with the recalculated verification figures, is shown in Figure 12. Some of the ECG strips where ground truth did not match with the expert's labels are shown in Figure 13.

5.7 Performance on CPSC 2018 dataset

On the CPSC 2018 dataset, we achieved an average f1-score of 0.92 and an average accuracy of 91%. In terms of sensitivity and specificity, the ECG-DinoV2 has performed better or comparable for almost all the classes. The detailed results are presented in Table 2 and compared with [22, 23]. The confusion matrix and ROC curves for all the labels are shown in Figures 14 and 15 respectively. We have achieved superior recall except for PVC, for which we have matched the performance, and achieved 100% recall for LBBB and STE. We achieved a better f1-score for all the classes, except for LBB, PVC, and Normal. The specificity of our model is better for all the classes and beats the SOTA by a good margin. We have achieved better precision for 1dAVb, AF, and RBBB. The ViT model outperformed all three models for STE, in terms of precision, with a value of 0.97. For RBBB, it achieved a precision of 0.98, beating other models, except for ECG-DinoV2. In terms of sensitivity, the performance of ViT is comparable with [22]. On the parameter of specificity, the model has outperformed [22] and is closer to ECG-DinoV2. The average f1-score of ViT model is also better than [22].

To understand the lower performance of ECG-DinoV2 for 3 classes viz PAC, PVC, and Normal, we performed an analysis of test samples for these classes. We found that it is due to the nature of PAV and PVC anomaly. Though the rest of the anomalies are present in all the heartbeats in ECG, the signatures of PAC and PVC are observed only in a few beats in the signal. Thus, when training and test data with a 2 s strip was prepared from the signal, the majority of these plots had normal sinus rhythm, but they were marked either PAC or PVC. Due to this, it confused the model at training time as normal samples were being presented as PAC, PVC, and Normal (from the Normal class), thus affecting the performance of all three classes, as evident from the confusion matrix in Figure 14. This problem can be solved by carefully selecting only those segments of PAC and PVC classes that have anomalies present for training/testing. For these two categories, a change is required in the augmentation strategy. The augmentation should be done mainly using noise addition and random sampling from a smaller window containing the defect. With this, the performance would improve for Normal, PAC, and PVC classes.

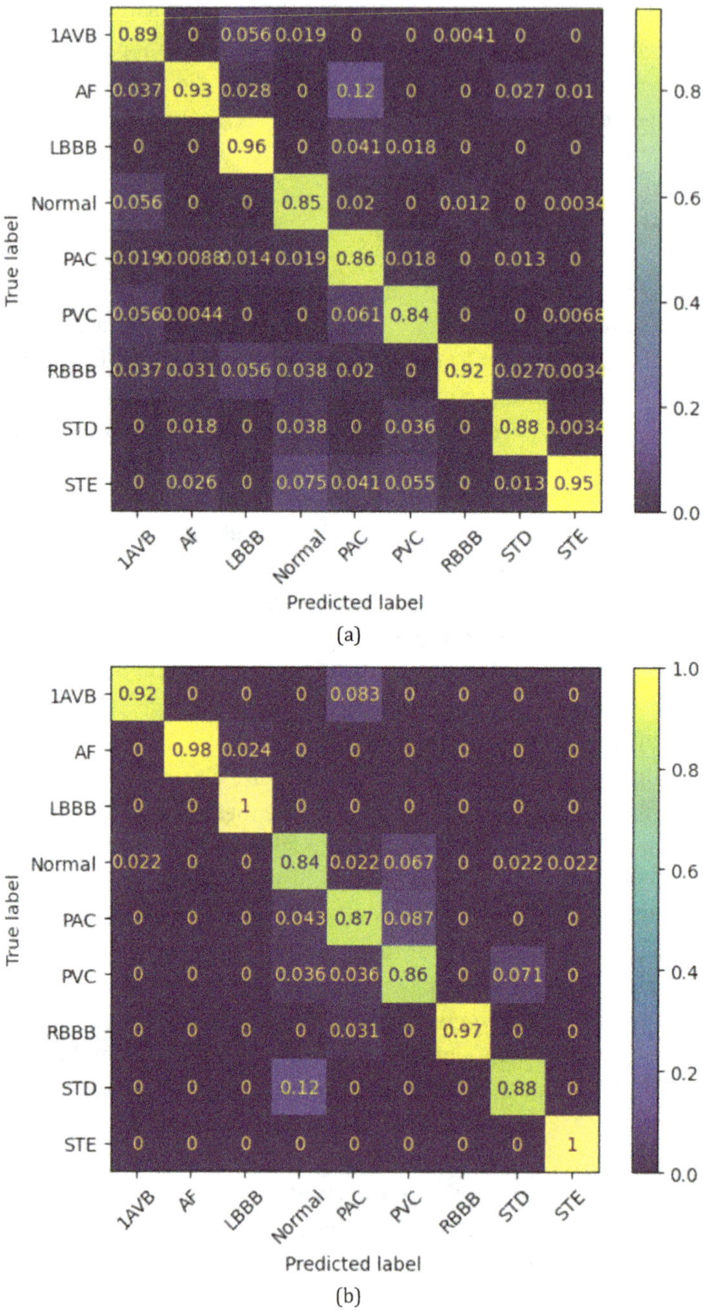

Figure 14: Confusion matrix for (a) ViT and (b) DINOv2 with CPSC 2018 dataset.

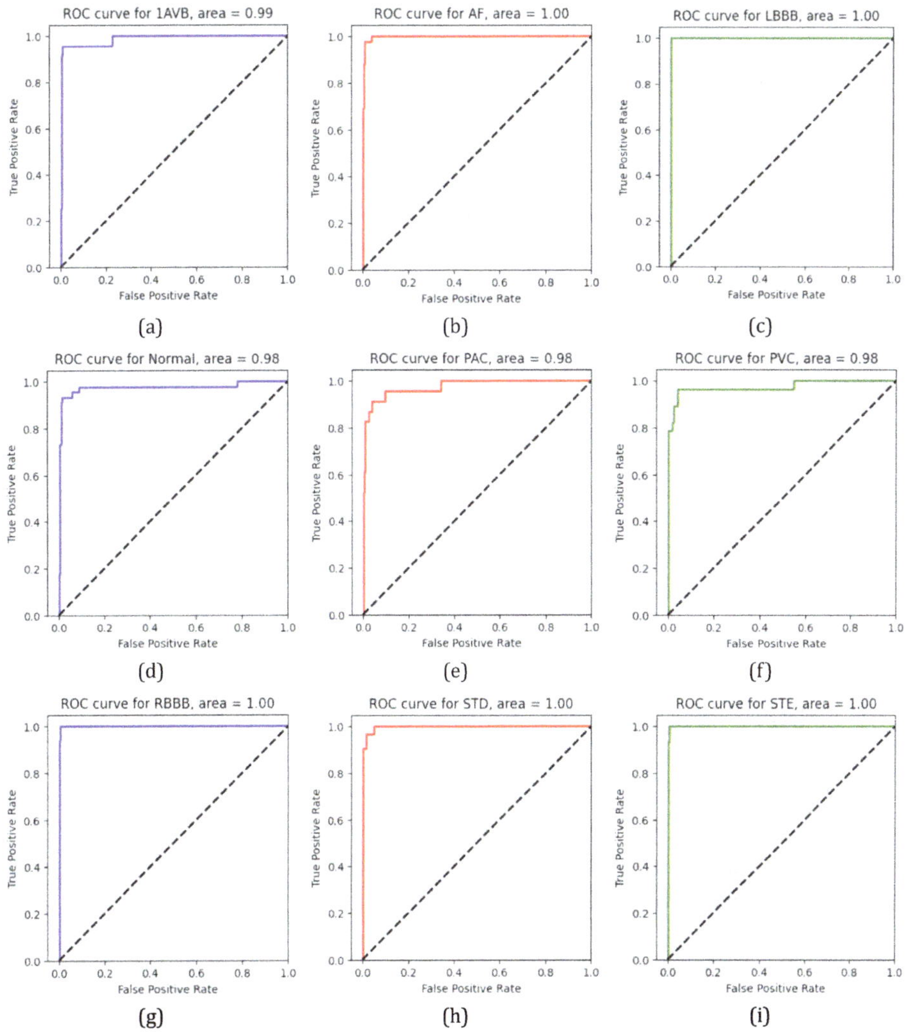

Figure 15: ROC curve for (a) 1dAVb, (b) AF, (c) LBBB, (d) Normal(SNR), (e) PAC, (f) PVC, (g) RBBB, (h) STD and (i) STE on CPSC2018 dataset.

Table 2: Performance comparison on CPSC 2018 dataset.

Metric	Method	1dAVB	AF	LBBB	RBBB	PAV	PVC	STD	STE	SNR
Precision	Yang	0.92	0.97	0.99	0.97	0.91	0.95	0.94	0.96	0.90
	Dhyani	0.91	0.89	0.97	0.97	0.81	0.97	0.78	0.93	0.96
	ViT	0.81	0.91	0.86	0.98	0.74	0.87	0.92	0.97	0.82
	ECG-DinoV2	0.96	1.00	0.92	1.00	0.77	0.83	0.90	0.94	0.86
Sensitivity	Yang	0.64	0.91	0.99	0.98	0.78	0.86	0.78	1.00	0.79
	Dhyani	0.88	1.00	0.97	0.94	0.79	0.83	0.85	1.00	0.84
	ViT	0.89	0.93	0.96	0.92	0.86	0.84	0.88	0.95	0.85
	ECG-DinoV2	0.92	0.98	1.00	0.97	0.87	0.86	0.88	1.00	0.84
Specificity	Yang	0.96	0.98	1.00	0.98	0.95	0.97	0.97	0.98	0.94
	Dhyani	–	–	–	-	–	–	–	–	–
	ViT	0.99	0.98	0.99	1.00	0.99	0.99	0.99	0.99	0.99
	ECG-DinoV2	1.00	1.00	1.00	1.00	0.98	0.98	0.99	1.00	0.97
F1-Score	Yang	0.57	0.91	0.92	0.96	0.60	0.82	0.77	0.59	0.66
	Dhyani	0.89	0.94	0.98	0.96	0.80	0.89	0.81	0.96	0.90
	ViT	0.85	0.92	0.91	0.95	0.79	0.85	0.90	0.96	0.83
	ECG-DinoV2	0.94	0.99	0.96	0.98	0.82	0.84	0.89	0.97	0.85

6 Conclusion & future scope

DINOv2 has impressive feature extraction capabilities, extracting good features from ECG images and identifying the six anomalies in 12 Lead-ECG data with increased accuracy. The algorithm (model) has been able to surpass the performance of the state-of-the-art by a good margin, with 2 s of ECG data only. This can be useful for real-time monitoring of the patients. As the method shows high sensitivity for the anomaly classes, it is a good candidate for first-level screening. In the future, the same approach can be expanded to more anomalies and even for multi-label classification of ECGs. A similar approach can be useful for other medical images like MRI, X-RAY, and skin scans. Another interesting area would be to learn a model for on-device deployments.

References

[1] A. H. Ribeiro, M. H. Ribeiro, G. M. M. Paixão, D. M. Oliveira, P. R. Gomes, J. A. Canazart, et al., "Automatic Diagnosis of the 12-lead ECG Using a Deep Neural Network," *Nature Communications*, vol. 11, pp. 1760, 2020, doi: 10.1038/s41467-020-15432-4.

[2] R. V. Andreao, B. Dorizzi and J. Boudy, "ECG Signal Analysis through Hidden Markov Models," *IEEE Transactions on Biomedical Engineering*, vol. 53, no. 8, pp. 1541–1549, Aug. 2006, doi: 10.1109/TBME.2006.877103.

[3] A. K. Mishra and S. Raghav, "Local Fractal Dimension Based ECG Arrhythmia Classification," *Biomedical Signal Processing and Control*, vol. 5, no. 2, pp. 114–123, 2010, doi: 10.1016/j.bspc.2010.01.002.

[4] Q. A. Rahman, L. G. Tereshchenko, M. Kongkatong, T. Abraham, M. R. Abraham and H. Shatkay, "Utilizing ECG-based Heartbeat Classification for Hypertrophic Cardiomyopathy Identification," *IEEE Transactions on NanoBioscience*, vol. 14, pp. 505–512, 2015.

[5] E. J. D. Luz, T. M. Nunes, V. H. C. De Albuquerque, J. P. Papa and D. Menotti, "ECG Arrhythmia Classification Based on Optimum-path Forest," *Expert Systems with Applications*, vol. 40, pp. 3561–3573, 2013.

[6] J. Zheng, H. Chu, D. Struppa, et al., "Optimal Multi-stage Arrhythmia Classification Approach," *Scientific Reports*, vol. 10, p. 2898, 2020, doi: 10.1038/s41598-020-59821-7.

[7] E. Butun, O. Yildirim, M. Talo, et al., "1D-CADCapsNet: One Dimensional Deep Capsule Networks for Coronary Artery Disease Detection Using ECG Signals," *Physica Medica*, vol. 70, pp. 39–48, 2020, doi: 10.1016/j.ejmp.2020.01.007.

[8] A. P. Shah and S. A. Rubin, "Errors in Computerized Electrocardiogram Interpretation of Cardiac Rhythm," *Journal of Electrocardiology*, vol. 40, pp. 385–390, 2007, doi: 10.1016/j.jelectrocard.2007.03.008.

[9] S. Sabour, N. Frosst and G. E. Hinton, "Dynamic Routing between Capsules," *Advances in Neural Information Processing Systems*, pp. 3856–3866, 2017.

[10] R. Avanzato and F. Beritelli, "Automatic ECG Diagnosis Using Convolutional Neural Network," *Electronics*, vol. 9, no. 6, p. 951, 2020, doi: 10.3390/electronics9060951.

[11] D. Zhang, S. Yang, X. Yuan, et el., "Interpretable Deep Learning for Automatic Diagnosis of 12-lead Electrocardiogram," *iScience*, vol. 24, no. 4, p. 102373, 2021, doi: 10.1016/j.isci.2021.102373.

[12] E. M. Lima, A. H. Ribeiro, et al., "Deep Neural Network Estimated Electrocardiographic-age as a Mortality Predictor," *MedRXiv*, 2021, doi: 10.1101/2021.02.19.21251232.

[13] A. Raymond and H. George, "Image Based Deep Learning in 12-lead ECG Diagnosis," *Frontiers in Artificial Intelligence*, vol. 5, p. 1087370, 2023, doi: 10.3389/frai.2022.1087370.

[14] A. Dosovitskiy, et al., "An Image Is Worth 16 × 16 Words: Transformers for Image Recognition at Scale," ICLR 2021, OpenReview.net.

[15] M. Caron, H. Touvron, I. Misra, et al., "Emerging Properties in Self-supervised Vision Transformers," Proceedings of the International Conference on Computer Vision (ICCV), 2021.

[16] M. Oquab, T. Darcet, T. Moutakanni, et al., *DINOv2: Learning Robust Visual Features without Supervision*, 2023, doi: 10.48550/arXiv.2304.07193.

[17] S. Śmigiel, K. Pałczyński and D. Ledziński, "ECG Signal Classification Using Deep Learning Techniques Based on the PTB-XL Dataset," *Entropy*, vol. 23, p. 1121, 2021, doi: 10.3390/e23091121.

[18] M. Hammad, S. A. Chelloug, R. Alkanhel, A. J. Prakash, A. Muthanna, I. A. Elgendy and P. Pławiak, "Automated Detection of Myocardial Infarction and Heart Conduction Disorders Based on Feature Selection and a Deep Learning Model," *Sensors*, vol. 22, p. 6503, 2022, doi: 10.3390/s22176503.

[19] Y. R. Jin, Z. Y. Li, Y. Q. Liu, et al., "Multi-class 12-lead ECG Automatic Diagnosis Based on a Novel Subdomain Adaptive Deep Network," *Science China Technological Sciences*, vol. 65, pp. 2617–2630, 2022, doi: 10.1007/s11431-022-2080-6.

[20] E. N. Marieb and K. Hoehn, "Human Anatomy and Physiology," *Chapter 18: The Cardiovascular System: The Heart*, 11th edn, 2019, ISBN: 97812922608.

[21] F. Liu, C. Liu, L. Zhao, et al., "An Open Access Database for Evaluating the Algorithms of Electrocardiogram Rhythm and Morphology Abnormality Detection," *Journal of Medical Imaging and Health Informatics*, vol. 8, pp. 1368–1373, 2018.

[22] Y. Xiuzhu, Z. Xinyue, et al., "12-Lead ECG Arrhythmia Classification Using Cascaded Convolutional Neural Network and Expert Feature," *Journal of Electrocardiology*, vol. 67, pp. 56–62, 2021, doi: https://doi.org/10.1016/j.jelectrocard.2021.04.016.

[23] D. Shikha, K. Adesh and C. Sushabhan, "Arrhythmia Disease Classification Utilizing ResRNN," *Biomedical Signal Processing and Control*, vol. 79, Part 2, 2023, doi: https://doi.org/10.1016/j.bspc.2022.104160.

Large language model (LLM)

Sean Choi and Jinyoung Jo

Leveraging linguistic features to improve machine learning models for detecting ChatGPT usage on exams

Abstract: This work presents a case study, linguistic analyses, and the results of experiments on using linguistic features to improve the detection mechanism of the use of large language models (LLM) to generate solutions for exams that require domain-specific knowledge. The study involves analyzing the responses of three groups of students: a group who verbatim copied outputs of ChatGPT to plagiarize solutions, another group who referred to external non-LLM resources (e.g., web search) to plagiarize solutions, a control group who did not plagiarize. Linguistic analyses show that solutions from groups that participated in plagiarism tend to be longer, use uncommon words, and are similar to each other compared to solutions that were not plagiarized. In addition, utilizing these characteristics as features improves F1 score of machine learning models that detect plagiarism as much as 5.5%. This study shows that certain linguistic features can be utilized in machine learning models to detect use of LLMs, ultimately improving academic integrity by deterring the unethical use of AI in academic settings.

Keywords: large language model, academic integrity, computer science education, plagiarism detection, machine learning

1 Introduction

Large Language Model (LLM) is a term used for deep neural network (a class of highly complex and compute-heavy machine learning models built for performing complex tasks), designed to represent the human language, often consisting of billions of parameters and are trained on a huge amount of textual data. Performances of various LLMs have improved vastly over the last decade, now excelling in complex tasks such as question answering, semantic search, and summarizing text. In particular, a company called OpenAI recently released an LLM-based product called ChatGPT [1, 2] that has gained worldwide popularity due to its ease of use and excellent performance in text/code generation, summarization, and question answering. To put the popularity into perspective, ChatGPT now holds the record for fastest-growing user base of any application in history with 100 million users in just two months [3].

Sean Choi, Department of Computer Science and Engineering, Santa Clara University, Santa Clara, CA, USA, e-mail: sean.choi@scu.edu
Jinyoung Jo, Department of Linguistics, University of California, Los Angeles, e-mail: jinyoungjo@ucla.edu

https://doi.org/10.1515/9783111344126-013

However, ChatGPT's sudden gain in popularity came with a severe side effect that the academic community is struggling to react – blatant violation of academic integrity. Since ChatGPT can adeptly generate a seemingly unique and sophisticated text based on user query, many academic institutions are reporting violations of academic integrity by using ChatGPT to generate solutions for homework, essays, and even timed exams. Making things worse, text generated from LLMs are hard to distinguish from human-generated text, as the goal of a high performing LLM is to replicate human language as closely as possible. Therefore, addressing the illegitimate use of LLM is quickly becoming a mainstream research topic.

To aid in such research efforts, this study presents a set of findings from analyzing students' exam responses, either generated illegitimately via LLM or legitimately by humans, in a real-world class setting. The goal of this study is to understand the effectiveness of ChatGPT in generating exam solutions and to recommend features that can be used to detect the use of ChatGPT. In particular, the analysis is based on a computer science class that requires deep background knowledge, where the teaching staff discovered that a portion of the students resorted to using ChatGPT and/or online search to plagiarize exam solutions to multiple exam questions. Based on such data set, this study compares the linguistic properties of the responses from a group of students from that class who used ChatGPT to generate answers and those of a group of students who used online search, with those of a control group who wrote their answers without relying on external sources. Finally, linguistic properties that significantly distinguished the three groups were used as input features to various plagiarism detection machine learning models to study the effects of adding such features.

This work is a first step toward understanding the linguistic characteristics of advanced LLMs, as compared to human language, and providing suggestions for potential features to discern text generated by a human versus text generated by LLMs. In addition, this work reports findings on teaching machine learning algorithms to better distinguish LLM-generated and human-generated text by providing the features. Finally, the goal of this work is to improve the performance of detecting text generated by LLMs to minimize the impact of illegitimate use of LLMs, which can ultimately encourage academic integrity, and prevent AI tools from reducing the effectiveness of the computer science education.

In summary, the main contributions of this work are as follows:

- Investigating feasibility and effectiveness of using ChatGPT in generating solutions for questions that require deep background knowledge
- Understanding the linguistic characteristics of text generated by LLMs
- Providing potential signals for plagiarism using ChatGPT and violation of academic integrity
- Studying and reporting the findings on the feasibility of improving plagiarism detection machine learning models by using the signals of plagiarism as input features to the machine learning algorithms

The paper is structured as follows. First, the paper presents the background on LLMs, models for text classification, cloud computing concepts, metrics used for language characterization, and scope of plagiarism (Section 2), followed by an overview of the course in which plagiarism was found, examination given to the students, and the types of plagiarisms observed (Section 3). Then, the paper presents the outcome of the evaluations (Section 4) and related work (Section 5). It concludes by discussing planned future directions (Section 6).

2 Background

This section provides relevant background information and the context of this work, which are LLMs, cloud computing, and language-characteristic metrics.

2.1 Large language model and ChatGPT

Language models have become an essential tool for a wide range of natural language processing (NLP) tasks, such as machine translation, text summarization, and question answering. Recent advancements in deep learning techniques have led to the development of large-scale language models that can generate human-like text and perform a wide range of language-related tasks. One such model is ChatGPT [1, 2], a large-scale generative language model trained by OpenAI. ChatGPT is based on the transformer architecture, which was introduced as an alternative to traditional recurrent neural networks (RNNs) for sequence modeling tasks. The transformer architecture consists of an encoder and a decoder, each composed of multiple layers of self-attention and feed-forward neural networks. The encoder and decoder work together to process an input sequence and generate an output sequence. The encoder first processes the input sequence and produces a set of context vectors, which are then used by the decoder to generate the output sequence. The self-attention mechanism allows the model to capture long-range dependencies between words in the input sequence, making it particularly effective for language modeling tasks.

ChatGPT has been trained on a massive corpus of text data, consisting of over 8 million unique documents, or 45 terabytes in size, from a diverse range of sources, including web pages, books, and online forums. In addition to using massive amount of training data, ChatGPT is one of the largest language models. ChatGPT version 3.5 consists of 175 billion parameters and ChatGPT version 4 consists of 100 trillion parameters, which translates to about 800GB and 500TB of model size, respectively. ChatGPT has been shown to be effective for a wide range of NLP tasks, attributed to the large size, including language modeling, which involves the model predicting the next word in a sequence (given the previous words), question answering, as well as generation of text, code, and dialogue. In particular, ChatGPT is renowned as a highly sophisticated chatbot that engages in human-like conversations with users. Figure 1 shows an example of ChatGPT user interface that solves an exam question, which involves domain-specific knowledge.

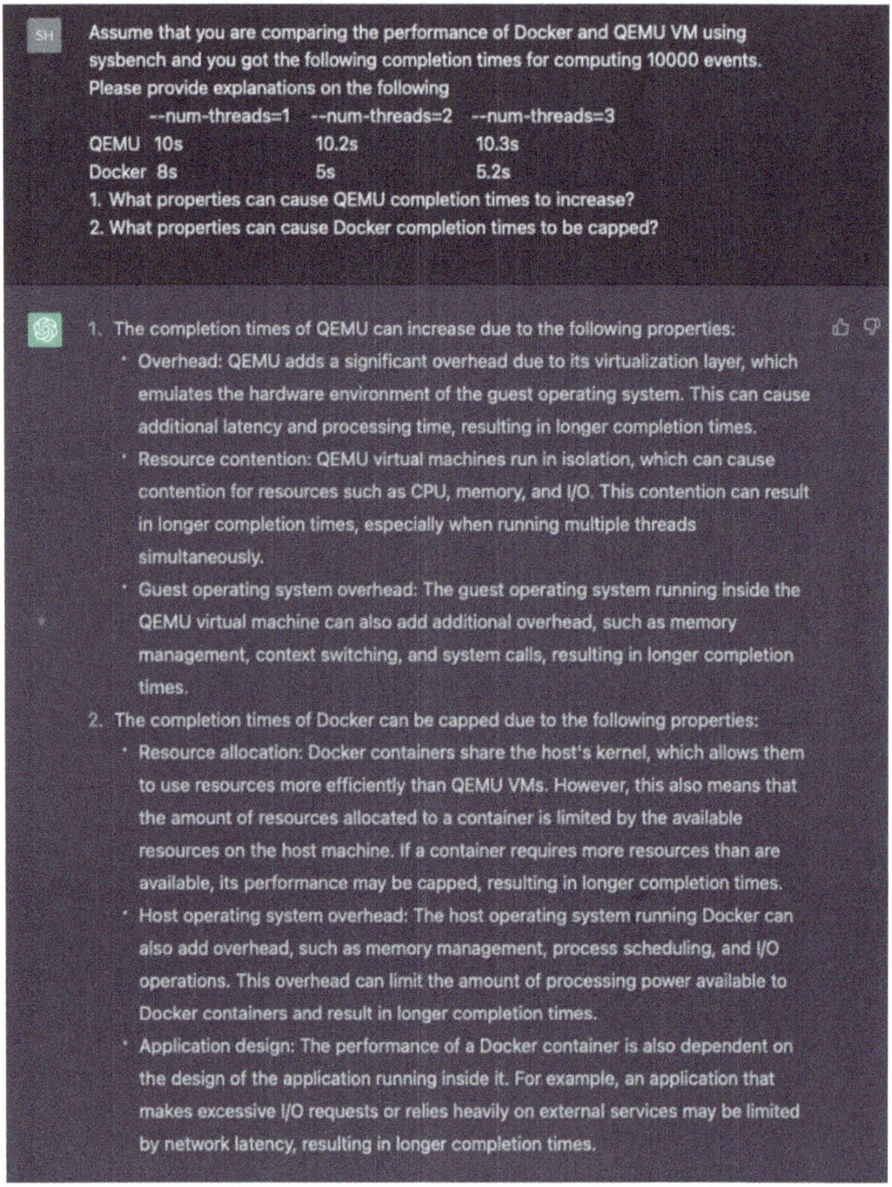

Figure 1: ChatGPT interface solving an exam question.

2.2 Machine learning models for classifying texts

With the advent of LLMs, the study on detecting LLM-generated text is also gaining significant interest. The problem of detecting LLM-generated text boils down to solving a high-level problem of text classification. To provide a brief context on machine learning-driven text classification, this section discusses two main topics: (1) types of text classification methods, (2) machine learning algorithms for classification, specifically boosted decision trees and feedforward neural network.

Text classification is a class of tasks that assigns a label (or a class), from a predefined set of labels, to a set of open-ended text data. Text classification can be used to classify just about any types of text, ranging from typical documents, books, web pages, user reviews, and even short sentences. At a high level, there are two types of text classification methods: (1) Rule-based and (2) Machine learning-based. A rule-based classification method utilizes a set of linguistic rules to classify text as shown in Figure 2. For example, a group of distinct words can be associated with a given label and whenever a text contains one or more of the words in the group, the label is assigned to the text. Rule-based classification method is often hierarchical and naturally forms a tree structure, consisting of a root node (the input to the tree), branches (the arrows), internal nodes (conditionals), and leaf nodes (final classification decision). A rule-based method has an advantage when it comes to human comprehensibility of the methods and result, but the main disadvantage of the rule-based method is that it requires deep domain knowledge, thus takes a long time to develop, and is quite difficult to scale. In comparison, machine learning-based classification method involves training a statistical model by feeding a set of labeled training data to a set of machine learning algorithms, as shown in Figure 3. The idea is that, given a set of labeled training data, ideally with sizable quantity and fair distribution of labels, a set of features per data point is generated using a customized feature generator. The features of a model can range from simple rules like the ones used in the rule-based classification to a hidden set of features that the algorithm figures out on its own. Crucially, the type of features are very flexible and can be mathematical to capture varying patterns within the data. Then, with the set of features for the training data generated from the feature extractor, a preselected machine learning algorithm is used to learn the patterns within the features, finally producing a machine learning model. Given the machine learning model, any unlabeled data point can be classified by asking the model what the label should be.

There are numerous types of machine learning algorithms used to create a classifier, but it is beyond the scope of this work to discuss all types of the algorithms. For brevity, this section discusses two relevant machine learning algorithms used for classification: boosted decision trees and feedforward neural network.

Decision tree is quite similar to the rule-based classification shown in Figure 2, but each internal node is mathematically selected using a node selection algorithm. An example of node selection algorithm is to form a decision tree greedily by selecting

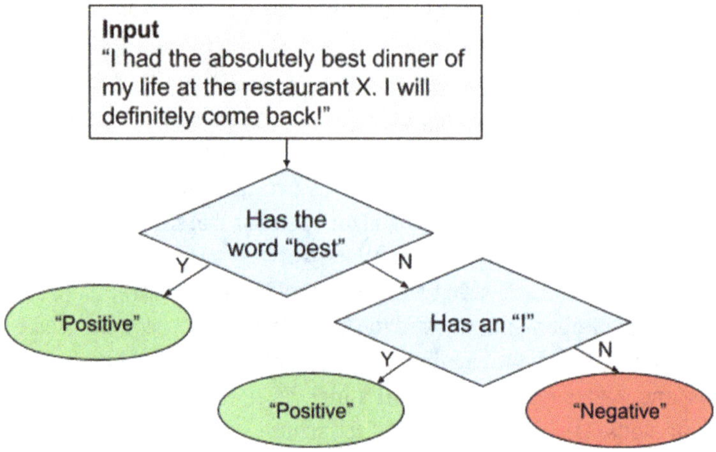

Figure 2: Example of rule-based text classification.

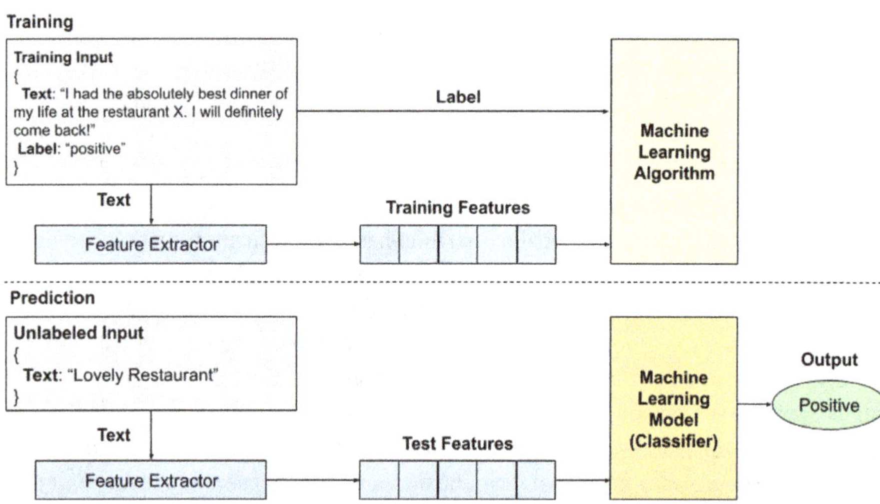

Figure 3: Example of machine learning-based text classification.

the best feature to split on at each level of the tree, rather than trying to find the optimal order of splits that gives you the best model. The "goodness" of a feature is generally computed using a mathematical formula. Two of the most popular formulas used are: (1) Entropy and Information Gain and (2) Gini Index. First, entropy can be seen as a measure of disorder in a node, computed as

$$E = -\sum_{i=1}^{N} p_i log_2(p_i)$$

where p_i is the probability of selecting an example in class i. Maximum value of entropy is 1 and minimum value is 0 and, at a high level, minimum entropy is achieved when the data only consists of a single class i, thus $N = 1$, $p_i = 1$ and the maximum entropy is achieved when the number of data point are divided equally. Given an entropy, the information gain is computed as

$$Gain = E_{parent} - E_{child}$$

What this means is that as the decision is made for the parent node, the entropy decreases due to filtering, thus we have a gain of information. The idea of greedy node selection, using information gain, is to compute the information gain for each potential split and greedily select the best one.

Second, Gini index is focused on quantifying the probability of a data point being misclassified; the lower the Gini index, the lower the likelihood of misclassification. Gini index is computed by the following formula,

$$Gini = 1 - \sum_{i=1}^{j} P(i)^2$$

where $P(i)$ is the probability of choosing the class i and j is the possible number of classes. So, greedy node selection using the Gini index would try to find the lowest Gini index for a particular class and use that class to split the conditional. Now, since greedily built decision tree may not be optimal and can be prone to overfitting, boosted decision tree is a way to combine multiple decision trees to build a better one. The details of the boosting methods are omitted for brevity as it is beyond the topic of this work, but some examples of boosted decision trees are XGBoost [4] and LightGBM [5].

Second, a neural network is a class of machine learning algorithms where the algorithm learns a set of hidden features from the relation of input features to the output value. An example of a neural network is schematically presented in Figure 4. There are multiple terms that make up a neural network:

- **Neurons**: It is the most fundamental unit of processing, or a mathematical model, that computes the weighted average of its inputs with an added bias using an activation function. Each neuron has a weight associated with it and the training phase of the neural network learns these weights based on the data. Each weight is normally a value between 0 and 1 due to the use of an activation function. The most widely used activation functions are: sigmoid, Tanh, and Rectified Linear Unit (ReLU).
- **Input layer**: The neurons of this layer represent the features of the data set, thus the number of neurons in this layer must match the number of features. The job of the input layer is to receive input and pass it on to the other layers of the network.

- **Output layer**: This layer represents the place that generates the final prediction. Depending on the type of prediction, you can have one or more neurons in the output layer.
- **Hidden layer**: This layer is the main purpose of using a neural network. Between the input and output layers, there are a number of hidden layers that can be configured as one of the hyperparameters of the training algorithm. The hidden layers can be seen as a set of features or patterns in the data that the algorithm automatically discovers during model training, which makes neural networks very powerful. The hidden layers can feed the weights it learned to the next set of layers or pass them backward, creating a loop.

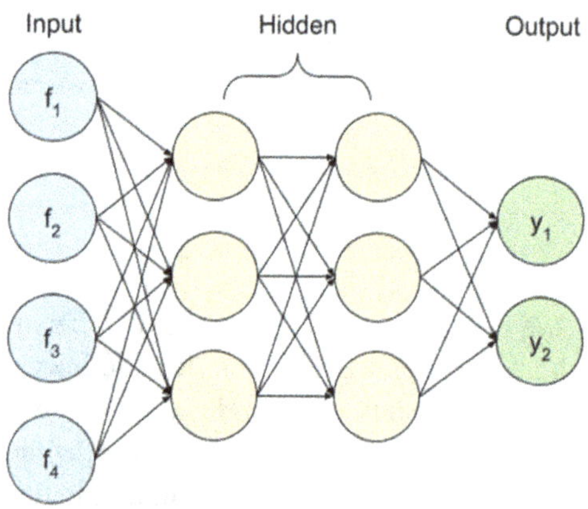

Figure 4: An example architecture of a feedforward neural network with two hidden layers.

Now, the main idea of a feedforward neural network is that each layer passes on the weights to the next layer, only in a unidirectional way. It is the simplest case of a neural network, but widely used due to its simplicity and low computational overhead, in comparison to other more complex neural networks.

2.3 Cloud computing and virtualization

The main topic of the course analyzed in this study is cloud computing. This section provides high-level context into cloud computing to help understand the exam questions and solutions. Cloud computing refers to the practice of using remote servers, accessed over the internet, to store, manage, and process data, instead of using local servers or personal computers. Cloud computing enables businesses and individuals to leverage the power of large-scale computing resources, which can be quickly scaled up or down, based on demand, to meet their computing needs. It also allows for greater flexibility and reliability, compared to traditional computing models.

The main driving technology behind cloud computing is virtualization. Virtualization is a technology to create virtual representations of servers, storage, networks, and operating systems using software [6]. Virtualization has revolutionized how software applications are managed and deployed, making it easier and more efficient to provision, scale, and manage applications at a large scale. Two pivotal virtualization technologies are virtual machines (VMs) and containers.

A VM is an emulation of a physical machine that runs on top of a hypervisor, a piece of software that allows multiple VMs to share the same physical resources. Each VM has its own operating system, software stack, and resources, including CPU, memory, and disk. VMs are typically isolated from each other and from the underlying host system, providing a high degree of security and flexibility. Some examples of VMs are QEMU [7] and VirtualBox [8]. VMs offer several advantages, including the ability to run multiple operating systems and applications on a single physical machine, the ability to easily migrate VMs between different hosts, and the ability to provision and scale resources dynamically.

A container is a lightweight and portable package that includes an application and its dependencies but shares the same kernel and resources as the host system. Containers are isolated from each other using namespaces and cgroups [9], which provide a similar level of isolation as VMs, but with lower overhead and faster startup times. Some examples of containers are Docker [10] and Kata [11]. Containers offer several advantages, including improved performance and scalability, faster deployment and startup times, and easier management and orchestration. Containers are also portable and can be easily moved between different hosts and environments.

In terms of architecture and performance, VMs are more complex and heavyweight than containers, requiring a full operating system and virtual hardware emulation. Containers, on the other hand, are lightweight and share the same kernel and resources as the host system, as mentioned earlier. This means that containers have lower memory and computational overhead and faster startup times than VMs, making them more feasible for deploying and scaling applications.

2.4 Metrics for language characterization

This study analyzes and compares the linguistic properties of three types of responses to the exam questions, ChatGPT-aided solutions, search engine-aided solutions, and "valid" answers that were generated without any aid from external references. The linguistic properties under investigation include: the proportion of stop words, length of responses as measured by the number of characters, words, and sentences, average sentence length, type-token ratio as a proxy for lexical diversity, average word frequency, the proportion of *I*, Automated Readability Index, Jaccard index [12], and cosine similarity of SBERT [13] encodings. The two text-similarity measures, namely the Jaccard index and cosine similarity, are calculated for each pair of responses that serve as answers to the same question. All other measures are calculated for each response.

- Proportion of stop words: Stop words are a set of words that carry little meaning; for example, in English, *the*, *is* and *and* can be classified as stop words. The evaluation uses stop words corpus from the Natural Language Toolkit package [14] of Python. The proportion of stop words is calculated as the number of stop words divided by the total number of words of a response. Stop words are excluded when calculating the number of characters and words, as well as sentence length, type-token ratio, and mean word frequency.
- Length of answers: As measures of answer length, the following metrics are calculated: (i) the number of characters, (ii) the number of words and (iii) the number of sentences contained in an answer. As will be discussed in Section 4, it is observed that increased length of answers was one of the signals of plagiarism.
- Length of sentences: Sentence length is calculated as the number of words contained in a sentence.
- Lexical diversity: Type-token ratio (TTR) was used as a proxy for lexical diversity. TTR is calculated as type frequency of a response (the number of unique words) divided by token frequency (the total number of words). A higher TTR indicates that the text has more diverse vocabulary.
- Word frequency: For each response, average word frequency is calculated. The frequency of each word is obtained from the SUBTLEX-us [15], frequency data based on a corpus of American English film subtitles. Higher word frequency means that the word is more commonly used. As will be noted in Section 4, use of sophisticated vocabulary was found to be a signal of plagiarism.
- Proportion of *I*: One of our informal observations was that valid answers more frequently use phrases that include the first person singular pronoun *I*, e.g., *I think*, *I would*. Thus, another metric used is the number of times *I* appears in each answer divided by the total number of words.
- Automated Readability Index (ARI): ARI [16] is an index of readability or understandability of a text. It is calculated as $ARI = 0.5 * ASL + 4.71 * AWL - 21.43$, where ASL stands for average sentence length (average number of words in a sentence) and AWL stands for average word length (average number of characters in a word).
- Text similarity: It is plausible to think that solutions that violated academic integrity, either aided by ChatGPT or a search engine, should share many words in common. To test this hypothesis, two measures of text similarity were collected: the Jaccard index and the cosine similarity of an embedding generated from a language model. The Jaccard index is calculated as the number of unique words common to two texts, divided by the total number of unique words in both texts. The cosine similarity is a measure of similarity between two vectors, as it holds a unique property where the cosine similarity only considers the angle between the vectors, not their magnitude. Given this definition, two orthogonal vectors

have a similarity of 0 and two proportional vectors have a similarity of 1. The cosine similarity between two vectors A, B is computed as:

$$cos(\theta) = \frac{A \cdot B}{||A||_2||B||_2}$$

In order to compute the cosine similarity, SBERT model is used to generate an embedding, in which individual words are represented as real-valued vectors. Then, the similarity values reported in Section 4 are computed between two embedding vectors generated from two distinct sets of words. Within each group of students, the text-similarity measures are obtained for every pair of responses for the same question, as calculating similarity between responses for different questions would trivially result in low similarity and holds little scientific value.

2.5 Definition and scope of plagiarism

This section provides a definition of plagiarism and delineates the scope of plagiarism that this work deals with. First of all, plagiarism is a broad term encompassing the use of contents, ideas or structures, without acknowledging the original source. This definition does not restrict the form of content, which means that plagiarism can come from any documents, regardless of their presentation, e.g., text, images, mathematical formulae, and more. This work focuses only on analyzing plagiarism based on textual contents. Second, the forms of plagiarism can vary substantially, based on their source. For example, Mozgovoy et. al. [17] presents a typology that categorizes plagiarism into five different forms, namely:

(1) Verbatim copying: e.g., copy-paste from an electronic source
(2) Hiding the instance of plagiarism by paraphrasing: e.g., adding, replacing or removing characters or words
(3) Technical tricks, exploiting weaknesses of current plagiarism detection systems: e.g., the insertion of similar-looking characters from foreign alphabets
(4) Deliberate inaccurate use of references
(5) "Tough plagiarism", i.e., types of plagiarism that are particularly difficult to detect for both humans and computers

This work mainly focuses on items (1), (2), (5), since most instances of plagiarism present in the current data set were an outcome of verbatim copying or paraphrasing of AI-generated text. In addition, the problem this work encounters is an instance of "Tough plagiarism" as well, since detecting AI-generated text is reported to be very difficult, as shown by a very low true-positive accuracy of 26% from a detection tool created by OpenAI.

3 Overview

This section provides an overview of the course and examination structure, focusing on the various types of questions and responses that are incorporated within the exam. In addition, this section delves into the intricacies of plagiarism as it pertains to the aforementioned responses.

3.1 Course and exam format

The course in question is entitled "Introduction to Cloud Computing", offered at Santa Clara University under the course code COEN 241 [19]. COEN 241 is a graduate-level course that presupposes completion of an undergraduate curriculum, with prerequisites in computer networks and operating systems. The course typically enrolls between 35 and 40 students, with the average final grade of the class being a B+. The course emphasizes project-based learning, with the final project constituting 50% of the total grade, whereas a single exam accounts for 25% of the total grade.

The exam, which is the main topic of the study, is given during the sixth week of the ten-week curriculum. The exam is administered via an online platform known as Canvas [20] and it is expected that the students taking the exam are also present on a Zoom [21] meeting throughout the exam. An example view of the exam in Canvas from the student's point of view is shown in Figure 5. The exam is composed of 50 to 60 questions, comprising both multiple-choice and two open-ended essay questions. The topics covered within the exam are predominantly concerned with various forms of virtualization, such as system, operating system, and application virtualization, as provided in Section 2. The exam has a duration of two hours, and once submitted, no further modifications to the solutions are accepted.

As shown in Figure 5, the exam format is quite generic and does not enforce special rules or usage of tools that place restrictions on students' computers while they are taking the exam. For example, there are tools such as LockDown Browser [22] that prohibit students from navigating away from the page while they are taking the exam. The rationale behind such an approach is to promote an open-book assessment environment that empowers the students to refer to any class materials they may require. Also, the students are expected to comply with the academic integrity pledge outlined in their institution's honor code. This unbound nature, however, creates an opportunity for students to easily and freely seek solutions from online resources and language models, which in turn significantly increases instances of plagiarism, as we discuss in the subsequent section.

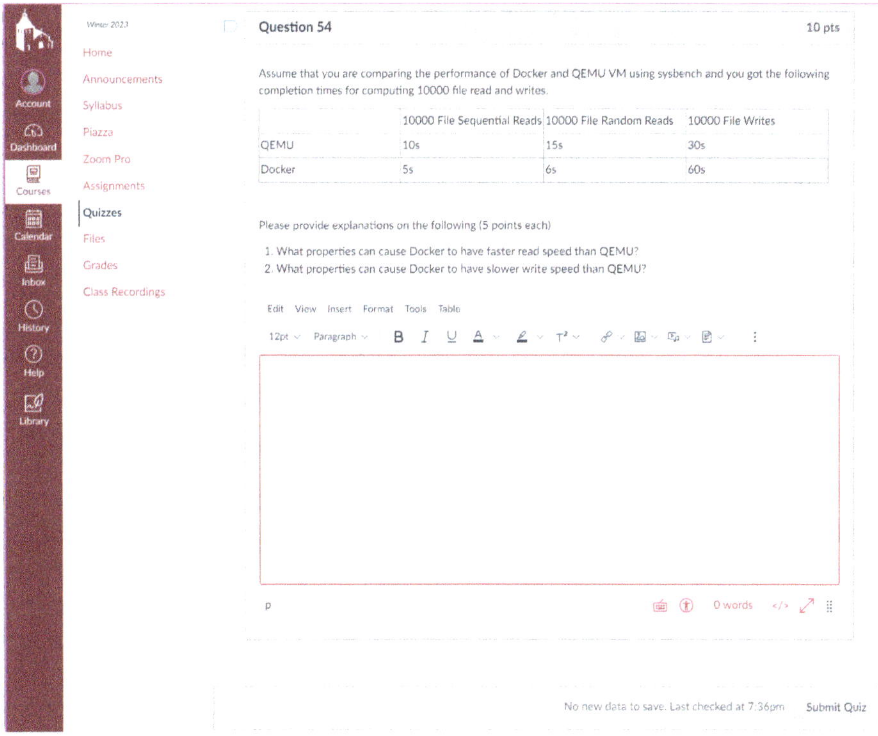

Figure 5: Student's view of the exam on Canvas.

3.2 Types of responses and plagiarism

Detecting instances of plagiarism in an academic setting remains a critical concern, particularly with regard to the free-form essay section of exams (as the multiple-choice format may not present clear indications of duplication.) Nevertheless, there exist certain informal indicators that may assist in the identification of plagiarism, specifically verbatim copying and paraphrasing. Such indications, while seemingly trivial, can prove instrumental in discerning instances of academic dishonesty. A few examples of these signals include:

- Similarities in the contents between solutions
- Solutions discussing topics, often incorrectly, that were not taught in the class
- Hard-to-replicate errors, such as spelling, that are shared in multiple solutions

In addition to these signals widely used to detect plagiarism in prior classes, more recent iteration of the course presented a new set of interesting signals of plagiarism that were found in a subset of the solutions, which are:

- Increased sophistication of the language
- Increased length of the solutions
- Increased correctness of the solutions that discuss topics not covered in the class
- Similarities in solution formats, such as bullet points and paragraphs/line breaks

To facilitate a comparative analysis, we can contrast the solutions that received no help (1) versus solutions that were plagiarized via ChatGPT (2) with a sample question: ***"What properties can cause Docker to have slower write speed than QEMU?"*** Also, note that both solutions received full credit on the question.

(1) *One potential reason is that Docker has write policies of AUFS, which means that Docker needs to copy up those files into its container layer for the first write to some files. In this case, there is an overheating for Docker writing.*

(2) *QEMU employs a full system virtualization approach, where a complete virtual machine is created with its own operating system. This approach provides better isolation and security but also incurs additional overhead, particularly when handling I/O operations. To improve I/O performance, QEMU can leverage hardware acceleration techniques such as using the VirtIO interface to directly access the hardware. This can lead to faster write speeds when compared to Docker, which employs a copy-on-write mechanism and may require more I/O operations for write operations.*

Even at a glance, we can easily see that solution (1) is far simpler and error-prone than solution (2) in terms of both vocabulary and language structure. Such *discovery* (or observation) of rather trivial signals from distinct groups incited the teaching staff to delve deeper into how the students came about these groups of solutions. Through an investigation, the teaching staff was able to identify a case of academic plagiarism and determined that the perpetrator received assistance from one of two sources: (1) a language model such as ChatGPT, or (2) a search engine, such as Google. The solution sets from each group were subsequently collated and subjected to a detailed analysis of their linguistic characteristics, as expounded in Section 2.4. The findings from evaluating the data are presented in the ensuing section (Section 4).

3.3 Statistical analysis of linguistic features

In order to investigate whether the numerical differences among the three groups (i.e., solutions that consulted ChatGPT, those aided by an online search, and "valid" answers that consulted no outside sources) in each language measure are statistically significant, we established mixed effects linear regression models using *lmerTest* [23] package in R [24]. The response variables were each language-measure presented in Section 2.4 and the fixed effect was GROUP with three levels (ChatGPT, Online and Valid). In all models except the ones for text-similarity measures, random intercepts for STUDENT and QUESTION were also included. We used the *anova* function in R to test whether the factor GROUP significantly increases the model fit to the data by comparing two models that are in a subset relationship, i.e., a model with GROUP and one without. The results of this likelihood ratio tests are reported as chi-squares. Post hoc pairwise comparisons of all levels of GROUP were conducted using the *emmeans* function of the *emmeans* package [25], with *p*-values adjusted for multiple comparisons using the Tukey method.

3.4 Improving plagiarism detection machine learning model

To understand the effects of adding linguistic features to the performance of plagiarism detection models, this work builds two types of classification algorithm, boosted decision trees and feedforward neural network, each with and without the linguistic feature inputs. A high-level overview of boosted decision trees and feedforward neural network is provided in Section 2. For each of these algorithms, we build three sets of models: a baseline model, a model with all linguistic properties that we investigated, and a model with only linguistic features that are found to significantly distinguish between the three groups of students' responses. The steps for building the sets of models are as follows:

1) **Baseline features and labels**: Baseline feature consist of only term frequency–inverse document frequency (TF-IDF) [26], a score calculated from unique words per data point after removing stop words. The main reason for using TF-IDF score as the baseline feature is to provide a simple and generic baseline that can potentially produce features to distinguish between classes, since TF-IDF gives high scores to words that do not appear often across data points. Numeric class labels, 0 for Valid, 1 for ChatGPT, and 2 for Online, are assigned to each data point for ease of modeling.
2) **Baseline models**: A gradient boosted tree and a feedforward neural network are built using TF-IDF as features and the numeric labels. We perform a grid search to find the best hyperparameter for the baseline models. The details of the hyperparameters selected for each algorithm is discussed in Section 4.
3) **Models with significant linguistic features**: Build two additional models by adding the linguistic measures that significantly differentiate among three groups of responses as features in addition to the TF-IDF scores. This step aims to investigate whether adding the significant measures as features aid in training a better model.
4) **Models with all linguistic features**: Build the largest model by adding all the linguistic features discussed in Section 4.2 as features, in addition to the TF-IDF scores. This step examines whether machine learning algorithms are capable of finding hidden relations between features that can further aid in detecting plagiarism.

Based on these three sets of models, we compare the accuracy and F1 scores to determine the performance differences between the types of models. These results are reported in Section 4.

4 Evaluations

To analyze the differences in responses that are generated by language model versus humans, we collected the set of data described in Section 4.1 and obtained the results presented in the following sections.

4.1 Data overview

There were 10 different questions that students provided solutions for. In total, the dataset consisted of approximately 150 samples of student responses that are categorized into three classes: (1) Valid: human-generated, without any external references, (2) Online: human-generated, from contents retrieved via a search engine, (3) ChatGPT: LLM-generated and copied over. Valid class consists of solutions that are not plagiarized, whereas the other two classes are determined to be plagiarized. The scores of solutions from class (1) range from 60% to 100% to reflect a true distribution of student grades, whereas responses from class (2) and (3) mostly received 90 +%. Most of the deductions in responses in class (3) came from discussion of topics that were correct, but out of the scope of the class, which may or may not be a legitimate deduction, depending on the classroom settings.

4.2 Per class characteristics

We investigated whether ChatGPT, Online, and Valid class significantly differed from each other in terms of each of the linguistic measures presented in Section 2.4. Results of the comparison between the three classes, using mixed effects linear regression models as outlined in Section 3.3, are presented in Table 1. First, the effect of CLASS did not significantly increase the model fit to the data of the proportion of stop words ($\chi^2(2) = 5.3$, $p = 0.07$), indicating that the three classes had a comparable proportion of stop words.

Next, solution length as measured by the number of characters, words, and sentences generally suggest that solutions aided by ChatGPT are longest, and Valid solutions are shortest. As for the number of characters, CLASS significantly improved the model fit to the data ($\chi^2(2) = 8.5$, $p < 0.05$). As can be seen in Figure 6, a violin plot that shows the distribution of the number of characters contained in a response, a post hoc analysis revealed that the difference in the number of characters between ChatGPT and Valid was significant ($\beta = 176.4$, SE = 59.6, t = 2.96, $p < 0.05$), while the difference between ChatGPT and Online ($\beta = 60.5$, SE = 42.8, t = 1.41, $p = 0.34$), and that between Online and Valid ($\beta = 115.9$, SE = 61.1, t = 1.90, $p = 0.16$) were nonsignificant. At the word level, there was a numerical trend in which ChatGPT had a greater number of words than Online, which in turn had a greater number of words than Valid. However, none of the pairwise comparisons were significant: the difference between ChatGPT and Valid missed significance ($\beta = 21.4$, SE = 8.65, t = 2.47, $p = 0.056$), and the difference between ChatGPT and Online was not significant ($\beta = 10.6$, SE = 6.34, t = 1.67, $p = 0.22$) nor was the difference between Online and Valid ($\beta = 10.8$, SE = 8.88, t = 1.21, $p = 0.46$). In terms of the number of sentences, CLASS significantly improved the model fit to the data ($\chi^2(2) = 14.3$, $p < 0.001$). ChatGPT had a significantly greater number of sentences than both Online ($\beta = 1.58$, SE = 0.48, t = 3.27, $p < 0.01$) and Valid ($\beta = 1.88$, SE = 0.66, t = 2.84, $p < 0.05$), while Online and Valid did not significantly differ from each other ($\beta = 0.30$, SE = 0.68, t = 0.45, $p = 0.90$).

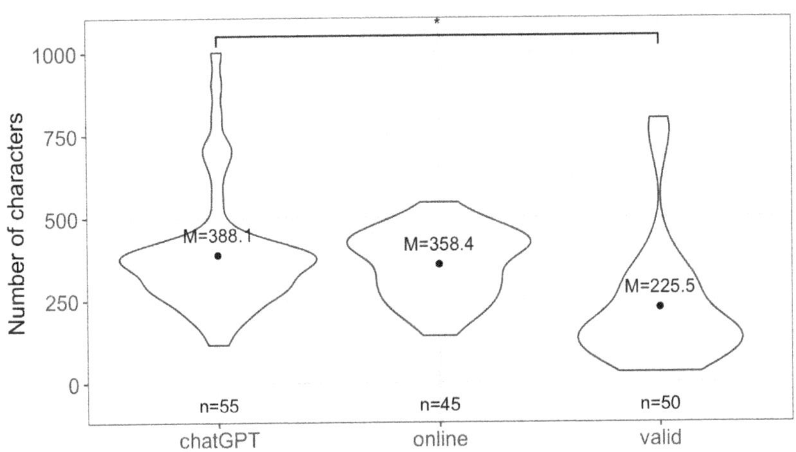

Figure 6: The distribution of the number of characters per response. *M* represents the mean value, and *n* is the number of samples included in each class. Stop words and white spaces are excluded in the count. (* indicates $p < 0.05$, ** $p < 0.01$, *** $p < 0.001$).

One of our informal observations was that sentences in the Valid class are shorter than those of the ChatGPT and the Online class. However, a statistical analysis of the average number of words contained in a sentence shows that while this is numerically true, adding CLASS to the model did not significantly improve the model fit to the data ($\chi^2(2) = 2.8$, $p = 0.24$).

We also examined word-level characteristics of the responses using type-token ratio (TTR), mean word frequency, and the proportion of the first person pronoun *I*. The model fit to the TTR data was significantly improved by adding CLASS ($\chi^2(2) = 6.1$, $p < 0.05$). However, none of the pairwise comparisons were significant. ChatGPT had a lower TTR than Online, but the difference was not statistically significant ($\beta = -0.04$,

SE = 0.03, t = −1.51, $p = 0.29$). Online had a lower TTR than Valid, which was not significant, either ($\beta = -0.03$, SE = 0.03, t = −0.94, $p = 0.62$). The difference between ChatGPT and Valid missed significance ($\beta = -0.07$, SE = 0.03, t = −2.42, $p = 0.06$). With respect to average word frequency, based on SUBTLEX-us, CLASS significantly improved the model fit ($\chi^2(2) = 8.1$, $p < 0.05$). As can be seen in Figure 7, a post hoc analysis showed that Valid had a significantly higher mean word frequency than both ChatGPT (β = 5,088, SE = 1,840, t = 2.77, $p < 0.05$) and Online (β = 4,926, SE = 1,921, t = 2.57, $p < 0.05$). ChatGPT and Online did not significantly differ from each other (β = −162, SE = 1,740, t = −0.09, $p = 0.995$). However, CLASS did not significantly improve the model fit to the data for the proportion of *I* ($\chi^2(2) = 1.9$, $p = 0.40$).

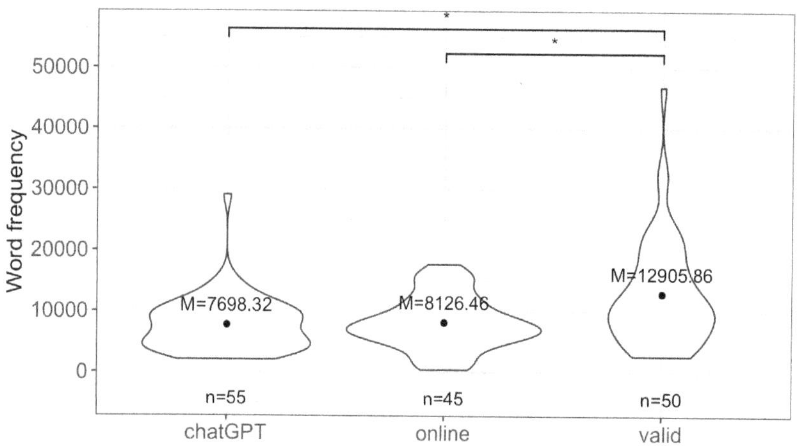

Figure 7: The distribution of the mean SUBTLEX-us frequency of words contained in a response. *M* represents the mean value, and *n* is the number of samples included in each class. (* indicates $p < 0.05$, ** $p < 0.01$, *** $p < 0.001$).

As for readability of texts, there was a numerical trend in which Valid responses had a lower ARI than both ChatGPT and Online responses, and adding CLASS to the model did significantly improve the fit to the data ($\chi^2(2) = 6.0$, $p < 0.05$). However, in the post hoc analysis, the difference between ChatGPT and Valid was not significant ($\beta = 2.05$, SE = 1.61, t = 1.27, $p = 0.42$) nor was the difference between Online and Valid reach significant ($\beta = 3.98$, SE = 1.67, t = 2.38, $p = 0.06$). The difference between ChatGPT and Online was also nonsignificant ($\beta = -1.93$, SE = 1.44, t = −1.34, $p = 0.38$).

4.3 Intra-class similarities

We examined text similarity among responses within each class using the Jaccard index and the cosine similarity of encodings using a distilled MiniLM [27] model called *all-MiniLM-L6-v2* [28]. As noted in Table 1, both measures show that the similarity values of Online are the highest, followed by ChatGPT, with Valid having the lowest similarity.

Table 1: Mean and standard deviation of each language measure for the three classes.

	ChatGPT	Online	Valid
Proportion of Stop Words	M = 0.41, SD = 0.1	M = 0.42, SD = 0.1	M = 0.44, SD = 0.1
Number of Characters	M = 388.1, SD = 177.9	M = 358.4, SD = 101.8	M = 225.5, SD = 186.1
Number of Words	M = 55.1, SD = 25.1	M = 51.2, SD = 14.7	M = 36.7, SD = 29.0
Number of Sentences	M = 4.1, SD = 1.9	M = 3.6, SD = 1.2	M = 2.7, SD = 2.0
Number of Words in a Sentence	M = 14.5, SD = 4.5	M = 14.9, SD = 3.9	M = 14.0, SD = 6.8
Type-token Ratio	M = 0.77, SD = 0.1	M = 0.81, SD = 0.1	M = 0.84, SD = 0.1
Word Frequency	M = 7698.3, SD = 4514.7	M = 8126.5, SD = 4317.9	M = 12905.9, SD = 8740.3
Proportion of *I*	M = 0.0017, SD = 0.003	M = 0.0004, SD = 0.002	M = 0.0022, SD = 0.005
Automated Readability Index	M = 16.08, SD = 4.8	M = 16.13, SD = 3.4	M = 13.19, SD = 6.3
Jaccard Index	M = 0.22, SD = 0.1	M = 0.39, SD = 0.2	M = 0.17, SD = 0.1
Cosine Similarity of SBERT Encoded Solutions	M = 0.73, SD = 0.1	M = 0.84, SD = 0.1	M = 0.64, SD = 0.2

For both measures, CLASS had a significant effect on text similarity. It was found that Online had a higher Jaccard similarity than ChatGPT (β = 0.16, SE = 0.01, t = 11.90, $p <$ 0.001), which in turn had a higher Jaccard similarity than Valid (β = 0.06, SE = 0.01, t = 4.35, $p < 0.001$; Figure 8). Similarly, Online had a higher cosine similarity than ChatGPT (β = 0.11, SE = 0.02, t = 5.88, $p < 0.001$), which in turn had a higher cosine similarity than Valid (β = 0.09, SE = 0.02, t = 5.11, $p < 0.001$; Figure 9).

The result is as expected, since the Online class consists of solutions simply copied from similar online sources, whereas ChatGPT tends to generate varying solutions of similar content. Finally, it is expected that responses in the Valid class show the lowest similarity, since students come up with their own sentences without relying on external sources.

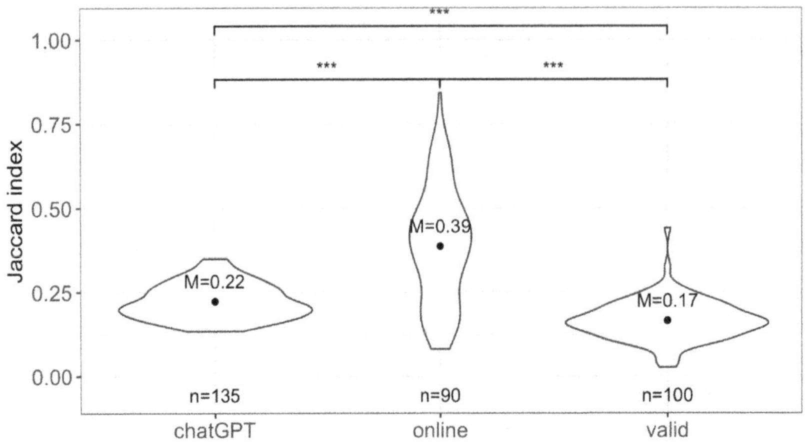

Figure 8: The distribution of Jaccard index. *M* represents the mean value and *n* is the number of samples included in each class. (* indicates $p < 0.05$, ** $p < 0.01$, *** $p < 0.001$).

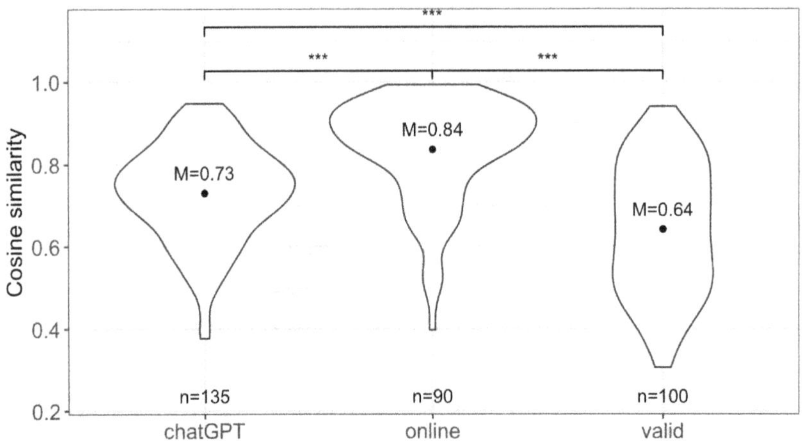

Figure 9: The distribution of cosine similarity values. *M* represents the mean value and *n* is the number of samples included in each class. (* indicates $p < 0.05$, ** $p < 0.01$, *** $p < 0.001$).

4.4 Performance comparison of machine learning models

As mentioned in Section 3.4, we evaluate the performance of two different machine learning algorithms on three types of models: Baseline Models (D_B, N_B), Models with Significant Features (D_1, N_1), and Models with All Features (D_2, N_2), where D_X represents the boosted decision tree model and N_X represents the feedforward neural network model.

The training data set is filtered to only classify between Valid and ChatGPT to simplify the problem as binary classification. The number of features for D_B and N_B is 1,014, which represents the number of unique words that exist across all the data points. As mentioned in Section 3.4, each feature of D_B and N_B corresponds to TF-IDF score of a particular word in the data set. Using count of word occurrence as the baseline feature, instead of TF-IDF, did not yield a promising result. The number of features for D_1 and N_1 is 1,017, which is the combination of TF-IDF scores with three significant linguistic features: number of characters, number of sentences, and mean word frequency. The similarity scores cannot be added as features, since one similarity value is calculated for a pair of responses rather than per response. The number of features for D_2 and N_2 is 1,023 with all the linguistic features, both significant and non-significant, added as features. For the added features, the values of the features are normalized using min-max normalization, which is

$$x_{scaled} = \frac{x - x_{min}}{x_{max} - x_{min}}$$

The hyperparameters for each type of models are as follows. For the boosted decision tree, the hyperparameters are: *Number of estimators = 500, Learning Rate = 0.1, Max Depth = 20*. For the feedforward neural network, the hyperparameters are: *Number of Hidden Layers = 3, Hidden Layer Sizes = [512, 128, 32], Max Iteration = 100, Activation Function = ReLU, Solver = Adam, Early Stopping = True*. The hyperparameters for the feedforward neural network model is selected via a grid search framework, which selected the above hyperparameters between different hidden layer sizes, activation functions (tanh, *ReLU*), and more.

After specifying the model parameter, the model was evaluated via 10-fold stratified cross validation, which basically is a 10-fold cross validation that makes sure each split has the same distribution of classes. The main program is written in Python, mainly using the scikit-learn framework [29] for model training and evaluation. For boosted decision trees, the program uses the GradientBoostingClassifier class and for feedforward neural network, the program uses MLPClassifier. The code runs on Google Colab [30], which is a cloud platform for running a notebook, to build and evaluate the models. The code for selecting the feature, training the models and performing cross validation can be found in the shared code [31].

The average of accuracy, precision, recall, and F1 scores are calculated, presented in Table 2. Note that since precision and recall are averaged separately, the mean value of F1 may differ from computing F1 scores from mean precision and mean recall.

The results showed that adding only the significant features improves F1 score by 3.281% for the boosted decision tree and 12.38% for the feedforward neural network. This result is promising in that the boosted decision tree model that adds only significant features outperforms the model that adds all features by up to 7.133%.

Table 2: Model Performance Metrics.

	Mean accuracy	Mean precision	Mean recall	Mean F1	% Δ Mean accuracy	% Δ Mean F1
D_B	0.724	0.800	0.680	0.701	NA	NA
D_1	0.723	0.771	0.700	0.724	−0.138%	**3.281%**
D_2	0.694	0.709	0.660	0.674	−4.144%	−3.852%
N_B	0.664	0.663	0.853	0.727	NA	NA
N_1	0.790	0.816	0.873	0.817	**18.98%**	**12.38%**
N_2	0.765	0.762	0.883	0.802	15.21%	10.32%

4.5 Feature importance for the boosted decision tree model

Lastly, in order to examine whether the significant features are the most important features for the trained model, each feature was given a feature importance score and indices of the top 10 features were found. To elaborate again, features up to indices 1,013 are TF-IDF scores, features with indices 1,014, 1,015, 1,016 are the three significant features, and the rest are nonsignificant features. The feature importance is only retrieved from the boosted decision tree models for simplicity.

The list of the top 10 feature indices are as follows:

- D_B: [643, 981, 7, 334, 516, 508, 970, 572, 579, 399]
- D_1: [**1,014**, **1,016**, 184, 411, 489, 687, 674, 981, 266, 25]
- D_2: [**1,015**, 1,017, 184, 411, 212, 489, 674, 687, 523, 2]

Unsurprisingly, for D_1, two of the significant linguistic features added to the model are found to be the most important features. One of the significant linguistic features is found to be the most important feature for D_2 as well. This shows great promise in using the linguistic features as input features for training machine learning algorithms.

5 Related works

At a high level, this work combines the background knowledge of three topics of study: 1) detection of language model usage, 2) detection and prevention of plagiarism, and 3) study of linguistic properties of language models. This section provides a short summary of prior works in each of these fields and shows how each work is related to the present study.

Detection of language model usage: The ubiquitous use of LLMs, particularly in academic environments, is a recent phenomenon that has stimulated considerable interest in detecting their usage [32]. Thus, detecting the usage of language models is a very active field of study, as of writing [33]. The development of tools for identifying language model usage has been an active area of research, with noteworthy contributions from the GPTZero framework [34, 35], which is gaining traction in both academia and industry [36]. Furthermore, OpenAI, the creators of ChatGPT, has released a tool to identify the presence of AI-generated text [18]. However, these tools are still in their early stages of development, and OpenAI acknowledges that its model has a true positive rate of only 26% and a false positive rate of 9%, with performance deteriorating as text length decreases. These findings underscore the significant challenges associated with detecting language model usage, and highlight the ongoing efforts in this area.

Detection and prevention of plagiarism: Detection and prevention of plagiarism has been a topic of study for multiple decades. Some notable works, specific to computer science education, include Donalson et al. [37], which is one of the earliest attempts at detecting plagiarism in computer science programming assignments, and MOSS, discussed in [38], which is widely used in various institutions to detect plagiarism. Furthermore, there are tools such as Turnitin [39] to detect plagiarism, but these tools rely on similarity to existing documents of databases, which can easily be avoided by a language model if it is optimized to generate new sequence of text that has not been submitted to the Turnitin database. There are also recent works on fooling MOSS via a language model [40]. These findings show that while the study of plagiarism detection has a long history, detecting LLM-aided plagiarism is a relatively recent challenge. Research on detecting plagiarism will continue for a long period of time, and this paper attempts to provide new insights to this field of study.

Study of linguistic properties of language models: This study also involves analyzing linguistic properties of text generated by language models and compares them against linguistic characteristics of text generated by humans. Studying linguistic features of AI-generated text is still at a beginning stage, e.g., [41], and the present work is one of the first efforts to understand whether and how AI-aided text differs from human-generated text in an academic and educational setting.

6 Discussion and future works

This section summarizes findings of this study and discusses its potential use cases and future directions.

Discussion of the results: The present study investigates which metrics of linguistic characteristics, if any, distinguish between three types of solutions to essay questions in an exam of a cloud computing class, i.e., solutions aided by ChatGPT, those aided by a search engine, and honest solutions. We found that solutions that consulted ChatGPT are longer than honest solutions, as measured by the number of characters contained in a response. Similarly, an analysis of the number of sentences within a response shows that ChatGPT-aided responses are longer than both search engine-aided solutions and honest ones. Together, the findings suggest that students write in a concise manner when they come up with solutions on their own. We also found that words used in honest solutions have a higher frequency, on average, than those used in ChatGPT- and search engine-aided solutions, suggesting that the former group contains more common and familiar words than the latter groups. Finally, based on the Jaccard index and the cosine similarity, we found that solutions that consulted search engine had the highest text similarity, and honest ones had the lowest similarity.

As for using the significant linguistic features as input features to machine learning models, the experiments show great promise in improving the machine learning models. By employing linguistic domain-specific knowledge to find the measures that effectively distinguish between plagiarized and not plagiarized solutions, we enable models to utilize deeper insights that are not found in the mere ordering or frequency of words that most machine learning algorithms use. Thus, this work shows promising first steps in building LLM detection models by using linguistic features. However, given that this is an early work in this field, additional work should be done to identify additional linguistic features and/or to improve training of the models, especially regarding building different types of models and more complex/deeper neural network models.

Additional signals via language models: A planned future work is to use LLMs to generate additional signals for violation of academic integrity. For example, ChatGPT is known to excel at summarizing text. Therefore, it is possible to feature-engineer additional signals using the summaries of each solution, further providing additional context in detecting academic integrity violations.

Data error, data bias, and sample size: Another planned future work is to increase the number of samples to reduce bias in data. A potential issue that can be seen in this work is that certain metrics can easily be disturbed by one or two data points with different characteristics. One way to resolve this issue is to add more samples to diminish the effects of such bias. Another way is to add more logic into processing the data. Both are planned future tasks for this work.

Complementing existing plagiarism detectors: Since even the most advanced detector from OpenAI [18] is showing a low true positive rate in detecting the use of LLMs, this paper can aid in providing features to use in improving such detectors. Furthermore, this work shows that success in detecting use of LLMs can increase greatly if there is a context to refer to. For example, by having the exam question and the class content as the context, one can easily generate another solution with similar linguistic characteristics to compare against other potential violations. This means that instructors can prepare their own LLM-generated solution to use as a basis for comparison, which is shown to increase the probability of detecting the use of LLM. While instructors should not rely solely or even heavily on these tools to automatically detect plagiarism, they can be used as a signal for further investigation.

7 Conclusion

The present study investigates the potential of AI tools such as ChatGPT to generate solutions to complex essay questions that require deep domain knowledge. While ChatGPT is able to generate coherent and relevant solutions, it provides solutions with certain linguistic features that are statistically different from the solutions from the control group. Furthermore, this study shows that using such statistically significant features as input features for training machine learning models are shown to improve the performance of such plagiarism detection models. In conclusion, this is a first step of many in encouraging academic integrity and minimizing the impact of AIs in reducing the effectiveness of the computer science education.

References

[1] T. B. Brown, B. Mann, N. Ryder et al., "Language Models are Few-shot Learners," *arXiv preprint arXiv:2005.14165*, 2020.

[2] A. Vaswani, N. Shazeer, N. Parmar et al., "Attention Is All You Need," *Advances in Neural Information Processing Systems*, pp. 5998–6008, 2017.

[3] K. Hu, "Chatgpt sets record for fastest-growing user base – analyst note," Feb. 2023. https://www.reuters.com/technology/chatgpt-sets-record-fastest-growing-user-base-analyst-note-2023-02-01

[4] T. Chen and C. Guestrin, "Xgboost: A Scalable Tree Boosting System," in *Proceedings of the 22nd acm sigkdd international conference on knowledge discovery and data mining*, pp. 785–794, 2016.

[5] G. Ke, Q. Meng, T. Finley et al., "Lightgbm: A Highly Efficient Gradient Boosting Decision Tree," *Advances in Neural Information Processing Systems*, vol. 30, 2017.

[6] AWS, "What Is Virtualization," 2023. https://aws.amazon.com/what-is/virtualization/#:~:text=Virtualization%20is%20technology%20that%20you,on%20a%20single%20physical%20machine.

[7] F. Bellard, "QEMU, a Fast and Portable Dynamic Translator," in *USENIX Annual Technical Conference, FREENIX Track*, pp. 41–46, 2005.

[8] J. Watson, "Virtualbox: Bits and Bytes Masquerading as Machines," *Linux Journal*, vol. 2008, no. 166, p. 1, 2008.

[9] R. Rosen. Resource management: Linux kernel namespaces and cgroups, 2013. https://sites.cs.ucsb.edu/~rich/class/old.cs290/papers/lxc-namespace.pdf

[10] C. Boettiger, "An Introduction to Docker for Reproducible Research," *ACM SIGOPS Operating Systems Review*, vol. 49, no. 1, pp. 71–79, 2015.

[11] A. Randazzo and I. Tinnirello, "Kata Containers: An Emerging Architecture for Enabling Mec Services in Fast and Secure Way," in *Proceedings of the Sixth International Conference on Internet of Things: Systems, Management and Security (IOTSMS)*. IEEE, pp. 209–214, 2019.

[12] S. Niwattanakul, J. Singthongchai, E. Naenudorn et al., "Using of Jaccard Coefficient for Keywords Similarity," in *Proceedings of the International MultiConference of Engineers and Computer Scientists*, vol. 1, pp. 380–384, 2013.
[13] N. Reimers and I. Gurevych, "Sentence-bert: Sentence Embeddings Using Siamese Bert-networks," *arXiv preprint arXiv:1908.10084*, 2019.
[14] S. Bird, E. Klein and E. Loper, *Natural Language Processing with Python: Analyzing Text with the Natural Language Toolkit*. O'Reilly Media, Inc., 2009.
[15] M. Brysbaert and B. New, "Moving beyond Kuc˘era and Francis: A Critical Evaluation of Current Word Frequency Norms and the Introduction of A New and Improved Word Frequency Measure for American English," *Behavior Research Methods*, vol. 41, pp. 977–990, 2009.
[16] R. J. Senter and E. A. Smith, "Automated Readability Index," Cincinnati University, Tech. Rep., 1967.
[17] M. Mozgovoy, T. Kakkonen and G. Cosma, "Automatic Student Plagiarism Detection: Future Perspectives," *Journal of Educational Computing Research*, vol. 43, no. 4, pp. 511–531, 2010, doi: https://doi.org/10.2190/EC.43.4.e.
[18] J. H. Kirchner, L. Ahmad, S. Aaronson et al., "New AI classifier for indicating AI-written text," Jan. 2023. https://openai.com/blog/new-ai-classifier-for-indicating-ai-written-text
[19] Santa Clara University, "Department of computer science and engineering." https://www.scu.edu/engineering/academic-programs/department-of-computer-engineering/graduate/course-descriptions/
[20] Instructure, "Canvas by instructure," https://www.instructure.com/canvas, 2023.
[21] Zoom Video Communications Inc., "Zoom," https://zoom.us, 2022.
[22] G. Cluskey Jr, C. R. Ehlen and M. H. Raiborn, "Thwarting Online Exam Cheating without Proctor Supervision," *Journal of Academic and Business Ethics*, vol. 4, no. 1, pp. 1–7, 2011.
[23] A. Kuznetsova, P. B. Brockhoff and R. H. B. Christensen, "lmerTest Package: Tests in Linear Mixed Effects Models," *Journal of Statistical Software*, vol. 82, no. 13, pp. 1–26, 2017.
[24] R Core Team, *R: A Language and Environment for Statistical Computing*. Vienna, Austria: R Foundation for Statistical Computing, 2021, https://www.R-project.org/
[25] R. V. Lenth, *emmeans: Estimated Marginal Means, aka Least- Squares Means*, 2023, r package version 1.8.4-1. https://CRAN.R-project.org/package=emmeans
[26] J. Leskovec, A. Rajaraman and J. D. Ullman, *Mining of Massive Data Sets*. Cambridge university press, 2020.
[27] W. Wang, F. Wei, L. Dong et al., "MiniLM: Deep Self-attention Distillation for Task-agnostic Compression of Pre-trained Transformers," in *Proceedings of the 34th International Conference on Neural Information Processing Systems*, pp. 5776–5788, 2020.
[28] N. Reimers, J. Gante and O. Espejel, "all-miniLM-L6-v2." https://huggingface.co/sentence-transformers/all-MiniLM-L6-v2
[29] F. Pedregosa, G. Varoquaux, A. Gramfort et al., "Scikit-learn: Machine Learning in Python," *Journal of Machine Learning Research*, vol. 12, pp. 2825–2830, 2011.
[30] E. Bisong, *Google Colaboratory*. Berkeley, CA: Apress, pp. 59–64, 2019, doi: https://doi.org/10.1007/978-1-4842-4470-87.
[31] Santa Clara University, "SCU CloudLab Github," https://github.com/The-Cloud-Lab/PlagiarismDetector, 2022.
[32] S. Barnett, "ChatGPT is making universities rethink plagiarism," Jan. 2023. https://www.wired.com/story/chatgpt-college-university-plagiarism/
[33] S. Mitrovic´, D. Andreoletti and O. Ayoub, "ChatGPT or Human? Detect and Explain. Explaining Decisions of Machine Learning Model for Detecting Short ChatGPT-generated Text," *arXiv Preprint arXiv:2301.13852*, 2023.
[34] E. Tian, "GPTZero: The World's No. 1 AI Detector with over 1 Million Users," https://gptzero.me/, 2023.

[35] S. E. Needleman, “ChatGPT creator releases tool to detect AI-generated text, calls it ‘unreliable’,” Feb. 2023. https://www.wsj.com/articles/chatgpt-creator-releases-tool-to-detect-ai-generated-text-calls-it-unreliable-11675204820

[36] T. H. Tran, “A college kid built an app that sniffs out text penned by AI,” Jan. 2023. https://www.thedailybeast.com/princeton-student-edward-tian-built-gptzero-to-detect-ai-written-essays

[37] J. L. Donaldson, A. M. Lancaster and P. H. Sposato, “A Plagiarism Detection System,” in *Proceedings of the twelfth SIGCSE technical symposium on Computer science education*, pp. 21–25, 1981.

[38] T. Lancaster and F. Culwin, “A Comparison of Source Code Plagiarism Detection Engines,” *Computer Science Education*, vol. 14, no. 2, pp. 101–112, 2004.

[39] T. Batane, “Turning to Turnitin to Fight Plagiarism among University Students,” *Journal of Educational Technology & Society*, vol. 13, no. 2, pp. 1–12, 2010.

[40] S. Biderman and E. Raff, “Fooling MOSS Detection with Pretrained Language Models,” in *Proceedings of the 31st ACM International Conference on Information & Knowledge Management*, pp. 2933–2943, 2022.

[41] D. M. Markowitz, J. Hancock and J. Bailenson, “Linguistic markers of AI-generated text versus human-generated text: Evidence from hotel reviews and news headlines,” Jan. 2023. http://psyarxiv.com/mnyz8

Michael Sandborn, Carlos Olea, Anwar Said, Mudassir Shabir,
Peter Volgyesi, Xenofon Koutsoukos, Jules White

Towards AI-augmented design space exploration pipelines for UAVs

Abstract: Design space exploration (DSE) is a key aspect of the engineering process, particularly for selecting components of a system and their parameters, subject to design goals and the physical constraints. Candidate designs that result from the design space exploration process must be evaluated to characterize their quality with respect to the desired performance. Recent advances in machine learning and artificial intelligence have introduced the potential for more rapid design space exploration in a target domain and continue to advance the process of rapidly producing and evaluating design candidates for a particular task. Accelerated and AI-augmented DSE will play a crucial role in developing future cyber-physical systems (CPSs). In this chapter, we focus on the domain of Unmanned Aerial Vehicles (UAVs) that pervade society and enable activities such as cargo transportation, rescue operations, and surveillance of large areas of land. A key consideration in the design and development process for UAVs is the selection and arrangement of components from a design space subject for an objective that is indicative of the vehicle's performance. We present a design pipeline to rapidly generate candidate UAV design topologies with a string-based design grammar and provide a design quality heuristic with a graph neural network (GNN)-based drag surrogate model [1]. The goal of this approach is to rapidly explore the design space of feasible UAV designs and provide an estimate of the quality of the design in terms of flight performance, based on the predicted drag force. Typical simulations for assessing the performance of a vehicle often rely on resource-intensive computational fluid dynamics (CFD) applications. Our goal with this work is to introduce a path toward circumventing this cumbersome requirement, to quickly generate design candidates that can be selected based on a heuristic performance metric and then filtered to reduce or remove the requirement of hefty simulation approaches. We also discuss the more recent advances in large language models (LLMs) for preliminary exploration and refinement of vehicle components and design space exploration.

Keywords: design space exploration, unmanned aerial vehicles, graph neural networks, design grammars, large language models

Michael Sandborn, Carlos Olea, Anwar Said, Mudassir Shabir, Peter Volgyesi, Xenofon Koutsoukos, Jules White, Department of Computer Science, Vanderbilt University, Nashville, TN, USA

https://doi.org/10.1515/9783111344126-014

1 Introduction

Recent advances in machine learning and artificial intelligence have unlocked a variety of applications in domains ranging from transportation to healthcare to engineering design. In this chapter, we examine the application of artificial intelligence to the design space exploration process of Unmanned Aerial Vehicles (UAVs), specifically for component selection and topology generation [1]. The key challenge to address is to reduce the dependency on heavyweight simulation pipelines that may require many computational resources, pay-walled software services, or substantial human effort. Broadly, there are four main stages to consider in the design process, which will be overviewed in the following subsections:

1. Design space constraints
2. Component selection
3. Topology selection
4. Design generation

1.1 Design space constraints

Design space constraints formalize the boundaries for a design domain and specify information such as the types of components that are allowed, the restrictions on the system size, layout, and performance requirements. In the UAV domain, example design space constraints for a candidate vehicle might require the presence of at least one propeller (so the vehicle can fly), an upper bound on the maximum width of the vehicle (so the vehicle is airworthy), and a symmetry constraint (so the vehicle can maintain stable flight). While these are simple example constraints, more intricate design constraints such as a limit on the number of propellers to be supported by a single battery or the total component volume that must be maintained for situating components can also be incorporated. A formalization of pertinent constraints in a design domain helps to characterize what constitutes a valid design and thereby limits the search space to realistic designs.

1.2 Component selection

Selecting components for a UAV must account for several factors, including material properties, component dimensions, component parameters, and component arrangement. Here, we consider fundamental components to a baseline UAV, such as batteries, connectors, propellers, and wings, rather than specialized components such as actuated arms or dynamic components. Each component of a design may be parameterized by its quantitative properties. For example, a wing may have a spanning length, a thickness, a profile that describes the shape of its cross-section, and an orientation (e.g., horizontal or vertical). A propeller, similarly, can be defined by its diameter, pitch, rotation direction (i.e., to indicate whether it is a pushing or pulling propeller), and orientation. A connector component can be characterized by its diameter, material type, and length. A battery component can be characterized by its electrical properties such as voltage and capacity as well as its physical size such as volume and mass.

Each component can be characterized by a tuple, indicating its associated design variables. These are primarily categorical (e.g., a NACA profile for a wing) and real-valued parameters (e.g., the length of a connector). Component properties span multiple domains, including aerodynamic, geometric, and electrical. Each of these properties subtly affects the overall performance of the vehicle, which motivates the use of a performance heuristic to indicate the overall performance of the vehicle. We do not extensively consider the electrical domain or the aerodynamic domain, outside of our selection of the drag force, as a performance surrogate for the generated designs. This is discussed further in the following sections.

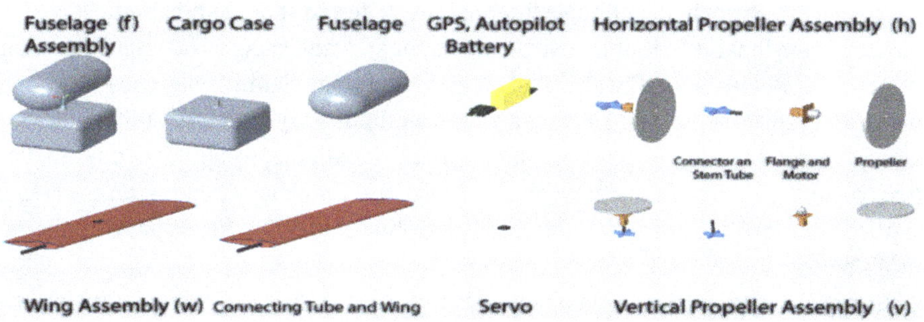

Figure 1: The component corpus [23] that we consider for constructing UAVs in this work. These components must be selected, parameterized, and assembled to represent a single UAV design for drag force analysis. Adapted from [2]. Used with permission.

1.3 Topology selection

We define the *topology* of a UAV as the arrangement of its selected components in 3D space. Additional constraints on the UAV designs space can be imposed here, for example, around the maximal height of the vehicle, the number of propeller groupings, and the front-to-back length of the vehicle. A vehicle's topology can be described by the connectivity between selection of the parameterized components. Typically, the connectors between components should be minimally viable, which means that a connector should be no longer than it needs to be. For example, a connecting rod holding two identical vertical propellers, each with a diameter of 4 inches and requiring 1 inch between them should be at minimum 9 inches in length. The topology selection is a crucial point in the design pipeline, and further constrains the space of possible designs. Both the component and topology selections of the design are combinatorial problems, with sparsity in the number of feasible designs. As an example, a UAV with four propellers may have its propellers arranged in a square (i.e., a quadcopter), a straight line, or with two pulling propellers at the top and two pushing propellers at the bottom. The constraint on viable vehicle topologies should be captured by the design grammar to represent how designs should be generated.

1.4 Design generation

The design generation process describes how design candidates from a target domain should be constructed such that the relevant components are captured in a structured manner, and that valid relations or connections between these components are implicitly represented or can be easily derived. Fundamentally, this problem refers to sampling from a design space, where a single point represents a candidate design, composed of components and their interactions. A design space can be represented as a multidimensional distribution over the components and their parameters. Common sampling strategies in this regard include Grid Sampling, Latin Hypercube Sampling, and Monte Carlo sampling. However, these methods may be more suitable once a baseline candidate design has been identified, for example, to optimize or fine-tune the structure of a fixed topology vehicle. Therefore, a topology generation process is required as a precondition to additional sampling strategies for improving the vehicle's performance.

A typical approach to topology generation is with design grammar, which is a grammar for a design domain that enumerates valid production rules to produce a design candidate from the target domain. In this case, a single design candidate is a UAV, represented by its selected components, their connections, and how they are arranged (the UAV topology). It is also possible to have variants of a design grammar, for example, one in which the specific components need not be specified a priori, or one in which the connections between the components can be implicitly derived, based on the component types (e.g., a propeller must be connected to a motor that must be wired to a battery using a connector of some length).

The design generation process then proceeds by sampling from the defined design grammar to collect a variety of candidate designs for assessment in a simulation or heuristic evaluation pipeline. Here, there is a trade-off between design diversity and design feasibility: one can either (1) modify a known good design that exists in the design space (e.g., increase the lengths of a quadcopter or introduce an additional propeller), or (2) trade-off flight feasibility for design novelty (e.g., generated designs must have a minimal "distance" between them, without consideration of how stable the vehicle might be in flight). The design space exploration process must reconcile these two conflicting objectives, as prioritizing one over the other drastically affects the size of the design space to consider. In the first case, sampling occurs more around a known good design in the design space, while in the second case, the design space is covered more comprehensively without additional certainty about the flight feasibility of the generated designs.

As a general guideline, the resources allocated for evaluating a design should be proportional to the effort placed in producing the design candidate. In other words, if a design is quickly generated with an automated process, it should require a similar level of effort to determine whether the design should proceed in the pipeline. As additional human modifications are invested into a promising design, more precise and resource-intensive evaluation methods should be employed.

2 Related work

As discussed previously, there are several key stages for efficient design space exploration, including design representation, design generation, and design evaluation. We describe design grammars for enhancing the design process, UAV design challenges, and Artificial Intelligence (AI) applications to design space exploration (DSE).

2.1 Design grammars

Fundamentally, a formal grammar is a set of production rules that defines a language over a domain. For example, a programming language can be defined by a formal grammar that captures valid syntactical relations between statements such as function calls and variable assignments. A set of grammar rules may contain terminal and nonterminal statements that correspond to different expansions, based on the available production rules.

Recent work in robotics [6] leveraged a recursive graph to generate locomotive robot topologies, given a target terrain. Taking as input the robot components to be connected as well as the terrain that must be traversed (e.g., steps, walls), a graph grammar produces small graphs from the expansion of nonterminals to represent connected components of the resulting robot design. Structural- and component-based rules are enforced by the production rules of the proposed graph grammar so that a proposed robot design has a physically realistic design. From a set of generated designs, a newly proposed graph heuristic search considers design, evaluation, and learning phases jointly to identify optimal designs from a set of design candidates. In comparison to this work, our approach relies on a string-based grammar that can be readily converted to a graph to represent a design. We rely on a graph neural network for heuristic estimation of UAV design quality using the drag force as our performance metric. Zhao et al. [8] presents a novel design space for the design of underwater vehicles using a graph grammar, combined with deformation cages. Each graph contains concrete components with edges indicating inter-component connections. The generated designs are then optimized, based on an objective that accounts for depth, mass, and ocean floor area surveyed in unit time. The work of Stöckli et al. [7] is inspired by the challenge in design space exploration whereby human or bio-inspired designs may perform well but may still be outmatched by unexplored alternatives. This work introduces a graph grammar that integrates dynamic simulation to evolve design candidates, focusing on the problem of brachiating robots. Sims [9] introduces an approach for generating morphologies of virtual creatures using genetic algorithms. A collection of fitness functions directs the evolution toward desired actions such as swimming and walking. The presented genetic language uses graphs to represent the shape and structure of simulated creatures. Mallozi et al. [5] proposes a context-sensitive grammar to systematically explore the design space of UAVs. A context-sensitive design grammar is formulated, based on a 3D grid, with either terminal or nonterminal symbols at each point. The context and state of a point refer to its neighbors in 3D space, and the rules determine how the nonterminal symbols are expanded, according to production rules. A heuristic is applied to select production rules, based on rules that match a given point. The process continues until only terminal symbols remain or no matching rules are present. The rules are derived from existing UAV designs and a specification satisfaction approach is discussed to encode design feasibility with SMT formulas.

2.2 UAV design

Unmanned Aerial Vehicles (UAVs) [4, 11] are becoming increasingly common in the society today. They reduce the burden of labor on humans for tasks such as watering crops and extinguishing fires. A key challenge in the advancement of UAV development and design for specific tasks is the sheer number of considerations in the design process. The UAV design process spans multiple disciplines, including electrical engineering, aerospace engineering, and control theory, and represents a complex and large set of optimization variables that interact across these domains. The design process typically begins with design constraints and mission objectives (e.g., hover for a certain amount of time or traverse a distance at an altitude within a certain time). These objectives guide the component selection process (e.g., a fast vehicle should be light and have smaller propellers while a vehicle that must carry a 100lb payload should have larger propellers). Varsha et al. [3] provide a conceptual overview of the process of the UAV design process, beginning with constraint generation and the design of the wings, propellers, and fuselage sizing. Papageorgiou et al. [10] provide in-depth discussion of the multidisciplinary nature of designing a UAV, beginning with mission specification and vehicle sizing to aerodynamic efficiency and alternative vehicle topologies for adequately maintaining components, while also maintaining flight stability.

In this work, we aim to address the complexity of the UAV design process by introducing a string-based grammar for design generation, coupled with a graph neural network surrogate model to heuristically estimate the flight performance of a candidate design, based on its drag force.

2.3 Artificial Intelligence in design space exploration

Artificial Intelligence in recent years has accelerated the design space exploration process, which includes both identification of feasible design candidates as well as heuristic performance estimation to determine which candidates should receive additional refinement with human intervention. As universal function approximators, neural networks can approximate functions that are typically costly to compute. One example of this is in physics-guided learning, where a model is trained to predict the behavior of a dynamical system such as a fluid dynamics simulation [18]. Rehman et al. [19] examines the use of learning drag coefficients of underwater obstacles using multilayer perceptrons (MLPs) and a hydrodynamics model. Viquerat et al. [20] explores the use of convolutional neural networks (CNNs) to predict drag forces on arbitrary 2D shapes in laminar low-Reynolds number flows across shapes generated with Bézier curves as well as NACA airfoils. Muralidhar et al. [21] develops a network architecture for predicting the drag forces on individual 3D particles, suspended in a moving fluid. A physics-guided loss function is introduced to ensure predictions from the proposed model do not violate known physical constraints. Sanchez-Gonzalez et al. [23] leverages graph neural networks (GNNs) to formulate "Graph Network-based Simulators" where the state of a physical system is represented by particles as nodes and learn the dynamics between them with message-passing. Ozdagli et al. [18] present a physics-guided learning architecture to develop an explainable surrogate model for evaluating the structural integrity of the hull of an underwater vehicle.

In UAV design, there are broadly two approaches for artificial intelligence in accelerating the design iteration process. The first is with design-level evaluation, which means that a design candidate has been produced and must be evaluated against its performance criteria (i.e., mission specification to move to a waypoint or deliver cargo). Results from system-level evaluations can inform iterative changes to be made to one or more components present on the existing candidate design, or otherwise help to identify additional directions to explore improvements in flight performance (e.g., use a larger number of lower capacity batteries, swap multiple smaller diameter propellers for a single larger diameter propeller, etc.). In general, component-level changes require fixed vehicle topology, while design level changes permit a change to the design topology. The challenge is to bridge the gap between the component- and system-level modifications. To this end, component-level design space exploration can allow for filtering from a potentially large space of possible components. Vardhan et al. [14, 17] pursues this direction for propellers by combining machine learning with a numerical simulation tool called OpenProp [40] to train a prediction model that can identify the geometry and efficiency of a propeller, given a performance requirement (e.g., desired RPM, required thrust force, etc.). A hybrid optimization approach is formulated to generate and evaluate baseline geometries of propellers from a geometric design space on the order of 10^{27} elements, describing the propeller chord profile, diameter, hub diameter, and a requirement space on the order of 10^{11} elements, describing the propeller thrust, vehicle velocity, and RPM using Random Forests and Decision Trees.

Additionally, Vardhan et al. [15, 16] proposes a surrogate model to estimate the drag forces on an underwater vehicle using a neural network, trained on outputs from a computational fluid dynamics (CFD) simulation. An end-to-end software pipeline is developed for a parametric underwater vehicle design to be synthesized to a 3D computer-aided design (CAD) model, which is then passed to an OpenFOAM [37] simulation to determine the drag forces of the vehicle. The results from this simulation then inform updates to an inner-loop optimization algorithm to improve the subsequent generated designs. This work closely parallels our goals in the UAV domain.

We seek a pipeline to rapidly generate designs that can be evaluated using a surrogate model trained on simulation data. From here, the simulation results inform changes to the vehicle at the component level (e.g. change a motor for a single propeller) or at the system level (alter the width of the vehicle, change propeller positioning), which require a human in the loop to iteratively make such changes.

To achieve this, we source data from the modeling software Creo [38], with the parameterized components shown in Figure 1. These components are combined to assemble into a variety of vehicles that are initially randomly generated to train the surrogate model predict the drag force on the vehicle. Additional designs are then generated from our string-based UAV design grammar and similarly evaluated. The learned surrogate model to predict the drag force is achieved using a deep graph convolutional neural network (DGCNN) in the graph regression setting, which takes as input a graph representing a vehicle and produces as output the affected drag areas (the *drag profile*) in the x, y, and z flight directions for the vehicle under analysis. In the next section, we describe our pipeline and problem formulation in detail.

3 UAV design pipeline

Figure 2 depicts an idealized end-to-end UAV design process pipeline consisting of two stages: Design generation and Heuristic evaluation. We first produce designs from the design grammar and convert them to the graph format to represent components and their connectivity. The CAD assembly pipeline consumes this graph and generates the 3D representation of this model, which is then fed through a drag simulation pipeline to generate training data. Once we have collected a dataset of pairs of generated vehi-

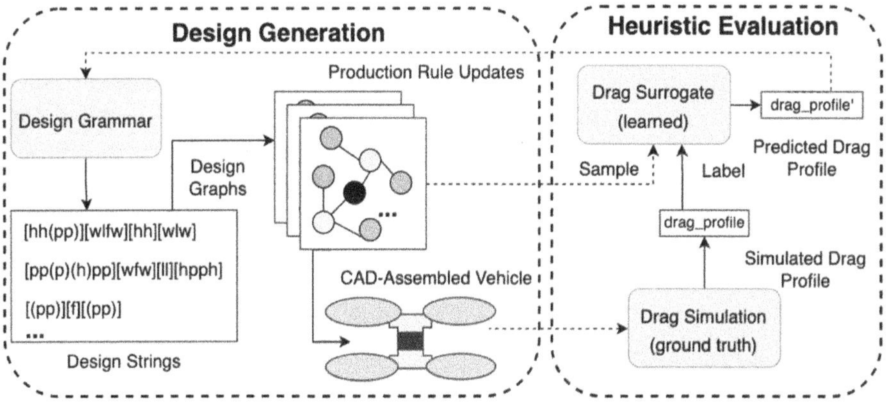

Figure 2: The proposed UAV design pipeline approach includes 2 main stages: (1) Design generation and (2) Heuristic evaluation. In stage (1), we produce designs from a design grammar and convert them into graphs to represent vehicle connectivity and component attributes. The assembled design (e.g., CAD model) is then provided as input to the drag simulation. In stage (2), the drag simulation computes the drag profile for the assembled UAV in the *x*, *y*, and *z* directions. The UAV design graph is the input sample and the ground truth drag profile is the label to train the GNN drag surrogate.

cle models and their drag profiles, we train a surrogate model that performs regression on the graph that represents the constructed vehicle. These predicted drag results can be further incorporated into the design generation step by feeding results back into the design grammar to modify production rules so that the generated vehicles adhere more closely to vehicles with a reduced drag profile. This can be further improved using a constrained optimization formulation or genetic algorithms, for example, to update the grammar rules from drag simulation results.

The space of possible UAVs represented by graphs with parameterized components as nodes and component connectivity as edges is very large. Moreover, the space of UAVs capable of stable flight is a small fraction of the possible designs in this graph-based space. Even without considering nongeometric parameters such as electrical or control parameters for manipulating the vehicle, the design space is combinatorial in the number of components that are present in the vehicle. For example, a vehicle containing n parameterized components has m inter-component connections (e.g., battery connected to fuselage, motor connected to propeller, etc.) as graph edges. Assume that each component must be connected to the vehicle, which gives a lower bound of $m_{min} = n - 1$ for the number of edges on the vehicle (e.g., some components must be connected to multiple others –a motor must be connected to a battery as well as a propeller, or a battery may be connected to multiple motors, etc.). Assume also that a UAV graph contains on average $n_{avg} = 20$ components, and that a UAV graph has at most one quarter of the number of possible edges connecting components, then $m_{max} \leq (n_{avg} * (n_{avg} - 1)) / 8$, which we argue is a reasonable assumption since vehicle components will be sparsely connected (i.e., have a few incident edges at most; the battery or fuselage will generally have the most connections because of the surrounding components or connection to multiple motors to obtain current from a single battery). The number of possible graphs that correspond to the described UAV graph space is given by $C(C(n_{avg}, 2), m_{max}) \approx 9.71^{44}$, where $C(x, y)$ represents the combinations given by x choose y. This space is even larger when considering the varying vehicle sizes, represented by the set of graphs with total number of edges ranging from m_{min} to m_{max}. The size of this space motivates our approach to rapidly identify feasible designs that are evaluated heuristically according to their drag force during flight.

3.1 Problem formulation

We consider UAVs that are represented by a graph converted from a string format in our design grammar. The graph contains nodes that encode properties of the represented components (e.g., a propeller node contains attributes indicating the propeller diameter, thrust, efficiency, propeller profile, etc.) with edges between components indicating how the components are connected. We note here that the component types are instantiated to concrete choices from our component corpus in the design graph, but they are abstractly represented in our design grammar. In other words, a design with two adjacent propellers at the front of the vehicle, according to the grammar interpretation rules, does not constrain which two propeller types must be used. However, when the design is converted to the graph format, a component selection (a concrete propeller instance, in this case) is required to assemble the vehicle. A designer could optionally randomly populate instances of components, based on the component type prescribed by the grammar, or heuristically identify component instances that fit with the design encoded by the design grammar.

Component-level studies to understand the relationships between a specific battery, motor, and propeller combinations can enhance the component selection process. These are collectively referred to as BMP (battery, motor, propeller) components and are often tested or analyzed in tandem. However, this analysis is out of the scope in this work and, here, we assume that the designer has already identified such groupings of these components. Naturally, one cannot pair any battery with any motor to current draw and motor rating requirements, and similarly a motor is constrained to specific propeller types, based on the torque that it can produce. These are design-level constraints, ideally addressed prior to the design generation phase. At the very least, a collection of batteries, motors, and propellers that pair well (i.e., are electrically and physically compatible) can drastically reduce the design space to be explored. However, it is worth noting that even with predefined collections of BMP components, there is still a combinatorial number of potential vehicle topologies to describe how these components are arranged in 3D space. An additional component to consider is the Electronic Speed Controller (ESC), which sends electrical signals to the motor from the battery, based on a required thrust force or available current. The ESC must interface with a controller that dictates how motors should behave under different conditions during flight. We do not consider the ESC component in our approach, but this component is often required to fine-tune the flight behavior of a selected vehicle, and introduces additional challenges in the design space exploration process.

$$\langle \textit{terminal-token} \rangle ::= \mathrm{h} \mid \mathrm{v} \mid \mathrm{l} \mid \mathrm{w} \mid \mathrm{f}$$

$$\langle \textit{non-terminal-token} \rangle ::= (\mid) \mid [\mid]$$

$$\langle \textit{comb-token} \rangle ::= \mathrm{h} \mid \mathrm{v} \mid \mathrm{l}$$

$$\langle \textit{fuselage-cluster} \rangle ::= [,\ \mathrm{w\ f,\ w,}\ ] \mid [,\ \{\langle \textit{comb-token} \rangle\},\ \mathrm{f},\ \{\langle \textit{comb-token} \rangle\},]$$

$$\langle \textit{sub-cluster} \rangle ::= (,\{\langle \textit{comb-token} \rangle\},\langle \textit{comb-token} \rangle,\ \{\langle \textit{comb-token} \rangle\},)$$

$$\langle \textit{wing-cluster} \rangle ::= \mathrm{w},\ \langle \textit{sub-cluster} \rangle,\ \mathrm{w} \mid \mathrm{w},\ [\mathrm{v}],\ \mathrm{w}$$

$$\langle \textit{comb-cluster} \rangle ::= [,\{\langle \textit{comb-token} \rangle\},\{\langle \textit{sub-cluster} \rangle\},\ \{\langle \textit{comb-token} \rangle\},\{\langle \textit{sub-cluster} \rangle\},\{\langle \textit{prop-cluster} \rangle\},]$$

$$\langle \textit{cluster} \rangle ::= \langle \textit{wing-cluster} \rangle \mid \langle \textit{comb-cluster} \rangle$$

$$\langle \textit{design} \rangle ::= \{\langle \textit{cluster} \rangle\},\langle \textit{fuselage-cluster} \rangle,\{\langle \textit{cluster} \rangle\}$$

Figure 3: The Extended Backus-Naur (EBNF) Form of our string-based UAV design grammar for generating UAV design candidates.

3.2 Design generation

Our design grammar operates over strings, whose production rules encode allowable component types, their implicit connection constraints, and the relative positioning of the components on the vehicle body. Figure 3 shows our string-based design grammar to generate UAV designs. Designs are constrained to be symmetric and can be interpreted as symmetric along the x, y, or z directions, which is a configurable parameter to the grammar interpreter, which produces a graph, given a design string that originates from the design grammar.

Given a generated UAV design string from the grammar, we form the corresponding graph by creating nodes for each of the n parameterized components in the design. The graph's m edges indicate which components are connected to which other components on the UAV.

In general, it is difficult to account for component overlap in generated designs: for example, how should the length of a connecting rod be adjusted if it interferes with the path of a spinning propeller? These unwanted component interactions are known as *interferences* and can prevent a vehicle from a successful flight during simulation. We mitigate these to the extent possible within our design grammar by providing a conservative length between each of the components of a generated vehicle. The optimization of these lengths can be completed at a further step in the design process during fine-tuning, when a promising or sufficient quality design has been identified.

We further note that the actuation of the UAVs is achieved with an Electronic Speed Controller (ESC) to coordinate the current required by the battery and translate the voltage into the appropriate electrical signals to alter the behavior of the motor of the rotating propellers. This is yet another additional step in the design process that we do not address in this work. The drag simulation that we use is provided by Walker et al. [24] as a Python library and requires a 3D model (e.g., STL file) of the vehicle as input. Again, the performance estimation of the generated designs is heuristic in nature and is motivated by the hypothesis that a low drag force on a vehicle correlates with relative ease of stable flight; we do not attempt to tune or optimize flight control parameters in the ESC in this work and instead focus on rapid design generation and evaluation with design grammars and surrogate models.

The design grammar aims to reconcile the conflicting objectives of maximizing the proportion of the design space explored, containing performant designs, and minimize the proportion of the design space explored, containing non-performant designs. For example, if a motor and propeller are known to be incompatible with one another, they should not be viable candidates for component selection in the generation of a candidate design. Similarly, if the total vertical thrust of a vehicle is larger than its normal force when the vehicle is at rest, the vehicle will not fly and should not be considered for construction. The exploration of performant and non-performant regions of the design space is directly affected by the constraints imposed on the design generation process. Here, we encode UAV designs as strings with relative positioning of symbols, corresponding to the component and how they are arranged and implicitly derived connections. For example, a propeller cannot exist without a motor, and a flange is required to seat a motor to hold a propeller. Therefore, a single propeller-motor-flange assembly can be indicated by a single symbol, either "*h*" or "*v*" for horizontal or vertical propeller and motor assemblies, respectively.

The design grammar shown in Figure 3 shows the production rules for our UAV design grammar. Broadly, the terminal and nonterminal tokens are organized into different cluster types, multiple of which comprise a single design. Large clusters are indicated with "["and "]", while subclusters contained within clusters are indicated with "("and ")". The leftmost cluster of any given string (i.e., the substring with matching "["and "]", closest to the beginning of the string) represents the front of the vehicle in 3D space. Similarly, the rightmost cluster represents the rear of the vehicle. Within a cluster, the subclusters can be interpreted from left to right, so that the leftmost subcluster within a cluster refers to the left side of that section of the vehicle (e.g., front, middle, or back). Components within a subcluster may extend either above or below the cluster in which they are contained. This notation allows us to specify forward and backward, side to side directions, and up and down locations for placing components on a vehicle. This also grants additional freedom in the interpretation of elements from the design grammar since adjacent components can be connected to either left or right connectors or top or bottom connectors as they are available.

Once a design string has been generated, it is augmented and transformed into a graph to be consumed by the assembly pipeline for drag simulation. The design interpreter consumes the generated strings and options for the component types to be used (e.g., a motor-propeller combination or a wing shape) as well as their metadata about connection ports and allowable connections (e.g., a wing can be connected from top to bottom or through its span; a propeller cannot have anything connect to its rotating blade; a battery must be contained in the fuselage housing to prevent damage during flight; each of the motors must be connected to the propellers they serve and the battery to obtain current). Where possible, spatial component attributes, such as length and width, are computed to obtain heuristic bounding boxes to prevent interferences when placing components in 3D space. There is a degree of flexibility inherent in the interpretation step, allowing for the placement of subclusters above or below their containing clusters, or grouped vs. alternating subclusters. Furthermore, a design string could be interpreted as having one, two, or more principal axes for connectors. This would allow for either more planar flat vehicle designs or more densely packed and taller designs. Examples of designs constructed from strings of our design grammar are shown in Figures 4–6.

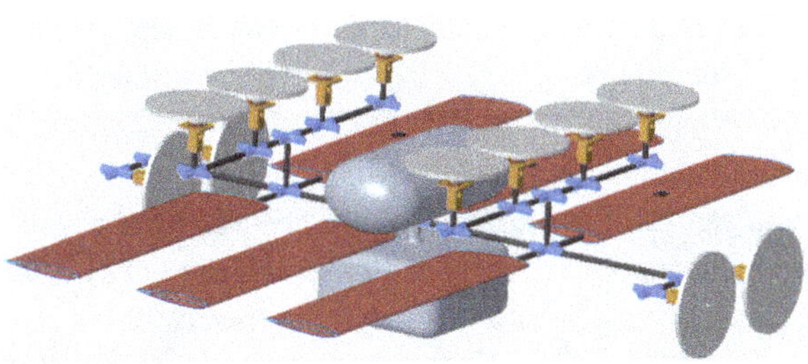

Figure 4: An example vehicle represented by the string "[hh][w(vvvv)w][wfw][w(vvvv)w][hh]". Adapted from [2]. Used with permission.

3.3 Drag force data collection

To train the surrogate drag model from drag simulations, we randomly generate UAV designs by iteratively adding a component to an open location that repeatedly jumps to a random connector that is open on the vehicle or otherwise, an available connector on the most recently added component. This assembly process is completed using Creo [38] along with a custom Python library that is built on the Creoson server tool [39]. Creo is required in the initial data generation phase to provide the inertial transforms and moments for the assembled vehicle, based on its components. Although

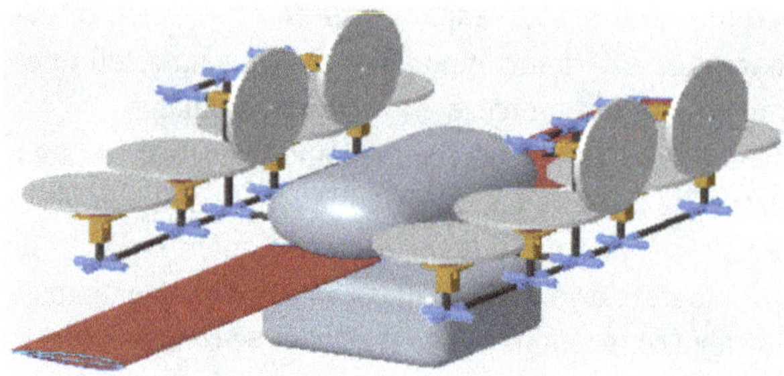

Figure 5: An example vehicle represented by the string "[vv(hh)vv][wfw][vv(hh)vv]". Adapted from [2]. Used with permission.

Figure 6: The process of assembling a UAV design, generated from our grammar. The fuselage and cargo container with wings are shown in the far-left image. The center image shows the vehicle after an initial propeller cluster is added to the vehicle, and the right image shows an identical cluster on the far side of the vehicle, to maintain symmetry. Adapted from [2]. Used with permission.

these designs are random, it can be argued that the model is forced to solver a harder problem than more structured and feasible designs: the irregularity in blockage areas of random designs can be more difficult to predict than uniformly constructed and more symmetric designs that are closer to being capable of flight.

The input to the drag model is the STL file representing the generated vehicle, its parameterized components, and their mass properties. The output of the drag model is the centers of drag acting on the input vehicle and the drag areas in each of the x, y, and z directions affected by the drag force, called the *drag profile*. The drag force on the vehicle is calculated as an intermediate quantity and can be derived from the simulation result. In essence, we aim to minimize the drag areas that are affected by drag force, given that the larger the area in each direction, larger will be the drag force acting on it. Note that to identify the blockage area for a UAV in 3D space is not straightforward, given that the overlapping components in the three axis directions create nontrivial overlapping geometry when projected into two dimensions.

$$F_D = \frac{1}{2}\rho v^2 C_D A$$

Equation 1: The drag force equation. We would like our model to learn the area A that represents the portion of the vehicle that experiences drag during flight, based on blockage areas derived from overlapping components along the axis of flight.

The goal of this pipeline is to identify and explore vehicles that have minimal drag areas in a direction of interest, guided by the intuition that lower drag force is exerted on a vehicle with a lower drag area in that direction. This is our motivation for the heuristic performance estimation of generated vehicles: vehicles that are both feasible and have minimal drag areas in the desired directions of flight should perform better than those with larger drag areas during flight. Examples of randomly generated designs and their blockage areas are shown in Figure 7.

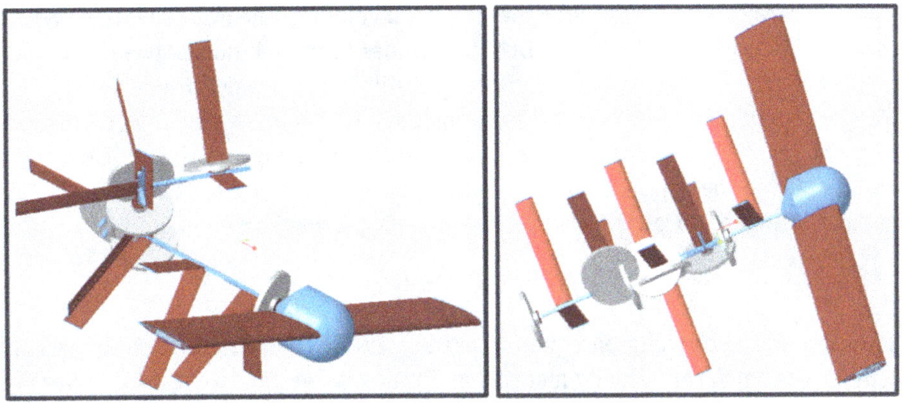

Figure 7: Randomly generated designs used to train the surrogate model on drag profile data, based on vehicle blockage areas.

3.4 Drag surrogate model

We learn a drag surrogate model of UAVs using graph learning, specifically, the Deep Graph Convolutional Network (DGCNN) [22]. This network was originally proposed to work in the graph classification setting, but here we adapt it to the regression setting for predicting drag profiles, given a representative graph of a UAV. Graph representation learning in recent years has gained momentum, owing to the powerful and general formalism for representing a wide range of real-world data and natural processes [13, 25]. Graph Neural Networks process graph structures to produce outputs for a problem domain, and work by learning graph structures through topology and node information. GNNs are commonly used in tasks such as graph classification, graph regression, link prediction, and community detection [26, 27]. We are concerned with the performance estimation of UAVs via their drag profiles which are real-valued vectors. Therefore, this is a graph regression problem.

Problem Statement: Let $G = \{G_1, \ldots, G_N\}$ be a set of graphs representing UAVs and $Y = \{y_1, \ldots, y_N\}$ be their corresponding drag profiles. Given G and Y, we aim to learn a representation vector h_G that helps in predicting $y_{G'}$ for an unseen graph G'.

The typical approach for graph-level learning involves aggregating extracted node-level features. However, this aggregation typically results in considerable information loss and, consequently, lower model performance. Especially in the graph regression problem, it is crucial to maintain as much node information as possible from the original graph. Retaining node-level information allows models to learn both local- and global-level information and achieve improved performance. Guided by this insight, Zhang et al. [22] proposes the Deep Graph Convolutional Neural Network (DGCNN), which includes a novel pooling layer to arrange the extracted node features in a consistent ordering, which allows for the use of convolutional neural networks (CNNs). Compared to traditional GNNs, the DGCNN includes 1D convolutional layers to extract expressive graph-level representations, resulting in impressive performance.
We leverage DGCNN in this work with a slight adaptation to for the graph regression task. The graph convolutional layers typically include message passing to learn-node features. Given the adjacency matrix of the input graph A and the feature matrix X containing node features, DGCNN uses the following form of convolution:

$$H = f\left(\tilde{D}^{-1}\tilde{A}XW\right)$$

Equation 2: The convolution equation used by DGCNN. Given a UAV design graph with its associated connectivity matrices and trainable weights, we learn a vector representation to predict drag profiles of the vehicle.

Where $\tilde{D}$ is a normalized degree matric for the input graph, $\tilde{A}$ is the adjacency matrix with self-loops, W is the matrix of trainable parameters, and H is the learned representation. Here, f can be any nonlinear function. As mentioned, DGCNN introduces a SortPooling layer, which imposes a consistent ordering of graph nodes so that traditional neural networks or CNNs can be applied. This is achieved using the Weisfeiler-Lehman (WL) coloring scheme to sort node features. Given H^l, where l is the last layer of GNN convolutions, the SortPooling layer first sorts the learned representation H row-wise in descending order and then sorts the graph nodes in the same order. To maintain scale invariance, the sorted features are further truncated, so only k percent of features are chosen to allow DGCNNs to impose ordering in feature space. After the SortPooling layer, the feature matrix is flattened and then two 1D convolutions are applied, followed by several MaxPooling layers. A fully connected layer with the desired activation is applied to produce the final output. We use GraphSAGE [28] instead of GCN [29], since it performs better in our case. We remove the Softmax operation from the final layer for the regression task.

4 Experimental results

We run experiments using the DGCNN architecture, described in the previous section, in the graph regression setting, providing UAV design graphs as input and predicting drag areas that affect the vehicle in flight. We use the publicly available DGCNN implementation in our experiments. Because of the limited data size, we consider a slim model for our experiments, with a 70:30 train-test split and L1 loss, with a learning rate of $1e^{-5}$. We use GraphSAGE convolution with three layers and 32 hidden channels. The number of neurons in the final fully connected layers is 416, 16, and 1, respectively. MinMax normalization is applied on the labels (i.e., predicted drag profiles) and the model is trained for 200 epochs. The performance of the model is described by its loss curve over the training epochs. We find decreasing loss in training and test sets, indicating the model can learn to predict drag profiles from the input UAV design graphs representing the vehicle design geometries. Figures 8–10 illustrate the differences in performance of the model when altering the number of hidden channels in the network, which represents the dimensionality of the learned feature vectors.

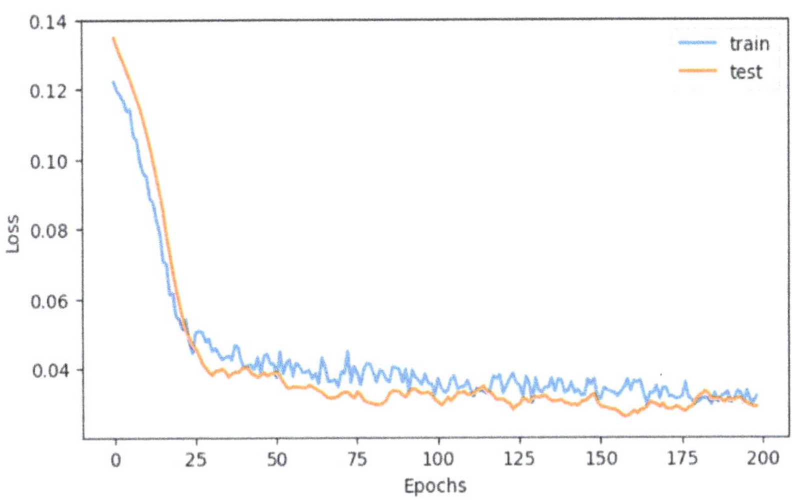

Figure 8: Train and test loss curve from training DGCNN with our UAV vehicle data for 200 training epochs, while retaining $k = 0.6$ of features, each represented by 16 hidden channels.

We observe that the model can successfully predict drag profiles from the provided UAV design graphs, indicated by the decreasing loss curves in the above figures. When adjusting the number of hidden channels that represent the dimensionality of the learned feature vectors, we find that the model overfits on the training data with 64 hidden channels and generalizes to the test data better with hidden channels of 32 and 16. Additional architectural and parameter variations may yield improved model

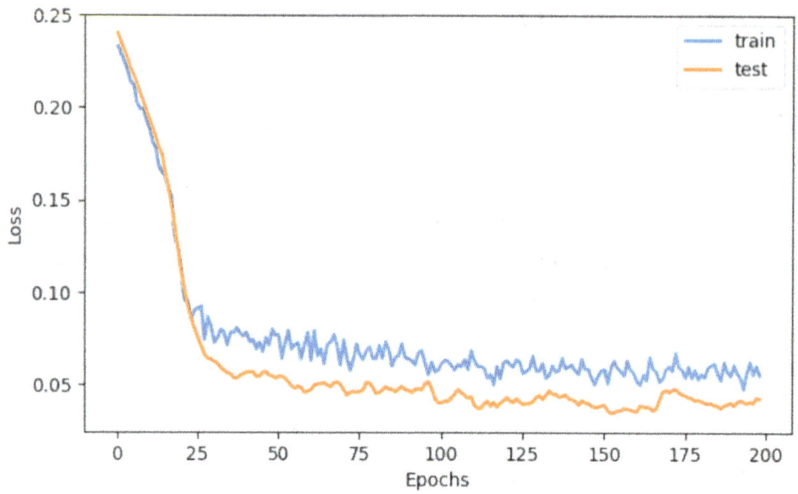

Figure 9: Train and test loss curve from training DGCNN with our UAV vehicle data for 200 training epochs, while retaining k = 0.6 of features, each represented by 32 hidden channels.

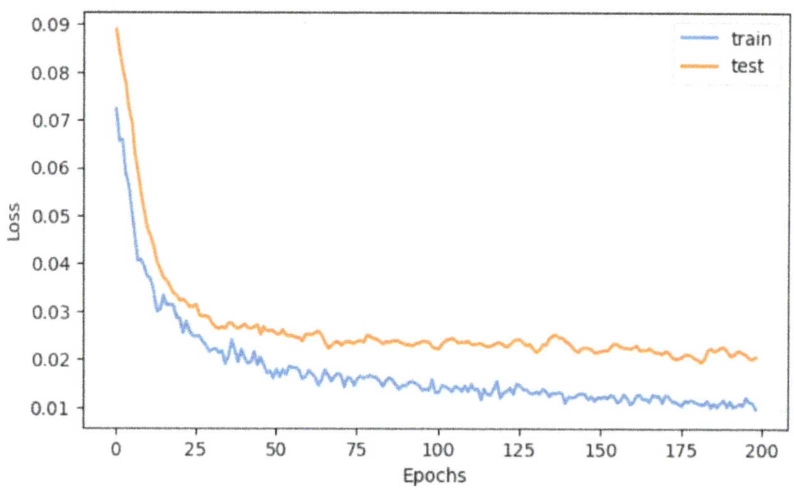

Figure 10: Train and test loss curve from training DGCNN with our UAV vehicle data for 200 training epochs, while retaining k = 0.6 of features, each represented by 64 hidden channels.

performance with increased data size, but we do not explore such modifications in this work. We now discuss additional approaches to UAV design generation and evaluation, based on recent advances in Large Language Models (LLMs).

5 Large language models in the design process

In recent months, Large Language Models (LLMs) have garnered substantial attention from the scientific community for their ability to predict the next word in a sequence, given a natural language input prompt [12, 30–32, 36]. Interaction with LLMs is achieved using an input *prompt*, which is the set of text instructions provided to a language model to steer output generation [33, 34]. A prompt is fed as input to an LLM to produce a *response*, which contains the continuation of the input sequence by iteratively predicting the next tokens in the sequence to construct an output. Typically, the prompt that is input to an LLM is tightly coupled to the quality of the response that is produced by the LLM. The process of refining input prompts given to an LLM to elicit the desired structural and qualitative characteristics is known as *prompt engineering* [33].

In addition to standard question answering and text completion tasks, language models have been shown to work well for domain-specific tasks such as medical knowledge retrieval [35] and 3D model generation [12]. Considering this recent progress, we introduce potential applications of LLMs in the design space exploration phase of the UAV domain.

5.1 Domain-specific language generation

A domain-specific language (DSL) is tailored to a specific application domain to describe and reason about elements of a design space or a specific class of problems. For example, regex is a domain-specific language for reasoning about the structure and patterns that can be found in text. Similarly, SQL is a domain-specific language for querying and presenting results from a tabular-structured database. In general, domain-specific languages enable a level of abstraction, consistent with the problem domain. The design grammar presented in Section 3.2 can be considered a domain-specific language for describing UAV designs with text. LLMs can also be prompted to produce or modify domain-specific languages or to write programs using a user-provided DSL.

5.2 Component generation

A UAV, like any cyber-physical system, requires the generation of constituent parameterized components (e.g., propellers, etc.) to be connected to represent an overall vehicle or design candidate. In this work, we leverage an existing component corpus provided by [23]. However, the need for novel components or refinement of existing ones for specific UAV use cases may be required. In this case, an LLM can provide feedback on how to generate or modify component models according to a performance specification.

5.3 Component positioning alternatives

Once a candidate design is identified, it is not always clear how to alter the design to improve its flight performance or there may be limited context about how to make a design modification at different stages of the design pipeline. An LLM can be prompted to suggest how a component position should be adjusted, relative to stated goals of the design under consideration. For example, a UAV designed to carry cargo might benefit from having a wing closer to the front of the cargo container to reduce drag. A drone may have propellers that should be closer together, by shortening a connecting rod to reduce turbulence. Even the slightest change to a component's position can drastically alter the aerodynamic profile of a design under consideration. In this case, an LLM may present advice about how to modify a component layout, given information about the vehicle performance with the current layout.

5.4 Design generation co-pilot

Given that the capabilities of LLMs continue to evolve rapidly, a final realistic application would be their use as a co-pilot in the design process. An LLM can be prompted with a human-crafted design grammar, with modifications requested based on a target task (e.g. "this vehicle should be able to take off vertically" or "this vehicle should have thrust generated from a small number of propellers") to steer the design generation process. Additionally, an LLM can be fed heuristic performance scores along with a representation of the vehicle topology and solicited for modifications, based on the heuristic performance and a stated improvement goal (e.g., "this vehicle scores 6/10 since the wing drag is too large, how can I minimally adjust the design to improve its score by only modifying the wing geometry?"). Once a desirable workflow has been identified, the LLM can be repeatedly queried for improvements to handcrafted or heuristically performant designs. This approach may reduce the burden on the human designer to produce design candidates by freeing up time that can be spent on improving the heuristic evaluation framework.

6 Conclusion

We investigate efficient design space exploration of Unmanned Aerial Vehicles (UAVs) using a UAV design grammar and a neural network-based drag surrogate model. We devise a string-based grammar for generating candidate UAV designs based on component positioning and vehicle topology. We curate a dataset of randomly generated designs and curate a dataset of 3D models of UAVs as well as their drag profiles, collected from simulation. We then train a graph neural network (GNN) on a dataset of representative UAV design graphs that incorporates the components of the UAV and how they are connected (training example) and the associated drag profiles that are derived from simulation that requires the 3D model of the vehicle as input (training label). We present results from training our drag surrogate model and discuss how it can be combined with our design grammar or other design generation processes to iteratively refine or modify candidate designs, based on the desired performance. Finally, given the recent developments in Large Language Models (LLMs), we discuss possible applications of LLMs in the design space exploration process for UAVs, specifically for the generation and refinement of domain-specific languages (DSLs), components for UAV construction, suggestions for component positioning of a candidate design, and as a co-pilot for a human designer and refining known-good designs that can be improved. We expect additional design generation process (e.g., genetic and evolutionary algorithms) as well as neural network-based surrogates to play a crucial role in accelerating the design exploration phase of UAVs and other domains, and to avoid requiring resource-intensive simulation pipelines and significant human effort at early stages of the design process, before feasible designs are identified.

References

[1] M. Sandborn *et al.*, "What a Drag! Streamlining the UAV Design Process with Design Grammars and Drag Surrogates," *2022 International Conference on Computational Science and Computational Intelligence (CSCI)*, pp. 279–283, 2022.

[2] C. Olea, M. Sandborn, P. Volgyesi and J. White, "String Grammars for Preliminary UAV Design Exploration," *International Conference on Mechanical and Aerospace Engineering (ICMAE)*, 2023.

[3] N. Varsha and V. Somashekar, "Conceptual Design of High-Performance Unmanned Aerial Vehicle," in *IOP Conference Series: Materials Science and Engineering*, vol. 376. IOP Publishing, p. 012056, June 2018, doi: 10.1088/1757-899x/376/1/012056.

[4] B. Song, N. F. Soria Zurita, H. Nolte, H. Singh, J. Cagan and C. McComb, "When Faced with Increasing Complexity: The Effectiveness of Artificial Intelligence Assistance for Drone Design," *ASME Journal of Mechanical Design*, vol. 144, no. 2, p. 021701, 9 September 2021, February 2022, https://doi.org/10.1115/1.4051871.

[5] P. Mallozzi *et al.*, "A Grammar for the Representation of Unmanned Aerial Vehicles with 3D Topologies," *arXiv [Cs.RO]*, 2023, http://arxiv.org/abs/2302.13980. ArXiv.

[6] A. Zhao, J. Xu, M. Konaković-Luković, J. Hughes, A. Spielberg, D. Rus and W. Matusik, "RoboGrammar: Graph Grammar for Terrain-optimized Robot Design," *ACM Transactions on Graphics*, vol. 39, no. 6, p. 16, Article 188 December 2020, https://doi.org/10.1145/3414685.3417831.

[7] F. R. Stöckli and K. Shea, "A Simulation-Driven Graph Grammar Method for the Automated Synthesis of Passive Dynamic Brachiating Robots," in *Proceedings of the ASME 2015 International Design Engineering Technical Conferences and Computers and Information in Engineering Conference. Volume 7: 27th International Conference on Design Theory and Methodology*. Boston, Massachusetts, USA, 2–5 August 2015, V007T06A017, ASME, https://doi.org/10.1115/DETC2015-47641.

[8] A. Zhao *et al.*, "Graph Grammar-Based Automatic Design for Heterogeneous Fleets of Underwater Robots," *2022 International Conference on Robotics and Automation (ICRA)*, pp. 3143–3149, 2022.

[9] K. Sims, “Evolving Virtual Creatures,” *Proceedings of the 21st Annual Conference on Computer Graphics and Interactive Techniques, Association for Computing Machinery*, pp. 15–22, 1994, https://doi.org10.1145/192161.192167. SIGGRAPH ’94.

[10] A. Papageorgiou, “Design Optimization of Unmanned Aerial Vehicles: A System of Systems Approach,” in *Linköping Studies in Science and Technology*. Dissertations, Linköping University Electronic Press, 6 Dec. 2019, doi: 10.3384/diss.diva-161915.

[11] S. A. H. Mohsan *et al.*, “Unmanned Aerial Vehicles (UAVs): Practical Aspects, Applications, Open Challenges, Security Issues, and Future Trends,” *Intelligent Service Robotics*, vol. 16, no. 1, pp. 109–137, 2023, doi: 10.1007/s11370-022-00452-4.

[12] Y. Siddiqui *et al.*, “MeshGPT: Generating Triangle Meshes with Decoder-Only Transformers,” *arXiv [Cs.CV]*, 2023, http://arxiv.org/abs/2311.15475. ArXiv.

[13] A. Merchant, S. Batzner, S. S. Schoenholz *et al.*, “Scaling Deep Learning for Materials Discovery,” *Nature*, 2023, https://doi.org/10.1038/s41586-023-06735-9.

[14] H. Vardhan *et al.*, “Fusion of ML with Numerical Simulation for Optimized Propeller Design,” *arXiv [Cs.LG]*, 2023, http://arxiv.org/abs/2302.14740. ArXiv.

[15] H. Vardhan *et al.*, “Sample-Efficient and Surrogate-Based Design Optimization of Underwater Vehicle Hulls,” *arXiv [Cs.LG]*, 2023, http://arxiv.org/abs/2304.12420. arXiv.

[16] H. Vardhan and J. Sztipanovits, “Search for Universal Minimum Drag Resistance Underwater Vehicle Hull Using CFD,” *arXiv [Cs.CE]*, 2023, http://arxiv.org/abs/2302.09441. ArXiv.

[17] H. Vardhan *et al.*, “Machine Learning Assisted Propeller Design,” *Proceedings of the ACM/IEEE 12th International Conference on Cyber-Physical Systems, Association for Computing Machinery*, pp. 227–228, 2021, https://doi.org10.1145/3450267.3452001. ICCPS ’21.

[18] A. Ozdagli *et al.*, “Surrogate Modeling Using Physics-Guided Learning,” *Proceedings of Cyber-Physical Systems and Internet of Things Week 2023, Association for Computing Machinery*, pp. 130–135, 2023, https://doi.org10.1145/3576914.3587532. CPS-IoT Week ’23.

[19] K. U. Rehman, A. B. C¸ olak and W. Shatanawi, “Artificial Neural Networking (ANN) Model for Drag Coefficient Optimization for Various Obstacles,” *Mathematics*, vol. 10, no. 14, 2022, [Online]. Available: https://www.mdpi.com/2227-7390/10/14/2450.

[20] J. Viquerat and E. Hachem, “A Supervised Neural Network for Drag Prediction of Arbitrary 2d Shapes in Laminar Flows at Low Reynolds Number,” *Computers and Fluids*, vol. 210, p. 104645, 2020, [Online]. Available: https://www.sciencedirect.com/science/article/pii/S0045793020302164.

[21] N. Muralidhar, J. Bu, Z. Cao, L. He, N. Ramakrishnan, D. Tafti and A. Karpatne, “Physics-guided Design and Learning of Neural Networks for Predicting Drag Force on Particle Suspensions in Moving Fluids,” 2019, [Online]. Available: https://arxiv.org/abs/1911.04240.

[22] W. Peng, Y. Zhang, E. Laurendeau and M. C. Desmarais, “Learning Aerodynamics with Neural Network,” *Scientific Reports*, vol. 12, no. 1, Apr. 2022, [Online]. Available: https://doi.org/10.1038/s41598-022-10737-4.

[23] A. Sanchez-Gonzalez, J. Godwin, T. Pfaff, R. Ying, J. Leskovec and P. W. Battaglia, “Learning to Simulate Complex Physics with Graph Networks,” 2020, [Online]. Available: https://arxiv.org/abs/2002.09405.

[24] J. D. Walker, F. Michael Heim, B. Surampudi, P. Bueno, A. Carpenter, S. Chocron, J. Cutshall, R. Lammons, T. Bapty, B. Swenson and S. Whittington, “A Flight Dynamics Model for Exploring the Distributed Electrical Evtol Cyber Physical Design Space,” *2022 IEEE Workshop on Design Automation for CPS and IoT (DESTION)*, pp. 7–12, 2022.

[25] M. Zhang, Z. Cui, M. Neumann and Y. Chen, “An End-to-end Deep Learning Architecture for Graph Classification,” *Proceedings of the AAAI Conference on Artificial Intelligence*, vol. 32, no. 1, 2018.

[26] Z. Wu, S. Pan, F. Chen, G. Long, C. Zhang and S. Y. Philip, “A Comprehensive Survey on Graph Neural Networks,” *IEEE Transactions on Neural Networks and Learning Systems*, vol. 32, no. 1, pp. 4–24, 2020.

[27] A. Said, S.-U. Hassan, W. Abbas and M. Shabbir, "Netki: A Kirchhoff Index Based Statistical Graph Embedding in Nearly Linear Time," *Neurocomputing*, vol. 433, pp. 108–118, 2021.
[28] W. L. Hamilton *et al.*, "Inductive Representation Learning on Large Graphs," *arXiv [Cs.SI]*, 2018, http://arxiv.org/abs/1706.02216. ArXiv.
[29] T. N. Kipf and M. Welling, "Semi-supervised Classification with Graph Convolutional Networks," *arXiv Preprint arXiv:1609.02907*, 2016.
[30] H. Touvron *et al.*, "LLaMA: Open and Efficient Foundation Language Models," *arXiv [Cs.CL]*, 2023, http://arxiv.org/abs/2302.13971. ArXiv.
[31] A. Chowdhery *et al.*, "PaLM: Scaling Language Modeling with Pathways," *arXiv [Cs.CL]*, 2022, http://arxiv.org/abs/2204.02311. ArXiv.
[32] B. Zoph *et al.*, *Emergent Abilities of Large Language Models*. TMLR, 2022.
[33] J. White *et al.* A Prompt Pattern Catalog to Enhance Prompt Engineering with ChatGPT, 2023, http://arxiv.org/abs/2302.11382. ArXiv.
[34] Y. Zhou *et al.*, "Large Language Models Are Human-Level Prompt Engineers," *arXiv [Cs.LG]*, 2023, http://arxiv.org/abs/2211.01910. ArXiv.
[35] T. Dave *et al.*, "ChatGPT in Medicine: An Overview of Its Applications, Advantages, Limitations, Future Prospects, and Ethical Considerations," *Frontiers in Artificial Intelligence*, vol. 6, p. 1169595, 4 May 2023, doi: 10.3389/frai.2023.1169595.
[36] L. De Angelis *et al.*, "ChatGPT and the Rise of Large Language Models: The New AI-driven Infodemic Threat in Public Health," *Frontiers in Public Health*, vol. 11, p. 1166120, 25 Apr. 2023, doi: 10.3389/fpubh.2023.1166120.
[37] H. G. Weller, G. Tabor, H. Jasak and C. Fureby, "A Tensorial Approach to Computational Continuum Mechanics Using Object-oriented Techniques," *Computers in Physics*, vol. 12, no. 6, Nov/Dec. 1998.
[38] https://www.ptc.com/en/products/creo.
[39] http://www.creoson.com/.
[40] https://openprop.engineering.dartmouth.edu/.

Paula Lauren

Improving subword embeddings in large language models using morphological information

Abstract: Subword embeddings are integral to Large language Models (LLMs), such as the family of LLMs made available from Generative Pre-trained Transformers (GPT). One of the challenges with subword tokenization is determining the optimal set of tokens for effectively representing words. A subword tokenization algorithm that can reduce the number of subtokens for a word and capture morphological information would be advantageous. However, fully implementing and evaluating such an algorithm poses many challenges, including the necessity for high-performance GPUs, a distributed computing infrastructure, extensive bandwidth and networking capabilities, significant storage capacity, substantial RAM, and considerable energy resources. This research circumvents the aforementioned computational challenges by utilizing already trained GPT subtoken embeddings, in proposing a refined tokenization approach for improving subtokens for words. The evaluation of the proposed approach is done using analogy datasets, with emphasis on inflectional and derivational morphology tasks. In comparison to the original GPT subword embeddings, the proposed approach shows an overall improvement in reducing subword tokens in words and an overall improvement in the analogy tasks.

Keywords: subword embeddings, GPT, analogy

1 Introduction

The advent of Large Language Models (LLMs) represents a significant milestone in Artificial Intelligence (AI) and Natural Language Processing (NLP). These models have brought about transformative changes and opened up new possibilities. LLMs are capable of understanding and generating human-like text, enabling them to perform a wide range of language tasks, such as translation, summarization, question-answering, and creative writing [1–6]. Two popular LLMs are Generative Pre-trained Transformer (GPT) [7] and Bidirectional Encoder Representations from Transformers (BERT) [8]. The core technological achievement of these models can be attributed to the primary architectural aspect of transformers, a concept referred to as *attention* [9]. Attention enables

Paula Lauren, Lawrence Technological University, Southfield, Michigan, USA, e-mail: plauren@ltu.edu

https://doi.org/10.1515/9783111344126-015

the neural network to learn contextualized word representations by permitting continuous input selection [10]. Parallel learning of contextual information is done using multi-head attention as shown in Figure 1 from previous work [11].

The GPT and BERT transformer architectures rely on token embeddings for representing words or segments of a word, also known as subword embeddings, which are foundational to LLMs [12]. Subword embeddings are word vectors constructed from parts of the word and their associated subword co-occurrences from a training corpus. The advent of subword embeddings resolves a shortcoming with word embeddings, known as the Out-of-Vocabulary (OOV) problem [13]. The OOV problem occurs when a word is encountered in a downstream NLP task but does not have a corresponding word vector, due to the word not encountered during training, which often occurs with rare words [14]. The decomposition of a word into parts or subwords addresses rare words since these unique words can be reconstructed from their individual word segments. BERT uses WordPiece to deconstruct words, and results in a vocabulary consisting of 30 thousand words and subwords [15]. WordPiece first initializes the vocabulary to include every character that is available in the training data and then learns a given number of merge rules. GPT is based on Byte Pair Encoding (BPE) [16], which was influenced by an early data compression algorithm for replacing common pairs of bytes by single bytes [17]. BPE builds an all-inclusive vocabulary for representing any word by using variable-length character sequences, resulting in a fixed-size vocabulary consisting of about 50 thousand words and word segments or subwords.

Both BPE and WordPiece are statistical tokenization methods that identify common patterns and frequent sequences from the training data. However, in capturing morphological information, both approaches are somewhat limited, especially when compared with methods that are explicitly designed for morphological analysis.

Providing morphological insights in NLP can offer several significant benefits, especially in tasks that require nuanced understanding of language structure and meaning by providing an enhanced understanding of word forms, such as with inflectional and derivational morphology. Inflectional morphology understands how words change form to convey different grammatical aspects (such as tense, case, number, and gender) and can improve the accuracy of language processing tasks like part-of-speech tagging, syntactic parsing, and semantic analysis. Derivational morphology recognizes the formation of words and how they are related to each other (through prefixes, suffixes, and roots), which could enhance the model's understanding of word meanings and relationships. This understanding could assist in tasks like word sense disambiguation and semantic similarity assessment.

Another challenge with tokenization of words is determining the optimal set of subtokens for effectively representing words. This challenge arises from the need to balance various factors to ensure efficient and effective language processing, especially in the context of training and using LLMs. However, there are challenges with full implementation of a refined tokenization algorithm for improving subwords, such as the need for high-performance GPUs, distributed computing infrastructure, high-bandwidth and networking, large storage capacity, substantial RAM, and energy resources. This research proposes the inclusion of morphological information, and circumvents the aforementioned computational challenges by utilizing already trained GPT subtoken embeddings. The proposed approach reconstructs words, with the objective of subtoken minimization and refinement of subword tokens in the reconstruction of the word.

In this chapter, a proposed tokenization algorithm for creating subword embeddings, incorporating morphological information and building upon GPT subword tokens is described. The literature survey of subword embeddings, along with limitations as well as subword information in word representations, are described in Section 2. Subword token analysis using GPT subword tokens is described in Section 3. The proposed work is explained in Section 4, which includes the core algorithms in this study. Analysis and evaluation of the proposed work is described in Section 5. Conclusion is in Section 6 and future work is proposed in Section 7.

2 Related work

Subword information can be discovered in subword embeddings with GPT and BERT subword embeddings but this occurrence is often unintentional. It is often the case that subword information is methodically incorporated, as opposed to accidentally discovered, especially in arriving at word representations. Both approaches are described below.

2.1 GPT subword embeddings

GPT models captures some level of subword information. However, their capabilities in this regard are more statistical and context-driven, rather than based on explicit linguistic rules or deep morphological understanding. Their performance in these areas largely depends on the nature of their training. Despite its effectiveness in NLP, the use of BPE for deriving GPT subword embeddings has several drawbacks [18, 19]. It is less efficient for languages with complex morphologies, such as agglutinative languages (e.g., Finnish, Turkish, and Arabic). Additionally, BPE's context-insensitive approach to creating subwords can lead to suboptimal segmentations that do not fully capture syntactic or semantic nuances. There are also issues with tokenization inconsistencies, where the same word might be tokenized differently in varying contexts, potentially undermining the consistency of a model's understanding. Lastly, BPE can result in longer sequences for some words, increasing the computational complexity and impacting the efficiency of models, especially those with fixed maximum sequence lengths.

These limitations highlight the need for careful implementation and possible modifications of BPE in diverse linguistic and domain-specific applications. But despite the challenges, BPE in GPT models performs, overall, remarkably well in text generation tasks.

2.2 BERT subword embeddings

BERT models use WordPiece for subword tokenization, an approach that shares some similarities with GPT's use of BPE but also has distinct differences. Like GPT, BERT's subword processing is more statistical and context-driven, rather over using explicit linguistic rules or deep morphological understanding. WordPiece considers frequency as well as the language model's performance when creating subwords. WordPiece also has drawbacks despite its effectiveness [20, 21]. It tends to also struggle with languages that have complex morphologies, such as agglutinative languages due to potential suboptimal word segmentation. This method may produce longer sequences for certain words, especially more complex ones, leading to increased computational demands, particularly in models with fixed maximum sequence lengths. The fixed vocabulary size set during training limits WordPiece's adaptability to new words or evolving language usage. Lastly, WordPiece's approach to handling rare words by breaking them into smaller parts might not always effectively represent their full meaning or context. Despite the challenges, BERT's WordPiece provides remarkable performance in tasks requiring an understanding of language context, such as question answering and sentence classification.

2.3 Reconstructing word embeddings using subword information

Subword information has been included in the reconstruction of word embeddings using various distance measures such as Manhattan and Euclidean distances [22] and characters [23]. Early work on morphologically infused word embeddings incorporate the integration of recursive neural networks (RNNs) with neural language models [24]. Character n-grams have also been useful in extending pretrained word embeddings, which also captures morphological information [25]. Modeling subword information using a probabilistic weighting scheme incorporating pretrained word embeddings has resulted in improved subword decompositions and subword ranking [26].

3 Analysis of GPT subword tokens

Subword tokenization involves breaking down words into smaller units, known as subwords. A subword represents a subset of a word, and a word itself can be decomposed into one or more such subwords, as well as individual characters. As stated previously, subword embeddings are vector representations of the subword, which is constructed form subword co-occurrences from the training data. This approach to tokenization is particularly useful for managing out-of-vocabulary words and further reducing the size of the vocabulary that a model needs to handle. An example of a result from a suboptimal merge with GPT can be understood with the words *turbo-charge* and *turbomachine*. The tokenization of the first word is represented as [*'tur'*, *'bo'*,*'charge'*]. The word segment *turbo* is further decomposed as the two subwords *tur* and *bo*. Exploring other words with the *turbo* prefix highlights the drawback of the greedy algorithmic approach in the merge rules for generating GPT subword tokens. The tokenization of the word *turbomachine* by GPT is [*'t'*, *'urb'*,*'omach'*,*'ine'*]. This is one of the limitations of the GPT tokenization algorithm, which can generate subwords based on suboptimal merges. The vocabulary for GPT contains 50,257 tokens, which includes words, subwords, and characters. The tokens that make it into the vocabulary are determined by the frequency of their subparts in the training corpus. Worth noting that the word *turbo* is presented 16 times in the NLTK word corpus and primarily used as a prefix to other words. It suffices to say that words beginning with *turbo* are a rather rare occurrence and typically would require further tokenization into subwords from GPT.

In this study, the *nltk.corpus.words* available from Natural Language ToolKit (NLTK) is used to distinguish the word list utilized in previous work [11]. The word list in this study is essentially a list of mostly dictionary words from NLTK, a popular Python library working with English language data. As a preliminary analysis of subtokens, the GPT tokenizer was applied to the list of 234,892 words. Table 1 reports the number of words associated with the number of tokens (up to 10 tokens) and shows that the majority of words are comprised of four or less tokens. About 40% of the words in the dictionary contain three tokens. There are 14,117 words in the dataset that require only one token, which is the word itself, meaning that the word existed in GPT's vocabulary and did not necessitate further word segmentation. This preliminary analysis on the NLTK word corpus will serve as a baseline for further comparison.

A sample of 10 words taken from the NLTK word corpus for evaluation of the subwords generated from the GPT tokenizer is reported in Table 2. The interesting aspect of the NLTK word corpus is that it contains a significant number of unusual words, which makes it especially interesting in conducting experimentation with different subword tokenization approaches. The tokenization of the word *monosymmetrical* by GPT contains no morphological information, though *mono* is a well-known prefix in the English language. The subword tokens for the word *superrationally* in Table 2 correctly identifies the *super* prefix as well as *post* in the word *postpathological*. It is in-

Table 1: Tokens generated for NLTK word corpus using GPT Tokenizer.

Number of tokens	Number of words
1 token	14,117
2 tokens	52,233
3 tokens	95,559
4 tokens	52,667
5 tokens	16,501
6 tokens	3,905
7 tokens	739
8 tokens	137
9 tokens	31
10 tokens	3

teresting that the GPT tokenizer did not tokenize the prefix *in* for the word *inextinct.* As noted earlier on limitations of BPE in GPT, the approach does not incorporate morphological information in the tokenization process and if subtokens appear to capture morphological information, it is due to frequency not intentionality. Another uncommon word is *cinnamonwood* and it appears that the word *cinnamon* is not present in GPT, necessitating a more granular breakdown of the word as two subtokens. However, the word *cinnamon* does exist in the GPT vocabulary, as later discovered in this chapter.

Table 2: Example of subwords generated for words using GPT Tokenizer.

Word	Subwords
monosymmetrical	['mon', 'os', 'ymm', 'etrical']
unliveableness	['un', 'live', 'abl', 'eness']
stereotelemeter	['st', 'ere', 'ote', 'lem', 'eter']
progressiveness	['progress', 'iveness']
emanational	['eman', 'ational']
cinnamonwood	['c', 'innamon', 'wood']
inextinct	['ine', 'xt', 'inct']
superrationally	['super', 'ration', 'ally']
postpathological	['post', 'path', 'ological']
lithophotography	['l', 'ith', 'oph', 'ot', 'ography']

4 Proposed work

4.1 Reconstructing word representations

The analysis of the subword tokens in Table 1 reports that most of the words contained in the NLTK word corpus consists of more than one token. For words with one token, the corresponding word vector from GPT is all that would be needed for the word representation and the corresponding word vector. For words containing more than one token, it would necessitate a reconstruction of the word representation. Reconstructing word embeddings from subword embeddings requires consolidating the individual subword embeddings to arrive at a singular embedding for the word. This is done by matching up each subtoken of the subword with the corresponding subword vector in GPT to arrive at a composite vector to represent the entire word. The composite vector can be achieved by summing up the individual subword vectors, or averaging the vectors. In this research, the latter is done, where each successive match results in an averaging of the existing subword vectors, to arrive at the word embedding for the entire word.

4.2 Proposed algorithm for decomposing a word into subwords

The described procedure, FindSubwords, is an algorithm designed to break down a given word into its subword components, giving priority to prefixes and suffixes. Suffixes and prefixes are global lists containing well-known affixes. GPTtokenDict is a global dictionary containing the GPT tokens, along with the associated index for subword vector identification in GPT.

```
FindSubwords algorithm: Decompose a word into subwords.
procedure findSubwords(word)
  if word in GPTtokenDict then
     return [(word, GPTtokenDict[word])]
  end if
  original_word ← word
  prefix_part ← ''
  suffix_part ← ''
  for prefix in prefixes do
    if word.startswith(prefix) then
       prefix_part ← prefix
       word ← word[len(prefix):]
       break
    end if
```

```
    end for
    for suffix in suffixes do
      if word.endswith(suffix) then
         suffix_part ← suffix
         word ← word[:-len(suffix)]
         break
      end if
    end for
    result ← [ ]
    if prefix_part then
       result.append((prefix_part, GPTtokenDict[prefix_part]))
    end if
    remaining_word ← original_word[len(prefix_part):]
    result.extend(reduce_and_lookup(remaining_word, original_word))
    if suffix_part and not result[-1][0].endswith(suffix_part) then
       result← result + reduce_and_lookup(suffix_part, original_word)
    end if
    return result
end procedure
```

The FindSubwords algorithm is designed to decompose a given word into its constituent parts, based on predefined lists of prefixes and suffixes, and a dictionary (GPTtokenDict). This decomposition helps in understanding the structure of the word. The following steps describes the details of the FindSubwords algorithm:

4.2.1 Initial check in dictionary

The algorithm begins by checking if the entire word is present in GPTtokenDict. If the word is found, it returns a list containing a single tuple with the word and its corresponding index from GPTtokenDict.

4.2.2 Store original word

The original form of the word is stored in original_word for later use.

4.2.3 Prefix extraction

The algorithm iterates through a list of predefined prefixes. If the word starts with a prefix, then the algorithm stores this prefix in prefix_part and then removes it from the beginning of the word.

4.2.4 Suffix extraction

A similar process is applied to identify any suffix in the modified word. Again, ProcessAffix is used to potentially further modify the word and extract the suffixPart. Similarly, it iterates through a list of predefined suffixes. If the word ends with a suffix, this suffix is stored in suffix_part, and the algorithm removes it from the end of the word.

4.2.5 Processing the remaining word

The algorithm initializes an empty list result to store the results. If a prefix was found and removed, the algorithm adds a tuple containing the prefix and its corresponding value from GPTtokenDict to the result list. The remaining portion of the word, after removing the prefix, is then processed using a secondary procedure, reduce_and_lookup. The results from this are extended to the result list.

4.2.6 Suffix addition

If a suffix was found and removed, the algorithm checks whether this suffix is already included in the last element of the result list. If not, it processes the suffix, using reduce_and_lookup and extends these results to the result list.

4.2.7 Return result

Finally, the algorithm returns the result list, which contains tuples of the subparts of the word and their corresponding values from GPTtokenDict.

The output of the FindSubwords algorithm is a list of tuples. Each tuple contains a subpart of the word (either a prefix, a part of the remaining word, or a suffix) and its corresponding index from GPTtokenDict. The index is needed to find the corresponding subword vector in GPT.

4.3 Proposed algorithm for recursively decomposing word into subwords

The supporting algorithm to the FindSubwords algorithm is the ReduceAndLookup algorithm, which also utilizes the global dictionary (GPTtokenDict). It is intended to recursively decompose a word or subword into known subwords, as defined in the dictionary (GPTtokenDict). It is useful in parsing complex or compound words into smaller, recognizable units.

ReduceAndLookup algorithm: Recursively, decompose a word into known subwords

```
procedure reduce_and_lookup(word, original_word)
  if not word then
     return []
  end if
  if word in GPTtokenDict then
     return [(word, GPTtokenDict[word])]
  end if
  for i ← len(word) downto 1 do
     segment ← word[:i]
     if segment in GPTtokenDict then
        remaining_segment ← word[i:]
        return [(segment, GPTtokenDict[segment])] +
               reduce_and_lookup(remaining_segment, original_word)
     end if
  end for
  return []
end procedure
```

The following steps describes the details of the ReduceAndLookup algorithm:

4.3.1 Empty string check

The algorithm first checks if the input word is empty. If it is, it immediately returns an empty list, as there are no subwords to find in an empty string.

4.3.2 Word check in dictionary

The algorithm checks if the entire word is present in GPTtokenDict. If found, it returns a list containing a tuple of the word and its corresponding value from the dictionary. This step efficiently handles cases where the word is a known subword.

4.3.3 Subword segmentation and recursive decomposition

The algorithm iteratively examines segments of the word, starting from the longest segment (the whole word) and gradually reducing the segment size by one character at each step. For each segment, it checks if the segment is in GPTtokenDict. If the segment is found, the algorithm stores this segment and its value from the dictionary in a tuple. It then calls itself recursively with the remaining part of the word (the part after the identified segment) and the original word as arguments. The recursive call continues this process, breaking down the remaining part of the word into known subwords. The combination of the current segment tuple and the list returned by the recursive call forms the complete list of subwords for this iteration.

4.3.4 Return of subword combinations

If the algorithm finds a valid subword, it returns a list of tuples, each containing a known subword from the word and its corresponding value from GPTtokenDict. This list represents the decomposition of the word into the largest possible known subwords.

4.3.5 Fallback for no subwords

If no segment of the word matches any entry in GPTtokenDict, indicating that no known subwords are found, the function returns an empty list.

The output of the ReduceAndLookup algorithm is a list of tuples, where each tuple contains a known subword segment and its corresponding index from GPTtokenDict. This list forms the decomposition of the input word into recognized subwords.

4.4 Building word embeddings from subwords

The output from the FindSubwords algorithm will consist of a list of tuples containing the new subword associated with the index to identify the subword embedding (vector) in GPT. A search is executed in GPT to find the corresponding subword embedding from the index. After all of the subword vectors are gathered for each subtoken, an averaging of all of the subword vectors is executed to arrive at one word embedding for each word.

4.5 Analysis of subword tokens using proposed approach

Table 3 reports on the results using the proposed approach in tokenizing the same 10 words from Table 2 using the original GPT subtokens. The tokenization of the word *monosymmetrical* using the proposed approach contains morphological information, in that the subword *mono* is extracted as a subword that exists as a subtoken in GPT. The tokenization from Table 2 *is ['mon', 'os', 'ymm','etrical'*] and in Table 3 is *['mono', 'sym','met','rical'*]. The proposed approach does provide a more morphological breakdown of this word, even if both approaches result in the same number of subtokens. Similarly, the word *unliveableness* is tokenized as *['un', 'live', 'able', 'ness']* using the proposed approach, in comparison to using original GPT tokens *['un', 'live', 'abl', 'eness'].* The subword tokens for the word *superrationally* in Table 2 correctly identifies the *super* prefix as well as *post* in the word *postpathological.* In fact, the tokenization is the same using the proposed approach and the original GPT tokenization for the word *postpathological.* As stated, there are 234,892 words in the NLTK word corpus. Applying the proposed algorithm and comparing with the subwords generated per word against the subwords generated for the original GPT tokenization algorithm resulted in 54,061 matching tokenized words. Approximately 23% of the tokenized words from the NLTK word corpus resulted in the same subword tokens using the proposed approach and the original GPT tokenization. Reported in Tables 2 and 3 are the words *postpathological* and *superrationally* that have identical tokens from both approaches.

Table 3: Example of subwords generated for words using the proposed approach.

Word	Subwords
monosymmetrical	['mono', 'sym','met', 'rical']
unliveableness	['un', 'live', 'able', 'ness']
stereotelemeter	['stereo', 'tele', 'met', 'er']
progressiveness	['pro', 'gressive', 'ness']
emanational	['em', 'ana', 'tion', 'al']
cinnamonwood	['cinnamon', 'word']
inextinct	['in', 'ext', 'inct']
superrationally	['super', 'ration', 'ally']
postpathological	['post', 'path', 'ological']
lithophotography	['lith','oph', 'oto', 'graph','y']

The word *stereotelemeter*, using the proposed approach, provides an ideal tokenization into the subwords *['stereo', 'tele', 'met', 'er']* vs. the original GPT tokenized words *['st', 'ere', 'ote', 'lem', 'eter']*. It is worth noting that the subtoken *stereo* is not in the prefix list used in the FindSubwords algorithm. The word *progressiveness* tokenized by GPT is *['progress', 'iveness']*, which certainly conveys morphological information with the first subtoken and using less tokens than the proposed approach, which gives the resulting tokens *['pro', 'gressive', 'ness'*]. The extraction of the first subtoken *pro* is determined by the prefix list having priority in redefining tokenized words in the proposed approach. This shows that improvement could be made with the proposed algorithm, which will be discussed in future work.

The word *emanational* is tokenized by the original GPT tokenizer as *['eman', 'ational']*, with the proposed approach resulting in *['em', 'ana', 'tion', 'al']*. The root word is *emation*, with the *al* turning the word into an adjective, making the proposed approach ideal in identifying at least the suffix. The root word does not exist as a GPT subword, but a more morphological breakdown would be *['ema', 'na', 'tion', 'al']*. The word *inextinct*, with the *in* negating the word *extinct*, is a single morphological unit in itself. The original GPT tokenizer generates *['ine', 'xt', 'inct']* and the proposed approach recognizes the *in* because it is listed as a common prefix. Another uncommon compound word *cinnamonwood* is tokenized as ['cinnamon', 'wood'] using the proposed approach. It is interesting that the original GPT tokenizer rendered the subtokens as *['c', 'innamon', 'wood']*, since the word *cinnamon* is in the GPT subtoken vocabulary. The word *lithophotography* gave an interesting tokenization using both the original and the proposed approach. Original GPT gives *['l', 'ith', 'oph', 'ot', 'ography']* and the proposed approach gives *['lith', 'oph', 'oto', 'graph','y']*. The ideal morphologically rich tokenization would be *['litho','photo','graphy']*, which is independent of the GPT tokenization process. The proposed approach has both *y* and *graphy* captured in the suffix list. The reason why *graphy* was not selected is because *y* is listed first. Potential for improvement will be discussed in future work. The subtoken *lith* was selected because it is the longest sequence that exists in GPT subtokens, from the left side of the word, which is incorporated in the proposed approach. Note that the original GPT subtokens further breaks this subtoken into two subtokens, *l* and *ith*.

From previous work [11], the word automobile, using the proposed approach, is ['auto', 'mobile'] but using the base GPT, the tokenizer is ['aut','om','obile']. As noted, this inconsistency issue with GPT will occur if there is not a high enough frequency of subwords during training of the tokenizer. This is a limitation of BPE in GPT, in that the approach does not incorporate morphological information in the tokenization process and if subtokens capture morphological, it is due to frequency not intentionality. It is not known how many words were encountered during the training of GPT's language model, since that information is not publicly available. In fact, understanding the specifics of how these LLMs work poses numerous challenges due to the lack of access to the internal structure of these models [27].

Table 1 reports the number of subtokens generated from the NLTK word corpus of 234,892 words using the original GPT tokenization algorithm. Figure 1 reports on the number of tokens generated from the NLTK word corpus for GPT subword tokens and the number of GPT subword tokens needed in the proposed approach. Figure 1 presents a graphical representation of Table 1 for the counts of subtokens for all of the words in the NLTK word corpus, along with the subtokens generated using the proposed approach. An observable difference is noted at the number of words using two subtokens, with the proposed approach using significantly more two tokens, in comparison with the original GPT tokenization process. As noted in the plot, as the number of tokens increases, the proposed approach uses less number of subtokens for a word than the original GPT tokenization method. The original GPT tokenization approach requires more tokens – starting at three and continuing through the maximum number of subtokens generated on the NLTK word corpus at 10 tokens. Not shown are the number of 8 to 10 subtokens, which are negligible at *{8: 47 subtokens, 9: 4 subtokens, 10: 0 subtokens}* for the proposed approach and {8: *137 subtokens, 9: 31 subtokens, 10: 3*} for the original GPT subtokens.

As previously stated, the total number of GPT subtokens are 50,257 and these are based on word parts that were determined using BPE on a large corpus. In this study, it was determined that not all of the GPT subtokens were needed after tokenizing the ~235k words from the NLTK word corpus. The tokenizer from GPT requires 22,529 sub-

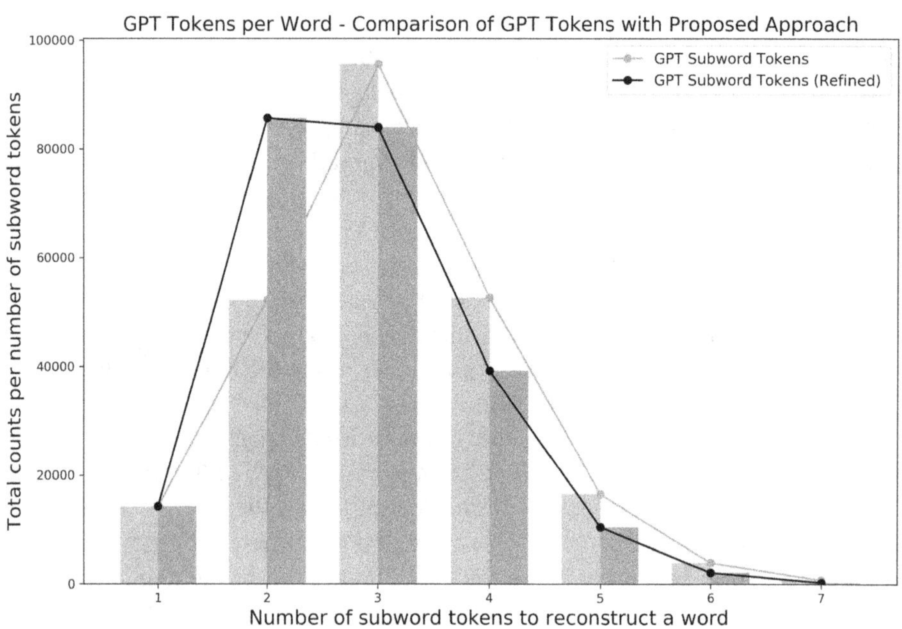

Figure 1: Tokens generated for NLTK word corpus using GPT subtokens and GPT subtokens (refined) using the proposed approach.

tokens out of the total 50,257 available from GPT. The proposed approach requires 18,433 subtokens out of the total 50,257 from GPT. The proposed approach requires less subtokens overall as well as less subtokens in the tokenization of a word, as noted in Figure 1.

5 Result analysis

This study involves additional analysis of the subtokens generated from the proposed approach, in comparison to the original GPT tokenization, as reported in the previous section. In this section, results are reported on an additional dataset that further differentiates this study from previous work []. The dataset involves linguistic relations pertaining to morphology, involving inflections and derivations, which are evaluated using the Bigger Analogy Test Set (BATS) [28]. Table 4 describes the 10 linguistic relations for inflectional morphology and Table 5 describes the 10 linguistic relations for the derivational morphology.

Table 4: Analogy dataset for inflectional morphology tasks [28].

Morphology	Subcategory	Relations
Inflections	nouns	**I01**: regular plurals (*book:books*) **I02**: plurals – orthographic changes (*wife:wives*)
	adjectives	**I03**: comparative degree (*strong:stronger*) **I04**: superlative degree (*strong:strongest*)
	verbs	**I05**: infinitive: 3ps.sg (*follow:follows*) **I06**: infinitive: participle (*follow:following*) **I07**: infinitive: past (*follow:followed*) **I08**: participle: 3ps.sg (*following:follows*) **I09**: participle: past (*following:followed*) **I10**: 3ps.sg: past (*follows:followed*)

An inflection refers to the modification of a word to express different grammatical categories such as tense, mood, voice, aspect, person, number, gender, and case. This process is a fundamental aspect of many languages, particularly those with rich inflectional systems. For example, in English, verbs are inflected for tense ("talk" becomes "talked" in the past tense), and nouns for number ("cat" becomes "cats" to indicate plurality). In other languages, like Russian or Arabic, words undergo more extensive inflections, modifying roots with a variety of prefixes, suffixes, infixes, or changes in the stem itself. These modifications convey precise grammatical and syntactic information and are integral to the structure and meaning of sentences in inflected languages. Inflections are distinct from derivations, another morphological

Table 5: Analogy dataset for derivational morphology tasks [28].

Morphology	Subcategory	Relations
Derivations	no stem change	**D01**: noun + less (*life:lifeless*) **D02**: un + adj. (*able:unable*) **D03**: adj. + ly (*usual:usually*) **D04**: over + adj./ved (*used:overused*) **D05**: adj. + ness (*same:sameness*) **D06**: re + verb (*create:recreate*) **D07**: verb + able (*allow:allowable*)
	stem change	**D08**: verb + er (*provide:provider*) **D09**: verb + ation (*continue:continuation*) **D10**: verb + ment (*argue:argument*)

process, where the addition of affixes creates new words with new meanings, rather than different forms of the same word.

Derivation in morphology refers to the process of creating new words from existing ones by adding prefixes, suffixes, or sometimes changing the word stem. Unlike inflection, which modifies a word to express grammatical relationships while keeping its core meaning intact, derivation often changes the word class and meaning. For instance, the English noun "kindness" is derived from the adjective "kind" by adding the suffix "-ness," which transforms an adjective into a noun. Similarly, "run" (a verb) can become "runner" (a noun) by adding "-er," or "happy" (an adjective) can become "unhappy" (another adjective, but with opposite meaning) by adding the prefix "un-." Derivation is a key mechanism in the expansion of vocabulary in a language, allowing for the expression of complex concepts and ideas by combining and modifying existing linguistic elements.

The evaluation on the inflectional and derivational datasets utilizes reconstructed word representations in the form of word embeddings. The details of the approach for evaluating the analogy task is found in previous work [11], which describes the 3COSADD [28] approach, along with Cosine similarity. The results reported for the inflectional morphology tasks are in Table 6. The proposed approach performed significantly better on several linguistic relations, with regular plurals at 75%, comparative at 42%, infinitive (3ps.sg) at 71%, infinitive (past) at 51%, and participle (3ps.sg) at 45%. In comparison, the values from the original GPT subtokens are, regular plurals at 45%, comparative at 18%, infinitive (3ps.sg) at 55%, infinitive (past) at 25%, and participle (3ps.sg) at 18%. There were also a few where the accuracy reported was less than five percent differential between the two approaches. The results reported for the derivational morphology tasks had a few tasks where the proposed approach did not perform as well, along with some tasks that were comparable to the results for inflectional morphology, as reported in Table 7.

Table 6: Results for inflectional morphology tasks.

Relations	**Original GPT sub- tokens**	**Refined GPT sub-tokens**	**Unique pair count**
regular plurals	0.45	0.76	50
plurals: orthographic changes	0.16	0.20	50
comparative	0.18	0.42	50
superlative	0.38	0.40	50
infinitive: 3ps.sg	0.55	0.71	50
infinitive: participle	0.35	0.42	50
infinitive: past	0.24	0.51	50
participle: 3ps.sg	0.18	0.45	50
participle: past	0.35	0.42	50
3ps.sg: past	0.30	0.38	50
Average Accuracy	**0.31**	**0.47**	400

Table 7: Results for inflectional morphology tasks.

Relations	**Original GPT sub-tokens**	**Refined GPT sub-tokens**	**Unique pair count**
noun + less	0.20	0.68	50
un + adj.	0.14	0.10	50
adj + ly	0.44	0.54	50
over + adj./ved	0.08	0.04	50
adj + ness	0.25	0.78	50
re + verb	0.22	0.04	50
verb + able	0.02	0.08	50
verb + er	0.16	0.63	50
verb + ation	0.18	0.44	50
verb + ment	0.33	0.77	50
Average Accuracy	**0.20**	**0.42**	400

6 Conclusion

This chapter reconstructs word representations with their associated word embeddings using a proposed tokenization method, incorporating pretrained subword embeddings from GPT. The proposed approach primarily focuses on the utilization of morphological information in the form of affixes (prefixes and suffixes) to reconstruct the tokenization process. The reconstruction of word representations and their associated word embeddings enables evaluation of the proposed subword tokenization algorithm on analogy datasets. An overall improvement in performance on inflectional and derivational morphological tasks is achieved. In addition, the proposed approach resulted in a reduction in the number of subword tokens needed in reconstructing a word. An interesting discovery pertains to the identification of the unique subtokens using the proposed approach, in comparison to the GPT tokenization method on the NLTK word corpus. The proposed approach requires 18,433 subtokens whereas GPT tokenization required 22,259 subtokens. This study shows that the proposed approach has the potential to enhance subtoken embeddings in LLMs in reducing computational resources and achieving more efficient tokenization.

7 Future work

The reduction of subtokens needed in this study for the tokenization of words is noteworthy but would need further study to show if it is effective in language understanding and generation. This study was done using GPT-2, which makes available the subtokens and which are presumably the same in GPT-3/4, but the newer GPT variants do not provide access to subtokens, though it would be preferable to evaluate using the latest GPT variant. This study required the use of GPT subtokens as a way to circumvent the computational challenges. It would also be interesting to implement the proposed tokenization approach without reliance on GPT subtokens. The proposed tokenization algorithm could also be further refined to capture more granularity as well as subword wholeness, as was noted with the tokenization of the word *progressiveness*. The tokenization of the word *lithophotography* revealed the importance in giving larger suffix words more weight. In addition, this study only works with the English language and would need adjustment for specific languages, especially agglutinative languages, where morphological information is especially useful.

References

[1] C. Lyu, J. Xu and L. Wang, “New Trends in Machine Translation Using Large Language Models: Case Examples with Chatgpt,” *arXiv preprint arXiv:2305.01181*, 2023.

[2] Y. Tan, D. Min, Y. Li, W. Li, N. Hu, Y. Chen and G. Qi, “Evaluation of ChatGPT as a Question Answering System for Answering Complex Questions,” *arXiv preprint arXiv:2303.07992*, 2023.

[3] T. Klein and M. Nabi, “Learning to Answer by Learning to Ask: Getting the Best of Gpt-2 and Bert Worlds,” arXiv preprint arXiv:1911.02365, 2019.

[4] T. Zhang, F. Ladhak, E. Durmus, P. Liang, K. McKeown and T. B. Hashimoto, “Benchmarking Large Language Models for News Summarization,” *arXiv preprint arXiv:2301.13848*, 2023.

[5] G. Franceschelli and M. Musolesi, “On the Creativity of Large Language Models,” *arXiv preprint arXiv:2304.00008*, 2023.

[6] C. Gómez-Rodríguez and P. Williams, “A Confederacy of Models: A Comprehensive Evaluation of LLMs on Creative Writing,” *arXiv preprint arXiv:2310.08433*, 2023.

[7] A. Radford, J. Wu, R. Child, D. Luan, D. Amodei, I. Sutskever., et al., “Language Models are Unsupervised Multitask Learners,” *OpenAI Blog*, vol. 1, no. 8, p. 9, 2019.

[8] J. Devlin, M.-W. Chang, K. Lee and K. Toutanova, "Bert: Pre-training of Deep Bidirectional Transformers for Language Understanding," In *Proceedings of the 2019 Conference of the North American Chapter of the Association for Computational Linguistics: Human Language Technologies*, vol. 1, pp. 4171–4186, 2018.

[9] A. Vaswani, N. Shazeer, N. Parmar, J. Uszkoreit, L. Jones, A. N. Gomez, Ł. Kaiser and I. Polosukhin, "Attention Is All You Need," *Advances in Neural Information Processing Systems*, vol. 30, 2017.

[10] K. Ethayarajh, "How Contextual are Contextualized Word Representations? Comparing the Geometry of Bert, Elmo, and Gpt-2 Embeddings," In *Proceedings of the 2019 Conference on Empirical Methods in Natural Language Processing and the 9th International Joint Conference on Natural Language Processing*, pp. 55–65, 2019.

[11] P. Lauren, "Reconstructing Word Representations from Pretrained Subword Embeddings," In *2022 International Conference on Computational Science and Computational Intelligence (CSCI)*, 2022.

[12] T. Lin, Y. Wang, X. Liu and X. Qiu, "A Survey of Transformers," arXiv preprint arXiv:2106.04554, 2021.

[13] D. Bahdanau, T. Bosc, S. Jastrzębski, E. Grefenstette, P. Vincent and Y. Bengio, "Learning to Compute Word Embeddings on the Fly," *arXiv preprint arXiv:1706.00286*, 2017.

[14] P. Bojanowski, E. Grave, A. Joulin and T. Mikolov, "Enriching Word Vectors with Subword Information," *Transactions of the Association for Computational Linguistics*, vol. 5, pp. 135–146, 2017.

[15] Y. Wu, M. Schuster, Z. Chen, Q. V. Le, M. Norouzi, W. Macherey, M. Krikun, Y. Cao, Q. Gao, K. Macherey., et al., "Google's Neural Machine Translation System: Bridging the Gap between Human and Machine Translation," arXiv preprint arXiv:1609.08144, 2016.

[16] R. Sennrich, B. Haddow and A. Birch, "Neural Machine Translation of Rare Words with Subword Units," arXiv preprint arXiv:1508.07909, 2015.

[17] P. Gage, "A New Algorithm for Data Compression," *C Users Journal*, vol. 12, no. 2, pp. 23–38, 1994.

[18] M. Podkorytov, D. Biś and X. Liu, "How Can the [Mask] Know? the Sources and Limitations of Knowledge in Bert," In *2021 International Joint Conference on Neural Networks (IJCNN)*. IEEE, pp. 1–8, 2021.

[19] S. Pokale, K. Taware, G. Fernandes, S. Kangane, P. Bhosale and L. Bewoor, "Text Summarization: GPT Perspective," In *2023 3rd Asian Conference on Innovation in Technology (ASIANCON)*. IEEE, pp. 1–7, 2023.

[20] K. Bostrom and G. Durrett, "Byte Pair Encoding Is Suboptimal for Language Model Pretraining," arXiv preprint arXiv:2004.03720, 2020.

[21] D. Vilar and M. Federico, "A Statistical Extension of Byte-pair Encoding," In *Proceedings of the 18th International Conference on Spoken Language Translation (IWSLT 2021)*, pp. 263–275, 2021.

[22] K. Stratos, "Reconstruction of Word Embeddings from Sub-word Parameters," arXiv preprint arXiv:1707.06957, 2017.

[23] Y. Kim, K.-M. Kim, J.-M. Lee and S. Lee, "Learning to Generate Word Representations Using Subword Information," In *Proceedings of the 27th International Conference on Computational Linguistics*, 2018.

[24] M.-T. Luong, R. Socher and C. D. Manning, "Better Word Representations with Recursive Neural Networks for Morphology," In *Proceedings of the seventeenth conference on computational natural language learning*, pp. 104–113, 2013.

[25] J. Zhao, S. Mudgal and Y. Liang, "Generalizing Word Embeddings Using Bag of Subwords," arXiv preprint arXiv:1809.04259, 2018.

[26] J. Jinman, S. Zhong, X. Zhang and Y. Liang, "Pbos: Probabilistic Bag-of-subwords for Generalizing Word Embedding," arXiv preprint arXiv:2010.10813, 2020.

[27] F. F. Xu, U. Alon, G. Neubig and V. J. Hellendoorn, "A Systematic Evaluation of Large Language Models of Code," In *Proceedings of the 6th ACM SIGPLAN International Symposium on Machine Programming*, pp. 1–10, 2022.

[28] A. Gladkova, A. Drozd and S. Matsuoka, "Analogy-based Detection of Morphological and Semantic Relations with Word Embeddings: What Works and What Doesn't," In *Proceedings of the NAACL Student Research Workshop*, pp. 8–15, 2016.

Massoud Alibakhsh

Swarm intelligence: a new software paradigm

A novel approach to integrating classical software with LLMs using Object Messaging and Intelligent Objects (OMIO)

Abstract: The recent advancements in AI, particularly Large Language Models (LLMs) and their applications in video and audio processing, have generated considerable excitement. These innovations have shown great promise in creating more interactive, engaging, and intelligent systems. However, despite these remarkable achievements, a fundamental gap remains in the application of AI within the business context. Businesses predominantly rely on software applications for automation and process management, but these traditional applications often operate in a structured, rigid manner that does not seamlessly incorporate the fluid, intuitive capabilities of modern AI. This disconnect presents a significant challenge: How can we meaningfully integrate AI into business applications to optimize communication, improve efficiency, and enhance decision-making?

The answer lies in adopting a new model and paradigm for AI integration in business, which is where the Object Messaging and Intelligent Objects (OMIO) model comes into play. This paper delves into this new paradigm, which is not only an innovative new approach to integrating AI with conventional software but a new way of thinking about AI and applications of LLM to create intelligent software, and give rise to collective intelligence. Some of the ramifications of this approach may present a significant challenge for the software industry and makes a compelling case as to the inevitability of adopting this approach. So, we discuss and review the history and evolution of software in general and its adoption in the business environment. In our opinion, software at work has arguably had the biggest impact on human activity and reshaping of the planet.

Keywords: Object Messaging, Intelligent Objects, Large Language Models

Note: This paper is an extended version of its original as presented at the CSCE 2023 conference.

Massoud Alibakhsh, Atlanta, GA, USA, e-mail: massoud@xeba.tech, www.omadeus.com

https://doi.org/10.1515/9783111344126-016

1 Introduction: it is all about communication

While LLMs excel in processing natural language, conventional software applications mostly communicate structured information in a rigid fashion. To bridge the gap between human usage of natural language and the structured data utilized by existing applications, communication tools such as email and 'Enterprise Social Network' tools such as Slack or Yammer emerged. While these tools filled the void by accommodating the informality of chat-like interactions, they struggled to accommodate the applications' structured data and more importantly incorporating the business nervous system, namely the workflow. The adoption and popularity of these tools such as email and Slack added to the proliferation of data in their arcane silos, in addition to the already existing structured data in the business applications databases. As these two separate models grew independent of each other, aside from occasional superficial interfaces to trigger additional notifications in email or channels, the gap continued to widen into a chasm. Even with the success and introduction of LLMs that can reliably process natural language, the prevalent form-based models employed by existing software applications present a fundamental obstacle for meaningful AI integration.

Examining some of the potential conflicts between these two computational models, we can observe that LLMs are probability-based machines with sometimes unreliable outputs, whereas conventional software is a deterministic state machine incorporating workflow, business logic and understanding of users and their roles.

We attempt to examine the challenges of integrating LLMs with form-based software, shedding light on the associated complexities, and introduce our new design paradigm, Object Messaging and Intelligent Objects (OMIO). This approach re-examines AI as a platform instead of its current popular use as simply a tool.

We also cover the evolution of computer programming models for the business world and the necessity of another paradigm shift to create fully intelligent software, composed of a swarm of intelligent objects with self-awareness, understanding of workflow, knowledge about various classes of stakeholders, and the ability to interact with each other and with users, with natural language.

Utilizing this model effectively merges natural language and structured data, facilitating a smooth integration of LLM functionalities into software applications. This paper sheds light on OMIO's capabilities and sets the stage for unleashing the transformative potential of LLMs, safely and reliably, thereby significantly boosting productivity in the modern business environment.

We will attempt to argue and demonstrate that the OMIO paradigm not only offers an elegant path for deep integration of LLMs and software applications with structured data, but also optimizes communication among humans [1] in the workplace, producing human and AI collective intelligence.

2 A quick background: in the beginning, there were the paper forms, then came the computers

Almost all business applications primarily serve as tools for communication. Initially, they followed a model of form automation, a concept derived from paper-based forms. When computers began to play a role in business, programmers collaborated with business managers to digitize these paper-based processes. This typically involved interviewing key staff and analyzing the paper forms used in current business operations for transferring information between individuals or groups. These paper forms, filled out by various people, often included headings and structured sections for data entry. The automation process generally started with flowchart creation, which evolved into data flow or control flow diagrams as the first step in designing automation.

The automated programs, created as a result, often mirrored the original paper forms on the computer screen, with each user accessing their specific section of the form to view or input data. The data entered was stored in a database, enabling later retrieval for analysis and report generation. This method marked a considerable improvement in information transfer speed and data collection accuracy, as computer programs could incorporate syntactical editing rules and enforce business regulations related to data entry. Furthermore, storing this data in a computer database enhanced the flexibility and efficiency of generating reports.

3 Two major paradigm shifts: first the GUI and then the Cloud

With the emergence of Personal Computers running on Local Area Networks (LANs) and the advent of Graphical User Interface (GUI) in the mid 80s and up to early 90s, most of these programs were upgraded or completely rewritten to take advantage of this new modern form of interface [9]. Software companies rushed to provide business applications offerings with the new GUI as this became the dominant competitive advantage in the business world [10]. Companies such as Apple ushered in the first GUI operating systems and later Microsoft followed suit. The GUI improved the basic human and machine interaction using the icon paradigm, along with new data entry gadgets such as the now ubiquitous mouse on every desktop. But the fundamental model underlying the process design for all these applications was still the paper form. Programmers continued to use the paper process as the starting point for creating business automation. In fact, the new colorful screens with icons and images further motivated the perpetuation of the paper process paradigm due to its apparent fidelity with the physical world. The unconscious assumption was that the existing familiarity of the physical forms reflected on the computer screens eased the users' anxiety about this new business tool, namely the computer, which had, up to that time, a formidable and intimidating reputation for many users in the business world, unfamiliar with such sophisticated tools.

The rise of the Internet and the emergence of cloud computing [2] forced most of these business applications to leave the world of LANs and simple client/server architecture and be redesigned for the new public network, the Internet or the Web [11, 12], and its emerging multi-tier distributed platform, the cloud [3, 13]. But the form-based paradigm of application design remained intact. These business applications take in structured data and operate like basic state machines. But humans communicate with natural language. That is where the shortcomings of the traditional form-based software applications lie. The natural language type of communication continued in the form of memos that were usually instructions from above, relayed by mid-level managers all the way down to the rank and file or summary reports compiled by mid-level managers to flow to the top informing high level executives of the goings on in the production environment.

4 Email and the promise of panacea: the first wave

Software applications' lack of support for natural language continued to make paper memos a necessity. Then, in the mid-1990s, a groundbreaking new "Killer Application" burst onto the business scene, generating much excitement and anticipation [4]. This heralded the end of traditional memos, with expectations that this new electronic version would enhance communication through natural language processing. Users were equipped with sophisticated inboxes, allowing them to compile a personalized information database, related to their specific tasks or projects. However, it soon became clear that this innovative tool had unintended consequences. The influx of both external spam and a growing volume of internal emails daily, revealed a fundamental flaw in the email model, which was initially overlooked. Features like 'cc' and 'reply all' only worsened the issue [5, 6]. This email system inadvertently placed the responsibility on each user to not only process and understand the information but also to determine the most effective way to distribute it within the organization, based on their individual grasp of its workflow and key players. While this might have been manageable in smaller organizations with fewer employees and simpler workflows, it proved to be an insurmountable challenge in larger entities, where timely and accurate information routing by individuals became impractical. Even in smaller settings, ensuring the prompt exchange of information was problematic, given its reliance on human intervention.

5 Enterprise social networks for business: the second wave

This is how enterprise social networking tools like Yammer [7, 8], Slack, and MS Teams gained prominence in the corporate world, offering solutions to the challenges posed by email. These platforms revolutionized communication by shifting focus from individual senders to topic-centric channels [1]. Managers, with a grasp of organizational workflows, would set up various channels, reflecting different aspects of these workflows. Employees could then subscribe to channels relevant to their roles, exchanging information within these focused streams. Although this method improved upon traditional email, particularly for internal communications, it did not completely resolve the underlying issues, and inadvertently introduced new complexities. Additionally, these channel-based communications were supplementary to existing business applications that processed structured data. Consequently, company information became fragmented, dispersed across emails, various Slack channels, and within traditional business applications.

In form-based applications, information flow is dictated by user roles and access permissions. Users inputting data typically have no control over how and where their data is routed; these pathways are preestablished by application designers, according to role-based workflows and data management plans. For instance, in an accounting system, a user might input data in the accounts receivable section, and later, another user with higher access privileges might access that data for generating financial statements or viewing the general ledger. However, the dialogue and discussions relevant to these data inputs often occur in emails and channel conversations, becoming isolated in information silos and frequently overlooked. This scenario highlights the apparent need for an integrated approach that combines the structured data from these applications with the fluid conversations happening around them, ensuring a more cohesive and efficient information management system.

6 Workflow-based communication: the 3rd wave – OMIO [patents pending]

Before exploring the solution and the features of the Object Messaging Model, let us first conceptualize the problem abstractly, using the metaphor of a biological cell, which in many ways mirrors actual organizational structures more accurately than one might think. Contemporary organizations consist of human elements and workflows, typically structured in a hierarchical, pyramid-like formation. These workflows are modern abstractions of the early twentieth-century assembly lines. Within these structures, human participants are grouped by specialized functions such as Production, Support, Sales, Marketing, and Administration, akin to organelles in a cell, all orbiting around the central workflow – the heart of the organization. These groups convene in a collective entity, or a 'superorganism,' with the unified goal of producing specific goods and services. Here, the hierarchical structure serves as the organization's skeleton, while the workflow is its pulsating heart.

Each group's capacity, much like organelles in a cell, hinges on the expertise and skill of its individual members. Crucially though, the overall effectiveness of the organization relies on the timely sharing of pertinent information and the harmonization of these human elements with the workflow, or the objects moving within this process. It is important to note that not every member is directly involved with every aspect of the workflow. Products and services, being the culmination of various components assembled within this workflow, suggest a web of intricate connections among these parts and the respective group members. The obvious conclusion is that the real glue that binds everyone in the organization are the components, parts, or the final product or products. And the conversations that matter in this setting are the ones related to the objects or virtual objects that are moving around within the workflows. And these conversations are innate attributes of the object itself. As described in the 3rd wave of corporate communication, these objects represent the new pivots of human communication in the workplace. The objects also have the responsibility of managing the variety of information about themselves as well as informing the stakeholders of significant events in a timely fashion [1].

7 A new software design approach for AI as a platform

Our innovative software application design method begins with an in-depth semantic analysis, tailored to the process set for automation. This methodology enables us to pinpoint key entities involved in the workflow, akin to components moving along an assembly line. These entities, encompassing both tangible and intangible objects, are subjects of potential discussions among humans. Subsequently, we allocate distinct, unique communication, and information channels to each entity, corresponding to different stakeholder categories within the organization. Consequently, all communication and information pertaining to an object are sorted by stakeholder class, each maintained within its dedicated channel and overseen by the entity itself. This process is akin to an individual managing her interactions, based on the nature of her relationships, such as differentiating between interactions with her mother and her boss. Importantly, these channels are inherent to the entity, and not user generated. They are either preconfigured by software designers or autonomously established by the entity, as necessary. Thus, these communication channels become intrinsic properties of the entities, much like the structured data elements that define and characterize them.

8 Self-aware intelligent objects

In the Object Messaging & Intelligent Objects model (OMIO), objects or entities are not only the pivot of connection for the relevant stakeholders but are always aware of their own states and are responsible to maintain all the information about themselves. When people interact with an object and affect any of its attributes or when the object changes state, moving within the workflow, it converts these events into messages and records them in their appropriate chat channels on behalf of the person or process responsible for the event. Messages or other types of information, such as video/audio files, links or documents, can be inserted into the object's channel for informing stakeholders and the object itself. The object then utilizes AI to read, understand, and remember the information from these files, as well as keeping them organized for later access by stakeholders. There is no need for users to come up with their own elaborate ways of creating folders for organizing files. That is the job of the object as the information belongs to the object and so it is processed, understood, remembered, and organized by the object itself. Any message insertion in the object's chat channel triggers the routing of the object to all the stakeholders' inboxes associated with this object. In this way, the stakeholders are informed in real time of important events affecting their objects of interest and can react and intervene, if necessary, by posting their reactions, thoughts, or concerns in the same channel. One can imagine that the object is calling for a virtual asynchronous meeting amongst the stakeholders for each significant event affecting the object.

In this fashion, it is the object's responsibility to communicate the event information to the stakeholders as soon as the event occurs. This also solves the synchronization problem with the workflow. All the stakeholders are thus tuned to the heartbeat of the organization, which is the tempo of the workflow. The events are usually changes in status or one of its attributes, or they can be messages inserted into one of its channels by stakeholders. All discussions and occurrences of events are stored in the appropriate channel of the object and are maintained by the object itself. And since the object is responsible for communicating any changes to its state in the form of natural language message insertion into its channels, the communication happens in real time and only to the appropriate stakeholders.

This is a significant feature of this model as it maintains a totally optimized communication in the form of relevant recipients only and precise timing for the transmission of these objects to human nodes within the organization. The precise timing is in synchronicity with the resonating pulse or tempo of the workflow as the heartbeat of the organization. The significance of this timing is usually overlooked in the current automated forms of communication in the workplace as the routing and timing of information transfer is dependent on the least reliable agents within the network, namely humans!

Object Messaging relegates the routing and the timing of all significant events about the object to the object itself. As a result, the emerging automated world is composed of a collection of intelligent objects, or super objects, as they get assembled, while they maintain all the conversations and event messages about themselves within appropriate channels as well as the list of all relevant stakeholders of various types. Even the insertion of a message is considered an event that causes the object to deliver a digital representation of itself along with its entire entourage (media, files, links, etc.) to the appropriate stakeholder's inbox.

In Object Messaging, the users' inboxes receive objects or super objects that have been associated with or self-identified themselves as relevant to a stakeholder or resource. The stakeholders can then examine the relevant chat channel to get informed of the latest events that have occurred to the object, since it was last examined.

9 Integration, deep integration, and AI fusion: meaningful integration with AI in one step

Some existing applications may find ways to do some integration with AI LLMs such as ChatGPT, which may provide some useful functions such as summarization of emails or writing emails, document generation for legal or other professions and code generation for programmers. All, if not most, business applications will struggle to find any place to plug in the AI in a useful or meaningful way. How would you connect ChatGPT to QuickBooks in its current form, for example? Where would you start?

Another example would be an Electronic Medical Record that has lots of text within each encounter. Perhaps you can summarize the text but that is of very limited value. And there are better ways that may sound more like science fiction. Imagine an intelligent Electronic Medical Record where at every point of user interaction, the care provider can call up the AI for assistance and the LLM is fully aware of the situation and provides meaningful information about some aspect of the patient such as an X-ray or some lab result or an illness. And all of that interaction is using natural language and recorded within the X-ray's private channel reserved just for care providers' discussions. We call that deep integration because the AI is passive and awaits patiently until it is called upon to review the records and then communicate its opinion to the provider.

One can also imagine yet another scenario where a patient's X-ray has just arrived in the provider's inbox and the AI within the X-ray itself has already examined it, read through the patient's history and relevant lab results, alerted the specialist, researched the latest on the issue on authorized predesignated sections of the Internet, sent a follow up appointment reminder email to the patient, inserted its comments into the appropriate private channel, and is ready to have a discussion with the care provider and the specialist in the private channel, all using natural language. We call that AI fusion. And that is what OMIO makes possible today.

10 A quick review of the mechanics

In the Object Messaging model, software designers identify key entities or objects and establish one or more dedicated communication channels for each. This approach integrates human communication as an inherent feature of the object. Through these channels, objects can store diverse formats of information, including videos, audio, and file attachments. As new data is added to these channels, it is simultaneously fed into the object's LLM, ensuring the object remains up-to-date with its evolving status. The LLM utilizes the object's structured data to better understand the context of stakeholder conversations within its chat channels and interpret any additional data or media files, like video call transcripts.

This model forms the backbone of the entire application, with each crucial object equipped with its own chat channel and structured data. The LLM is connected to the object at a singular point within the application, enabling deep integration. The LLM thus becomes an active participant in every chat room for all the objects. Applications built on the OMIO framework consist of a network of intelligent objects or "super components" (assembled objects), allowing the LLM to engage at every user interaction point. This represents, what we term, deep integration or AI fusion. Essentially, each object becomes a miniature LLM, trained on its specific data set that includes structured data, stakeholder dialogue, attachments, and a record of events and state changes across various channels. It analyzes channel conversations in real-time and serves as a knowledgeable advisor to stakeholders, in natural language.

Stakeholders are linked to these objects, based on their relevance or involvement in the workflow, forming myriad micro-networks centered on these objects. This new central point optimizes communication among stakeholders connected to a specific object. When different objects merge to form a super component, the stakeholder associations from the original objects are transferred to the new entity. Consequently, stakeholders are kept informed about the super component's status through these inherited connections.

11 Migrating to the new LLM AI platform

As with the previous transitional platforms, namely, GUI and the Cloud, the emergence of AI and the increasing popularity of LLMs such as ChatGPT represent a new virtual platform for most existing business applications to transition onto. Just like the previous transitions, this one requires not only a re-architecting of existing software applications but a new way of thinking about software design approach in general. In short, we need a new paradigm, and OMIO represents such a paradigm shift. The Object Messaging Model not only optimizes communication among human nodes but also facilitates a seamless and deep integration of Artificial Intelligence with any business software application. We believe it will prove to be the required architecture for a comprehensive and meaningful AI integration or an omnipresent AI at every point of human interaction with the application. An object containing people's chatter and any other information in any format about itself within its own appropriate channels can provide detailed context to an LLM for processing the conversation about the object. The structured data associated with the object is utilized as constraints for the processing of human chatter and other information about the object. The object's structured data may also be used in "Prompt Engineering" fashion to create the context for the LLM and a certain awareness as to its fundamental features before feeding it the information or channel chatters about the object. The object's chat channels include stakeholder's dialogs as well as the object's critical events and changes of state in the form of natural language. The OMIO provides the identification of all critical objects and extends one or more dialog channels as an innate attribute of the object. This streamlines a deep AI integration since there exists only a few localized points for this purpose within the application and those points of connections are the actual objects. And once the AI is connected to the object, it has access to the structured data and natural language in its chat channels as well as any attachments such as documents, media files, and hyperlinks. Traditional form-based applications combine multiple data elements in one screen (form) and were never provisioned to hold natural language dialogs for each critical element within the form! An LLM cannot be properly integrated to form-based applications since each screen or form contains several critical entities functionally associated together.

Furthermore, human dialog or additional information in various formats about the data fields do not exist within the application. Such dialogs are somewhat chaotically scattered all over emails and Slack or Teams' various channels, with arbitrary chosen names for channels, or email subjects. The OMIO model deals with these shortcomings in an elegant and effective way to create a system composed of lots of tiny bespoke LLMs representing critical entities in the system, wired together using business logic, aware of workflow and its stakeholders.

12 What LLMs are not suited for

Large Language Models (LLMs) have been at the forefront of AI's integration into various aspects of human life, from content creation to problem-solving. However, their application within the realm of automating and managing business workflows presents significant challenges that are not easily overcome. Business processes are fundamentally deterministic state machines, characterized by a series of predefined steps that lead to specific outcomes, based on certain inputs. These processes also involve understanding the intricate relationships between stakeholders and the critical entities within the workflow. The deterministic nature of business workflows and the nuanced understanding of roles and relationships within them are not within the innate capabilities of LLMs.

LLMs excel in processing and generating natural language, making them incredibly powerful tools for tasks that involve human language comprehension and production. However, business workflows require more than just language understanding; they necessitate a deep comprehension of the specific logic and rules that govern the workflow's progression. Deterministic state machines operate on the basis of clear, logical pathways that lead from one state to another, often involving conditional logic that LLMs are not inherently designed to manage. While LLMs can be trained to recognize patterns and even predict likely outcomes based on data, the deterministic and often binary decision-making process required in workflow automation is beyond their current scope. This limitation is not merely a technical hurdle but a fundamental mismatch between the capabilities of LLMs and the requirements of workflow automation.

Moreover, the complexity of business workflows extends beyond the mere sequence of actions. It encompasses an understanding of the roles and relationships between different stakeholders, as well as their interactions with various entities within the workflow. Each stakeholder has specific rights, responsibilities, and needs that must be accurately understood and managed throughout the workflow process. This level of understanding requires more than natural language processing; it demands contextual awareness and a dynamic adaptability to the evolving nature of the business environment. Current automation tools, designed specifically to handle such deterministic processes and relational dynamics, already effectively manage these aspects. They are built with the explicit purpose of capturing and enforcing the logic, rules, and roles that define business workflows.

This is where the Object Messaging and Intelligent Objects (OMIO) model offers a novel approach to integrating AI into workflow automation. Unlike LLMs, which primarily deal with language understanding and generation, the OMIO model embeds AI into every relevant object within the workflow. This means that instead of trying to manage the workflow through external analysis and intervention, intelligence is built directly into the workflow's components. Each intelligent object is aware of its role, state, and the relevant stakeholders, allowing for more nuanced and effective management of the workflow. This object-centric approach enables a dynamic and context-aware system that can adapt to changes within the workflow, understand the complex interdependencies between different entities, and make informed decisions, based on the embedded logic and data. In essence, the OMIO model recognizes that the solution to automating complex business workflows lies not in overarching language models but in imbuing the workflow itself with the intelligence necessary to navigate its intricacies.

In conclusion, while LLMs represent a significant advancement in AI's ability to process and generate human language, their application in automating deterministic business workflows is limited. The specific requirements of these workflows – logical progression, stakeholder relationships, and entity interdependencies – are better served by models designed to operate within these constraints. The OMIO model represents a compelling alternative, offering a way to integrate AI directly into the fabric of the workflow, thus providing a more effective and context-aware approach to automation.

13 A new kind of Operating System: introduction to OMIO OS

In the rapidly evolving landscape of technology, the integration of AI with operating systems represents a frontier ripe for innovation. The Object Messaging and Intelligent Objects (OMIO) model, a paradigm shift in software design, offers a groundbreaking approach to this integration. This essay presents a vision for a new Operating System (OS), dubbed "OMIO OS," which leverages the OMIO model to revolutionize user interaction with technology through natural language processing. The assumption is that all apps on this new OS follow the OMIO model of intelligent objects. Below, in Figures 1A and 1B, these objects are depicted visually.

OMIO OS is designed around the core principle of user interaction through natural language, eliminating traditional interfaces' complexity. By incorporating a master Large Language Model (LLM), OMIO OS understands user commands, intentions, and contexts, crafting strategies to fulfill requests by coordinating with applications built on the OMIO framework. These applications, inherently capable of natural language interaction, allow OMIO OS to seamlessly integrate various services, providing a cohesive and intuitive user experience.

13.1 The core components of OMIO OS

Master LLM: At the heart of OMIO OS is a sophisticated LLM that processes natural language inputs from users, interprets their needs, and devises strategies to address those needs using available applications and resources. Figure 1C depicts the master LLM along with all the OMIO-based apps and their dedicated LLMs that they use to interface with the master LLM using natural language.

OMIO-compliant applications: Applications developed using the OMIO model can understand and process natural language requests. They are designed to work in harmony within the ecosystem of OMIO OS, facilitating a broad range of tasks – from productivity to entertainment.

Natural language interface: Augmenting traditional GUIs, the OS features a natural language interface that allows users to communicate with their devices, as they would with a human assistant. The device uses its display as well as natural language to communicate.

Working example: "Hey OMI, I am hungry"

This example illustrates the seamless operation of OMIO OS in addressing a user's request:

User request interpretation: Upon receiving the user's statement of hunger, OMIO OS's master LLM interprets it as a request for meal options.

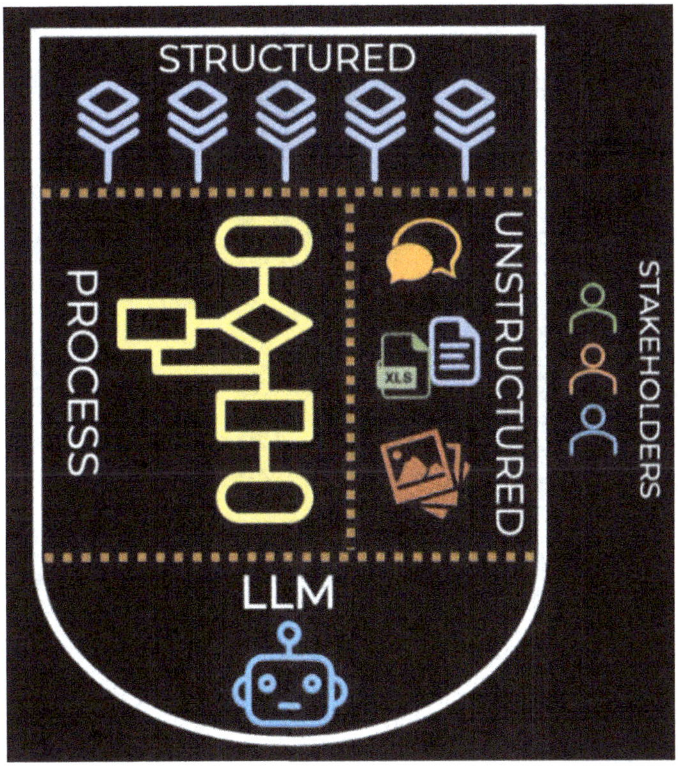

Figure 1A: The depiction of an intelligent object containing structured and not structured data, along with pointers to stakeholders.

Each object is a small model trained on all its available data to give it self awareness. This training continues as the object moves about the workflow and interacts with stakeholders and other objects. This information is transcripted into text and inserted into a predesignated channel, according to the application's business rules. The text is also used to update the object itself of its new state as part of its continuous awareness.

The business logic then dictates the transmission of the object to the appropriate stakeholders in real-time in order to keep them informed and seek feedback.

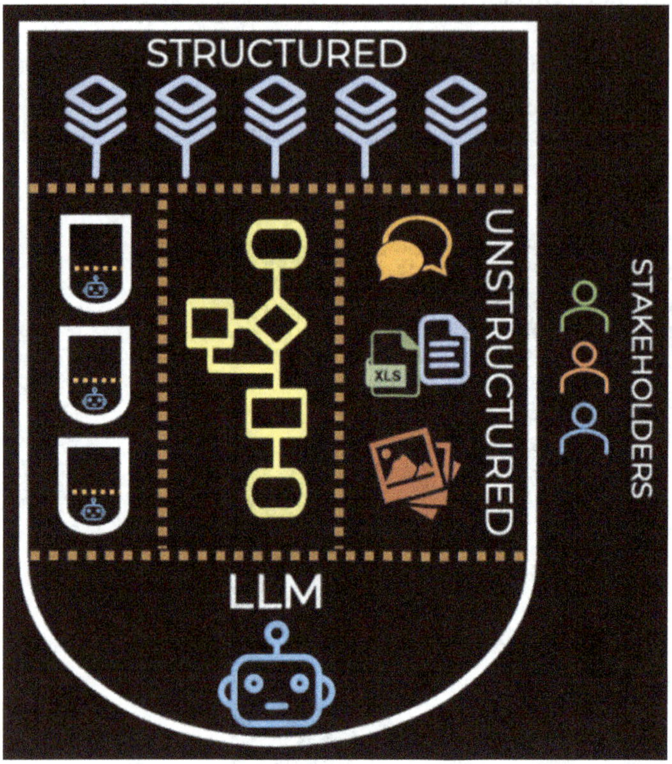

Figure 1B: The depiction of an intelligent super object, containing a group of related intelligent objects.

In OMIO, objects can be associated with other objects, based on some common attribute or state. In this fashion, they can form an intelligent super object. This can be akin to biology where a collection of cells can form an organ. The new super object is aware of its constituents and can communicate with them using LLM in a broadcast fashion or one at a time. The super object can have its own stakeholders and business logic, independent of its constituents. This also helps eliminate any duplicate messages, thus reducing noise in communication by aggregating similar messages and delivering it to the appropriate stakeholders only once. Again, an example from biology may make this clear. Cells can form organs by aggregating their efforts and communicate with each other within a framework or system. And organs form systems by connecting with each other, using arteries, veins, and the nervous system. OMIO is inspired by this biological model and uses LLMs as only one element to make all the constituents in a system intelligent, self-aware and capable of communicating using natural language.

Figure 1C: Depicts an OMIO OS with all OMIO-style apps interacting using natural language. The master LLM is the one at the center under the mic symbol. It is responsible to interact with the user.

Below is the execution of a planned strategy by the master LLM, as a response to "OMI, I am hungry". In the figure above, you can observe all the OMIO based apps installed on the device and being controlled by the master LLMs. The master becomes aware of all the apps during installation and an app informs the master LLM of its capabilities, its authentication requirements, etc. The master LLM then remembers this new capability and includes it in its future strategy to fulfill the user's requirements.

13.2 Strategic application interaction

Diet app consultation: OMIO OS queries the diet application, asking, "What is planned for lunch today?" The diet app, analyzing the user's meal history and nutritional plan, suggests "Fish for lunch," considering variety and health.

Budget app inquiry: Concurrently, OMIO OS checks with the budget app about the day's meal allocation, receiving a response of "$15" for lunch.

Restaurant app coordination: Armed with dietary recommendations and budget constraints, OMIO OS communicates with the restaurant app: "Find me the closest fish restaurants with a budget of $15." The restaurant app processes this request, identifying suitable options within the specified parameters.

User notification: OMIO OS consolidates the information and informs the user, "According to our budget and diet plan for today, here are a few fish restaurants nearby," displaying the options.

13.3 Advantages of OMIO OS

Simplification of user interaction: By leveraging natural language, OMIO OS drastically reduces the learning curve associated with new technology, making digital tools more accessible to a broader audience.

Seamless integration: The OMIO model ensures that all applications can communicate effectively, breaking down silos between different services and functionalities.

Personalization: The system tailors responses and suggestions to the user's unique preferences, history, and needs, enhancing the overall user experience.

Efficiency and productivity: By handling the coordination between different applications, OMIO OS streamlines tasks that would typically require multiple steps, saving users time and effort.

14 Conclusion

The Object Messaging Model, coupled with Intelligent Objects, adeptly integrates structured data and natural language communication, incorporating additional resources like videos, audio, internet links, and identifying information about relevant stakeholders and human resources within the object. Utilizing its LLM, the object continually updates itself with this plethora of information, thereby relieving humans from the tasks of organizing and directing information flow within the system. Understanding its stakeholders, their roles, and their connection to the object, it efficiently manages the distribution of pertinent events and information promptly. This task, critical in timely informing the right stakeholders, has traditionally seen human involvement as the least reliable factor. OMIO revolutionizes this aspect.

The end product of this innovation is an ensemble of these self-aware intelligent objects or "super objects," each embedded with an LLM that has thoroughly processed the object's structured data, associated attachments, and relevant discussions. These objects are primed to respond to any queries from human stakeholders as soon as they appear in their inboxes. They are adept at navigating through the workflow, autonomously determining their pace and timing at each stage, and conveying crucial information to their stakeholders in real-time. Encapsulated within a sophisticated messaging framework, these objects possess an advanced understanding of the application's nature and context, ensuring they are routed to the correct "inbox." named appropriately, according to the application's type and class. This overarching framework serves as the central "brain" of the application, orchestrating and interacting with these intelligent entities or super components.

15 Project OMADEUS

15.1 Project management and collaboration built with OMIO (patents pending)

To demonstrate the benefits and practical implementations of such a model, we are making available project management software designed with this approach. This is the first of its kind OMIO-based project management, collaboration, documentation, and communication too, unified in one application that guarantees optimal and real-time intelligent communication amongst the participants, including all the stakeholders. It is available for a free trial on www.omadeus.com to give you the experience and demonstrate the clear benefits of the OMIO model and the new world to come, a world without email frenzies and frivolous channel chatters. With this new model, intelligent objects, and not humans, manage the routing and on time delivery of events and messages to the right stakeholders, and AI is omnipresent at every point of user interaction, fully informed about the issue at hand and ready to assist, teach, and learn from you! Figure 2 below is a screenshot of Omadeus presenting all the Objects that have something to report to the stakeholder. These objects not only deliver comments from other stakeholders but report on significant events in their lives as they occur in real-time. When they show up in the inbox, they contain the object with all its components, including the chat channels, and can be examined from 360 degrees and queried via natural language. The object is self-aware and can inform other stakeholders of the interactions of the current stakeholder. As you can see, it is the object that is the pivot of communication between all the stakeholders. In this way, the object's intelligence and awareness of itself and its stakeholders guarantees the delivery of important messages and events in the form of messages to the appropriate stakeholders at the right time creating an optimal communication and completely eliminating noise (patents pending). In the screen below, the object highlighted is a feature being developed in a software project and has landed in the user's inbox to deliver the latest on its status. The user has opened up the LLM assistant (second column with the Omadeus icon) and requested a summary of all the conversations and events (contained in the third column) to this object. This way, the user is not forced to read all the messages and can quickly get up to date with the most significant events and can also provide feedback to the object and all the other stakeholders at the same time. All the files, including images, videos, dialogs, etc., are kept and managed by the object itself, removing humans from this burden. By way of analogy, similar to a neural net, optimization of communication between human nodes can bring about group collective intelligence. Add to that the timely assistance of an AI at every juncture for humans, that is the true recipe for humans and machine collective intelligence!

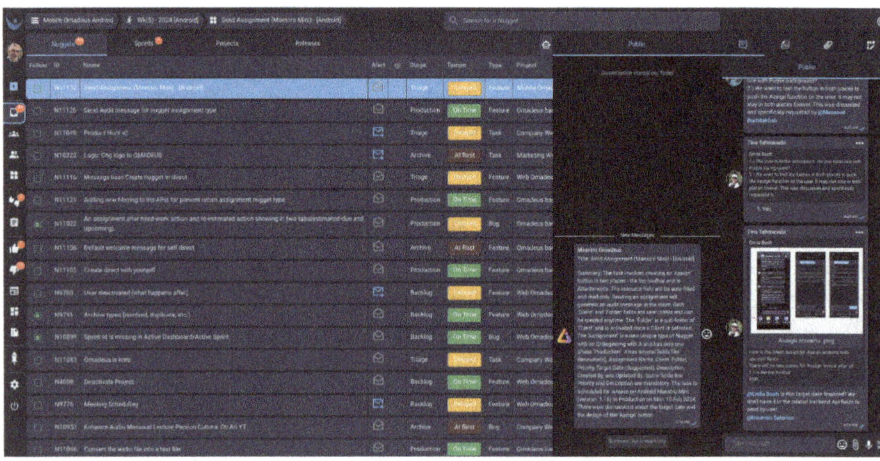

Figure 2: OMADEUS: The Inbox view depicting a "Feature Object," providing a summary to the stakeholder by its embedded LLM.

Disclosure: The author has an executive position with OMADEUS, with multiple patents pending for this technology and the manufacturer of OMADEUS software product based on OMIO.

References

[1] M. Alibakhsh, "System and Methods for Optimal and Synchronized Workflow-based Communication," in *2021 International Conference on Computational Science and Computational Intelligence (CSCI)*. Las Vegas, NV, USA, pp. 1451–1453, 2021.

[2] L. Qian *et al.*, "Cloud Computing: An Overview," in *Cloud Computing: First International Conference, CloudCom 2009, Beijing, China, December 1–4, 2009. Proceedings 1*. Berlin Heidelberg: Springer, 2009.
[3] B. Hayes, "Cloud Computing," *Communications of the ACM*, vol. 51, no. 7, pp. 9–11, 2008.
[4] G. Venolia *et al.* "Supporting Email Workflow." microsoft.com (2001).
[5] K. McMurtry, "Managing Email Overload in the Workplace," *Performance Improvement*, vol. 53, no. 7, pp. 31–37, 2014.
[6] L. A. Dabbish and R. E. Kraut, "Email Overload at Work: An Analysis of Factors Associated with Email Strain," *Proceedings of the 2006 20th Anniversary Conference on Computer Supported Cooperative Work*, 2006.
[7] K. Riemer and A. Tavakoli. "The role of groups as local context in large Enterprise Social Networks: A Case Study of Yammer at Deloitte Australia." http://hdl.handle.net/2123/9279 (2013).
[8] D. Wang *et al.* "Slack channels ecology in enterprises: How employees collaborate through group chat." (2021).
[9] B. Ives, "Graphical User Interfaces for Business Information Systems," *MIS Quarterly*, pp. 15–47, 1982.
[10] B. A. Myers "51. Graphical User Interface Programming." (2004).
[11] C. Standing, "Methodologies for Developing Web Applications," *Information and Software Technology*, vol. 44, no. 3, pp. 151–159, 2002.
[12] M. Lu and W. Yeung, "A Framework for Effective Commercial Web Application Development," *Internet Research*, vol. 8, no. 2, pp. 166–173, 1998.
[13] G. Rossi *et al.*, eds. *Web Engineering: Modeling and Implementing Web Applications*. Springer Science & Business Media, 2007.

Xiaowei Xu*, Bi T. Foua, Xingqiao Wang, Vivek Gunasekaran, John R. Talburt

Leveraging large language models for efficient representation learning for entity resolution

Abstract: In this paper, the authors propose TriBERTa, a supervised entity resolution system that utilizes a pre-trained large language model and a triplet loss function to learn representations for entity matching. The system consists of two steps: first, name entity records are fed into a Sentence Bidirectional Encoder Representations from Transformers (SBERT) model to generate vector representations, which are then fine-tuned using contrastive learning based on a triplet loss function. Fine-tuned representations are used as input for entity matching tasks, and the results show that the proposed approach outperforms state-of-the-art representations, including SBERT without fine-tuning and conventional Term Frequency-Inverse Document Frequency (TF-IDF), by a margin of 3–19%. Additionally, the representations generated by TriBERTa demonstrated increased robustness, maintaining consistently higher performance across a range of datasets. The authors also discussed the importance of entity resolution in today's data-driven landscape and the challenges that arise when identifying and reconciling duplicate data across different sources. They also described the ER process, which involves several crucial steps, including blocking, entity matching, and clustering.

Keywords: CCS CONCEPTS, representation learning, large language models, entity resolution

Additional Keywords and Phrases: contrastive learning, entity matching, cross-encoders models, TF-IDF

Acknowledgment: This research was supported in part by the US Census Bureau Cooperative Agreement CB21RMD0160002 for Record Linkage. This project used the facilities provided by the Arkansas High Performance Computing Center supported in part by grants from the National Science Foundation grants #0722625, #0959124, #0963249, #0918970 and a grant from the Arkansas Science and Technology Authority. This project was partially supported by the National Science Foundation under Award No. OIA-1946391.

***Corresponding author: Xiaowei Xu**, University of Arkansas, Little Rock, e-mail: xwxu@ualr.edu
Bi T. Foua, Xingqiao Wang, Vivek Gunasekaran, John R. Talburt, University of Arkansas, Little Rock

https://doi.org/10.1515/9783111344126-017

1 Introduction

The digital age has ushered in an era in which data are abundant, but the true challenge lies in understanding and processing these data effectively. Representation learning, with its ability to transform raw data into meaningful formats, has emerged as a beacon for this challenge, particularly in fields such as computer vision and information extraction. Automating the discovery of optimal data representations offers a fresh perspective on traditional tasks, thereby enhancing performance and efficiency.

Duplicate data, a pervasive issue in today's landscape, often incur significant costs in terms of money, time, and resources. Different organizations, from businesses to government agencies, grapple with the challenge of identifying and reconciling these duplicates, whether they originate from a single source or multiple disparate sources. This critical task is termed entity resolution (ER), which is a comprehensive process that addresses the challenge of identifying and linking records that refer to the same real-world entity across different data sources. ER becomes particularly complex when data are riddled with inconsistencies or when standardization is lacking. Consider, for example, an e-commerce database in which a single product, such as an "Apple MacBook Pro M2," can manifest in various ways. It may be represented as "Apple, M2 MacBook Pro" in one record or as "MacBook M2 Pro, Apple" in another, despite both referring to the same real-world entity. Similarly, a person named "John Tim Joe, 2022 Sunset Dr apt 217" in one record might appear as "John T Joe, 2022 Sunset Dr" in another record. Such discrepancies require rigorous identification, cleansing, and reconciliation. As described in [1], the ER process involves several crucial steps.

1. *Blocking (or indexing)*: Given the quadratic nature of the ER problem, in which every description should be compared to all others, blocking is applied as an initial step to reduce the number of comparisons. It groups similar descriptions into blocks based on certain criteria, ensuring that comparisons are executed only between descriptions co-occurring in at least one block. This step quickly segments the input-entity collection into blocks, approximating the final ER result.
2. *Entity matching*: This step involves applying a function that determines whether a pair of entity descriptions matches. Typically, a similarity function measures the similarity between two descriptions, with the aim of minimizing false-positive or false-negative matches.
3. *Clustering*: The final task in the ER workflow groups the identified matches together, ensuring that all descriptions within a cluster match. This step infers indirect matching relations among the detected pairs of matching descriptions, thereby overcoming the potential limitations of the employed similarity functions.

A common thread that runs through all these steps of ER, as described above, is the necessity of grouping duplicates together. Whether it is during blocking, entity matching, or clustering, the objective is to bring similar entities closer while pushing dissimilar ones apart. This underlying but important requirement forms the basis of our proposition. In light of this, we unveil TriBERTa, a groundbreaking representation-learning approach tailored for entity resolution. Through TriBERTa, we learn representations that inherently group similar entities together, while separating dissimilar ones, creating a foundational asset that can be leveraged across all steps of the ER process. The learned representation can be applied to each step of the entity resolution process, as shown in Figure 1. This approach harnesses the power of pre-trained language models and representation learning via a triplet loss function. It is important to mention that while our current evaluation focuses solely on entity matching, the potential of TriBERTa extends to other facets of entity resolution, including blocking and clustering, by virtue of its ability to learn representations that group similar entities together and separate dissimilar ones. To encapsulate our contributions:

- We pioneered a representation learning methodology for entity resolution with versatility to span the entire ER process, such as entity matching, data blocking, and data clustering.
- Our research rigorously evaluates the efficacy of learned representations using entity matching as a primary case study within the ER framework.
- A comparative analysis of dedicated end-to-end entity-matching models reveals the robustness and adaptability of our approach across diverse datasets.
- To ensure transparency and reproducibility, we offer extensive data preparation and analysis, enabling peers to replicate and build on our findings.

This chapter is an extension of work originally presented in the paper titled 'Large Language Model-Based Representation Learning for Entity Resolution.

Using Contrastive Learning' [24] which was accepted for presentation at the 2023 International Conference on Computational Science and Computational Intelligence (CSCI) held in Las Vegas, NV, USA.

The remainder of this paper is structured as follows: Section 2 discusses related work with a focus on the entity-matching task, followed by a detailed exposition of our proposed TriBERTa method in Section 3. Section 4 presents the evaluation results, and we conclude with prospective future directions in Section 5.

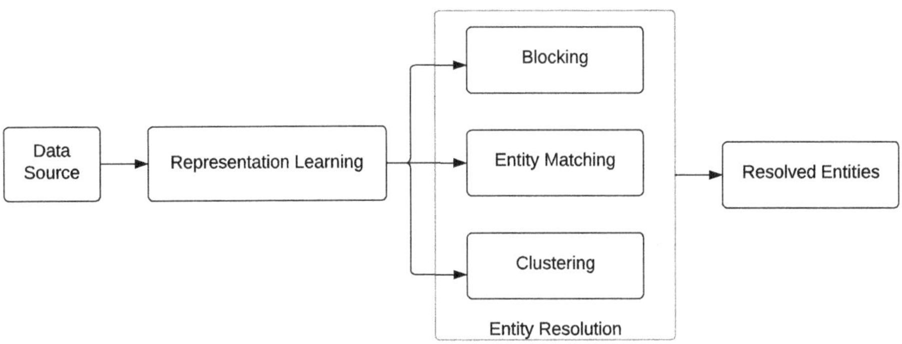

Figure 1: Our framework for entity resolution.

2 Related work

2.1 Entity resolution

Entity resolution has always been a subject of extensive research. While there is no dearth of research on ER, especially in domains like e-commerce, with papers leveraging benchmark datasets such as the "Amazon-Google," "Abt-Buy," "WDC products," "Google-Scholar," "iTunes-Amazon," and ACM datasets [2–6], a significant portion still leans on traditional methodologies.

A glaring gap in the current research landscape is the exploration of representation learning for ERs. Many early studies on ER relied on crowdsourcing approaches. Most crowdsourcing approaches rely heavily on human intervention for proper functioning. Some examples of these crowdsourcing platforms include Amazon Mechanical Turk (AMT) and Crowdflower, which benefited from simple tasks performed by people who were compensated for their efforts. However, crowdsourcing techniques are expensive and unsuitable for production environments [7]. The feasibility of human intervention diminishes as the size of the dataset increases because of the exponential growth in the number of required comparisons.

Subsequent research has helped develop ER-Systems such as Magellan [8, 9] and DeepMatcher [8, 9]. While these systems eliminate the need for human intervention, they often exhibit suboptimal performance (F1-scores), particularly when tested on unseen or noisy data [10]. For instance, on datasets with introduced typos or dropped tokens, Magellan and DeepMatcher yielded unsatisfactory results, rendering them unreliable for use in production.

Recently, the focus has shifted towards more deep learning (DL) approaches for ERs that solely focus on the entity matching task. Notable methods such as KAER[11], JoinBERT[10], SupCon[2], BERT[12], and Ditto[13] have all employed cross-encoders. Although these DL methods have shown promise in achieving good results for entity matching, they often fail to provide embedding for every input record. Having embedding for each record not only facilitates entity matching but also significantly eases the execution of other ER tasks such as clustering or blocking. Therefore, the provision of embedding emerges as a crucial requirement for a holistic and effective ER framework.

Our work with TriBERTa is rooted in the premise of representation learning, which we posit is a pivotal mechanism to bridge the existing gaps in ER tasks. The representations learned through TriBERTa are engineered to group similar entities while separating dissimilar ones, forming a foundational asset that can be leveraged across all steps of the ER process. Although our evaluation in this study is centered on entity matching, the essence of our approach is to demonstrate that a robust representation learning methodology can indeed be a game changer for all facets of ER, including blocking, entity matching, and entity clustering.

2.2 Representation learning: contrastive learning

Representation learning has witnessed resurgence in recent years, largely credited to groundbreaking advancements in computer vision and information extraction [14, 15]. In 2018, Wu et al. demonstrated by their experimental results that under unsupervised settings, contrastive representation learning results surpassed the state-of-the-art by a large margin [16]. In 2019, Henaff et al. developed a method that surpassed fully supervised pre-trained ImageNet results using unsupervised contrastive learning to improve transfer learning object detection on the PASCAL VOC dataset [17]. In 2020, Yonglong et al. achieved state-of-the-art results on unsupervised image and video learning benchmarks using contrastive learning [18].

Recently, representation learning, also known as contrastive learning, has been rapidly extended to Natural Language Processing (NLP) for a variety of tasks, including semantic text similarity, semantic search, translation sentence mining, product matching, and notably, entity matching [2, 3]. The fundamental principle of contrastive learning is straightforward: If two records are similar, they are pulled together in the embedding space[1]; conversely, if they are dissimilar, they are pushed apart.

This principle aligns seamlessly with the core objective of entity resolution, which necessitates the grouping of similar entities and separation of dissimilar entities across all its facets – blocking, entity matching, and entity clustering. Contrastive learning, as a DL approach, inherently facilitates the grouping of similar entities in the same embedding space while separating dissimilar entities. This characteristic is pivotal for the effective execution of all ER tasks, underscoring the potential of representation learning and, by extension, contrastive learning as a robust mechanism for advancing ER methodologies. Through this lens, contrastive learning emerges not merely as a technique for enhancing entity matching but as a comprehensive approach capable of significantly improving the broader spectrum of entity resolution tasks.

2.3 Triplet loss

Triplet loss is a loss function used in machine algorithms, in which positive and negative inputs are compared to a reference input called an anchor. Triplet loss is a specific type of contrastive learning loss function. The main idea is rooted in the context of nearest neighbor classification [19]. Given a triplet (anchor, positive, negative), the triplet loss function maximizes the difference between the anchor and negative inputs and minimizes the distance between the anchor and positive input. The loss function for one record can be calculated using the Euclidean distance function:

1 embedding space or vectors space is the k-dimensional space in which records are represented as vectors.

$$(A,P,N) = \max\left(||f(A) - f(P)||^2 - ||f(A) - f(N)||^2 + \alpha, 0 \right)$$

where A is an anchor input, P is a positive input of the same class as A, N is the negative input of a different class from A, α is the margin (distance) between the positive and negative pairs, and f is a default representation (an embedding).

Figure 2 illustrates the application of triplet loss in FaceNet [19]. In FaceNet, a convolutional neural network (CNN) is trained to optimize the embedding (representations) of the images [20]. A CNN is a type of artificial neural network that is widely used in computer vision and image recognition. As the first two images from the top left represent the same entity, the triplet loss function pulls them together (represented by the inward colliding arrows). For the two pictures in the bottom left, triplet loss increases their difference and pushes them apart (represented by arrows moving apart in opposite directions). This method allows for much greater representational efficiency and better identification of the same images.

The application of triplet loss extends beyond image recognition to the ER domain. By optimizing the representations of entities, triplet loss facilitates the crucial task of grouping similar entities together while separating dissimilar ones, which is a fundamental requirement across all facets of ER – blocking, entity matching, and entity clustering. Learned representations serve as a robust foundation that can be leveraged to enhance the efficiency and accuracy of the ER process.

2.4 Sentence BERT

Bidirectional Encoder Representations from Transformers (BERT) language model[2] is an open-source Transformers model[3] for Natural Language processing. It helps the computer understand the meaning of a text or word by using the words or text surrounding it. The BERT language model uses surrounding words or text to establish context. Sentence BERT (SBERT) was first introduced in 2019 by Nils Reimers and Iryma Gurevych as a modification of the pre-trained[4] BERT Language Model [21].

SBERT reduced the computation time used by BERT for finding the most similar pairs of sentences by a factor of 46,800 from 234,000 s (65 h) to 5 s while maintaining the accuracy of BERT [21]. It was mainly developed as a bi-encoder, as opposed to a cross-encoder. Bi-encoders are able to produce embedding for a given sentence. Figure 3

2 Language Model is probabilistic distribution over a sequence of words. It helps predict which word is more likely to appear next in a sentence.

3 Transformers is a novel deep learning architecture that solves sequence-to-sequence NLP tasks and can handle long-range dependencies in texts. Popular language models such as BERT and GPT come from transformers. For a detailed analysis, readers are encouraged to read "Attention is All You Need."

4 The model is already trained on unlabeled data over similar tasks.

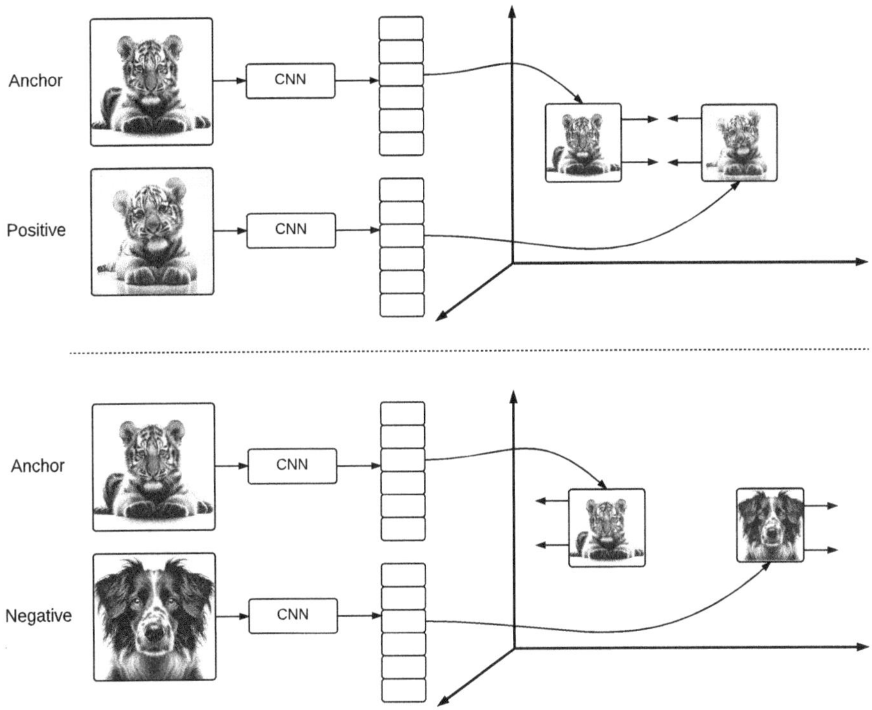

Figure 2: Triplet loss representation [19].

shows a bi-encoder and a cross-encoder. In the bi-encoder, two sentences were independently fed to the BERT. BERT then outputs two vectors (or embedding) u and v, which can be compared using a distance metric, such as cosine similarity, Manhattan distance, or Euclidean distance. In contrast, the cross-encoder did not provide any embedding. In the cross-encoder, both sentences are fed to BERT to produce a value between 0 and 1.

SBERT was fine-tuned[5] using Siamese and triplet structure networks to capture meaningful similarities between sentences. The innovation brought about by SBERT in generating embedding for sentences holds significant promise in the domain of entity resolution. By efficiently producing embedding for textual descriptions of entities, SBERT facilitates the grouping of similar entities while separating dissimilar entities across all facets of ER: blocking, entity matching, and entity clustering. The ability to generate meaningful embedding quickly and accurately is a cornerstone of effective representation learning, which, as previously discussed, is pivotal for advancing ER methodologies.

5 the model is first initialized with the pre-trained parameters; then all parameters are fine-tuned using labeled data from the downstream tasks.

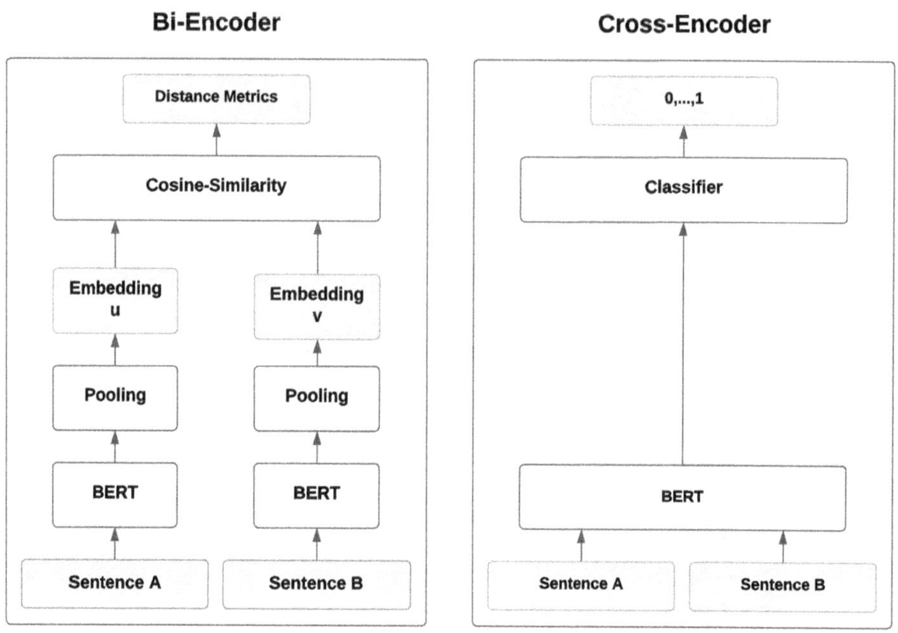

Figure 3: Bi-encoder and cross-encoder architecture [22].

3 Methodology

As mentioned earlier, although we developed an approach that is applicable to all ER tasks, we only evaluated it on the entity-matching task. Our approach, TriBERTa, comprises two steps:

1. First, the entity records are used as inputs to fine-tune a Language Model and get embedding for every record,
2. Second, a classification task (or entity matching task) is performed using these embedding to determine whether two entities are a match. As our application and evaluation are constrained to entity matching, the second step of this approach is classification. For other facets of the ER, the second step could be blocking or clustering. The embedding architecture we used is similar to the one used in FaceNet; however, instead of images as inputs, we are using text data and replacing the CNN model with a language model, SBERT in our case. In addition, we performed a simple classification task using the embedding obtained from the language model. A simple logistic regression[6] model

6 Logistic Regression is a machine learning algorithm used in classification task to analyze the relationship between some dependent variable and a set of independent variables.

was used to demonstrate the performance of this approach. Figure 4 shows the overall TriBERTa framework, which is further explained in the following sections.

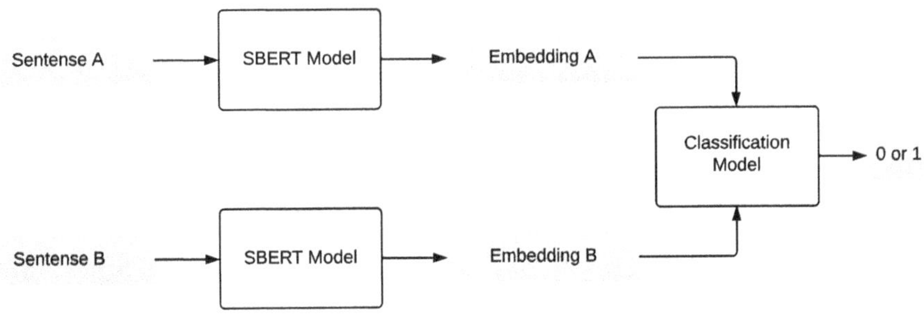

Figure 4: TriBERTa framework.

3.1 Entity resolution embedding framework using triplet loss

The first step in the overall approach is to determine the embedding (or vectors) for each record. As we know by now, all of the entity resolution steps require that same or duplicate entities be grouped together, so to make sure that every record is mapped correctly in the embedding space we chose triplet loss as a loss function. The goal is to pull together as close as possible vectors that are similar to one another, and push as far as possible vectors that are dissimilar.

As described in Section 2.3, one advantage of contrastive learning through triplet loss is pulling the anchor and positive as close as possible and pushing negative as far as possible from the anchor. Figure 5 shows the framework of the embedding phase. Instead of using a pair of records, an approach used in various studies [2, 3, 12], our embedding framework uses three records or triplets. The three sentences (or records) were fed independently to the SBERT model.

3.1.1 Data set preparation for the embedding framework

The embedding framework requires triplet records to be fed into SBERT independently. For this, we modified our datasets to anchor, positive, and negative. Therefore, for each instance (represented as anchor) in the dataset, an instance with the same entity label or id (represented as positive) and another instance with a different entity label (represented as negative) were randomly selected to generate the dataset.

Figure 6 illustrates the data preparation for the triplet loss function. The first table (or origin data) from the left shows that every record has id truth (or Truth ID). The ID truth corresponds to the class of each record. Two records with the same id truths were duplicate records (or the same). As we can observe from the table, the first two records

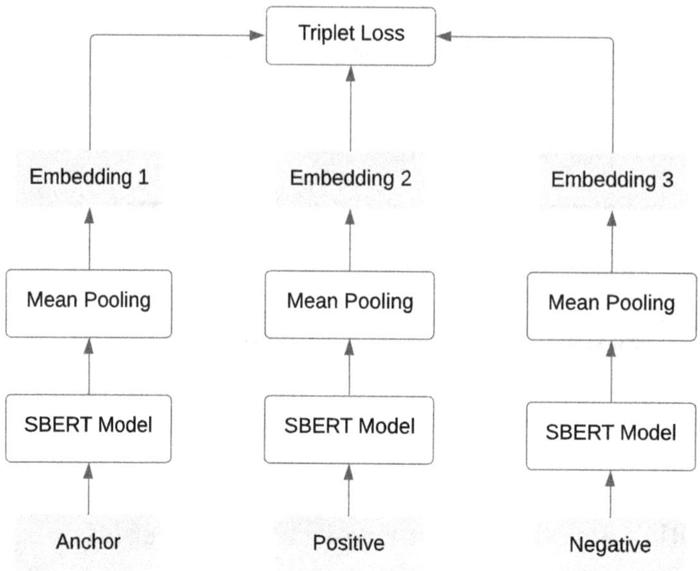

Figure 5: Embedding framework.

Jane Mary Doe and Jane M. Doe are duplicates; therefore, they have the same Truth ID. Now that we know duplicate records by their ID truth, we can easily create the second table and third table from the left. For instance, for the first record Jane Mary Doe (our first anchor), we randomly select Jane M. Doe (our first positive) and William P Smith (our first negative). In addition, we dropped rows with null values at the end of the modification.

It is important to note that every positive or negative has an equal probability of being selected. For example, if we were to have another duplicate record for the name Jane Mary Doe, Jane M. Doe and that duplicate would have an equal chance of being randomly selected as a positive.

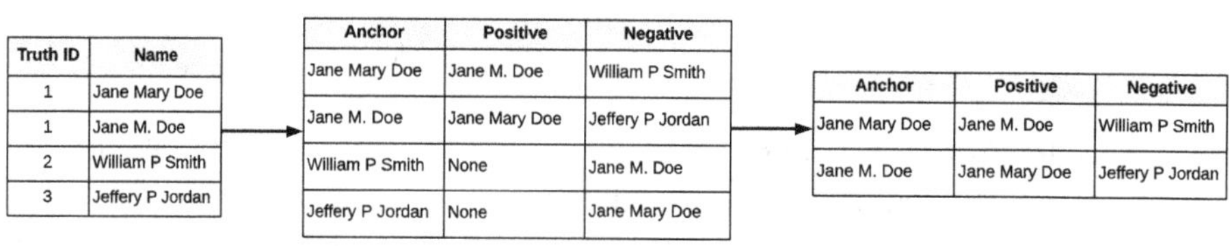

Truth ID	Name
1	Jane Mary Doe
1	Jane M. Doe
2	William P Smith
3	Jeffery P Jordan

Anchor	Positive	Negative
Jane Mary Doe	Jane M. Doe	William P Smith
Jane M. Doe	Jane Mary Doe	Jeffery P Jordan
William P Smith	None	Jane M. Doe
Jeffery P Jordan	None	Jane Mary Doe

Anchor	Positive	Negative
Jane Mary Doe	Jane M. Doe	William P Smith
Jane M. Doe	Jane Mary Doe	Jeffery P Jordan

Figure 6: Data preparation for triplet loss function.

3.1.2 Model selection and fine tuning

Different language models were considered for this task. To select the best model for our embedding task, we chose 14 language models and compared their performance for nearest neighbor search based on cosine similarity. The nearest neighbor should be a positive entity for each anchor.

To choose the best SBERT model, we fine-tuned the 14 language models[7] on a small sample dataset (restaurant dataset of 100 records), similar to our datasets, and compared their nearest neighbor searches for similar batch sizes and epochs. We used the *all-distill RoBERTa-v1,* a RoBERTa model, because it provided the highest accuracy (0.986) of all models tested, as shown in the Appendix table.

To produce a fixed-size output vector for each of our records (or inputs), we added a mean pooling layer (see Figure 5). The mean pooling layer provides the average of all embedding that *all-distill RoBERTa-v1* gives us. This provides us with a fixed embedding vector of 768 dimensions, regardless of the length of the input record.

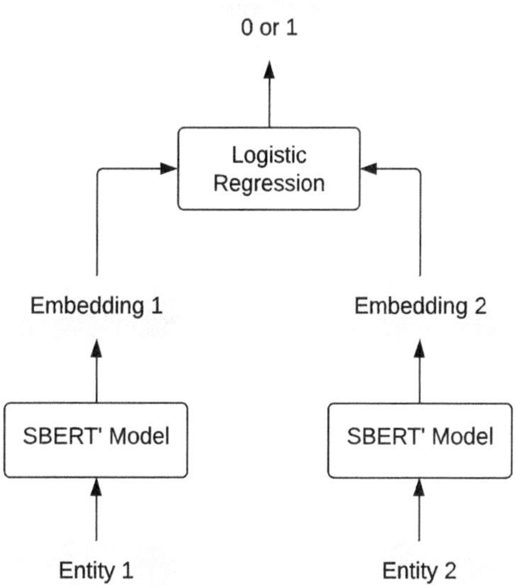

Figure 7: Pairwise classification framework.

7 Each model has a different parameters and architecture. See appendix table for the list of the models considered and their respective cosine similarity accuracies. More information on each model is available on https://huggingface.co/models or https://www.sbert.net/docs/pretrained_models.html.

3.2 Pairwise classification task (entity matching)

The second step in our approach is the application of entity matching, which is a pairwise classification task. As mentioned earlier, although we use entity matching as a second step to evaluate our approach, the second step after obtaining the embedding is blocking, entity matching, or clustering, which are all steps in the entity resolution.

To prove the efficacy of our approach, we used the basic logistic regression model in the second step to classify each pair of entities as a match or no match. Using the fine-tuned *all-distill RoBERTa-v1* model (referred to as SBERT hereafter for simplicity) chosen above from the SBERT package, we find the embedding of each record, which is then fed into the logistic regression model to classify every pair of records as match or no match, as shown in Figure 7.

For this purpose, we modified our datasets for a binary classification task. There are multiple ways to prepare datasets for binary classification tasks. One method is to use the original dataset and select for every record in a dataset a record that matches and another record that does not. Another method uses triplet datasets. For every record composed of anchor, positive, and negative in the triplet, the first two columns (anchor and positive) will represent a match and the first and last columns will represent a no match; therefore, each triplet generates two training samples: one is labeled 1 and the other is labeled 0. We chose the latter (triplet data to the classification dataset) because it maintains a 50/50 split in positive and negative labels, as shown by the following lemma:

$$\forall A \in D \exists P, N \in D$$

that is, for all Anchor A in Dataset D, there exists a positive sentence P and a negative sentence N in dataset D.

This modification is illustrated in Figure 8. Our final classification dataset has twice as many records as the triplet dataset. For the first record, Jane Mary Doe (our first anchor) and Jane M. Doe (our first positive) were selected as positive match with label 1. Jane Mary Doe (our first anchor) and William P Smith (our first negative) are selected as a negative match with label 0.

4 Experimental results

In our comprehensive evaluation, we rigorously tested our framework on three widely recognized datasets, benchmarking our results against state-of-the-art representations [21], including the original SBERT model devoid of triplet loss (referred to as RoBERTa) and the conventional TF-IDF method (referred to as TF-IDF). Notably, the underlying model for the SBERT versions remained consistent: the all-distill RoBERTa-v1.

This distinction arose from the fine-tuning process, with our approach leveraging the datasets for this purpose. Both TF-IDF and non-fine-tuned SBERT were employed

Anchor	Positive	Negative
Jane Mary Doe	Jane M. Doe	William P Smith
Jane M. Doe	Jane Mary Doe	William Peters Smith
William P Smith	William Peters Smith	Jane M. Doe
William Peters Smith	William P Smith	Jane Mary Doe

Record 1	Record 2	Label
Jane Mary Doe	Jane M. Doe	1
Jane Mary Doe	William P Smith	0
Jane M. Doe	Jane Mary Doe	1
Jane M. Doe	William Peters Smith	0
William P Smith	William Peters Smith	1
William P Smith	Jane M. Doe	0
William Peters Smith	William P Smith	1
William Peters Smith	Jane Mary Doe	0

Figure 8: Data preparation for the classification task using triplets.

to derive distinct embedding. These embedding were then processed through one of the simplest yet most effective machine learning models, Logistic Regression, for pairwise classification. Our empirical evaluations underscored the prowess of learned representations in entity matching by setting new benchmarks against established representations.

Building on this foundation, we further embarked on a comparative analysis against dedicated end-to-end entity-matching models across various datasets, each presenting unique challenges. This broader evaluation illuminated the robustness and adaptability of our approach, especially when juxtaposed against models tailored specifically for the entity-matching task, reinforcing our belief in the versatility and efficacy of our representation learning approach.

4.1 Datasets

The datasets used to test our framework were predominantly sourced from the public domain, but we also explored other datasets, as mentioned in references [11, 13].

Initially, our primary datasets included the GeCo census dataset [23], Cora dataset [23], and restaurant dataset [23]. The GeCo census dataset, which contains address records of people living in the US, was synthetically modified by us to introduce more duplicates. It encompasses 19,993 records with details such as name, address, zip code, city, state, and SSN. The Cora dataset details scientific publications across different topics, with 1,295 records. The restaurant dataset has 868 records, each detailing aspects such as name, address, city, state, zip code, phone number, and other associated data. Table 1 summarizes the initial datasets used in this study. Notably, the GeCo census dataset had the highest duplicate count primarily because of its synthetic nature. This was followed by the Cora dataset and then the restaurant dataset.

To further validate and compare our methods against state-of-the-art cross-encoders dedicated to entity matching only, we employed three additional datasets: GoogleScho-

Table 1: Summary of the first three datasets.

Dataset	Total number of records	Total number of unique records
GeCo	19,993	9,996
Cora	1,295	112
Restaurant	868	756

lar, iTunes-Amazon, and ACM, all of which are detailed in references [11, 13]. These datasets, chosen for their unique characteristics and challenges, provided a more comprehensive evaluation platform for our framework. The datasets we used to further test our learning representation framework are structured or "dirty" (or unstructured.) Unstructured data contain noises (missing values, missing characters, etc.) that pose a challenge to any entity-matching model.

4.2 Metrics

We considered five metrics to assess the performance of our framework, including cosine similarity for the embedding step, and accuracy, precision, recall, and F1-scores for the classification step.

The metrics used were the common metrics used for classification and embedding tasks. The evaluation metrics are presented in Appendix 6.2 Evaluation Metrics. Cosine similarity is a measure of similarity between two sentences. The closer the value is to 1, the more similar the two sentences are. Accuracy measures the total number of values correctly predicted over the total number of records. Precision is the proportion of actual true positives correctly predicted by the classifier. This is a measure of the true positive rate in the dataset. Recall measures the total number of positive cases from all true positives.

4.3 Design of the evaluation

We designed an evaluation to test the performance of both parts of our framework: embedding and classification. We split each dataset into three: training, testing, and validation. For Step 1, the embedding part, we used the validation and training data to fine-tune our SBERT (i.e., *all-distill RoBERTa-v1*) model. For Step 2, the classification task, we used the training and test data to classify pairs of data. Figure 9 illustrates this phenomenon.

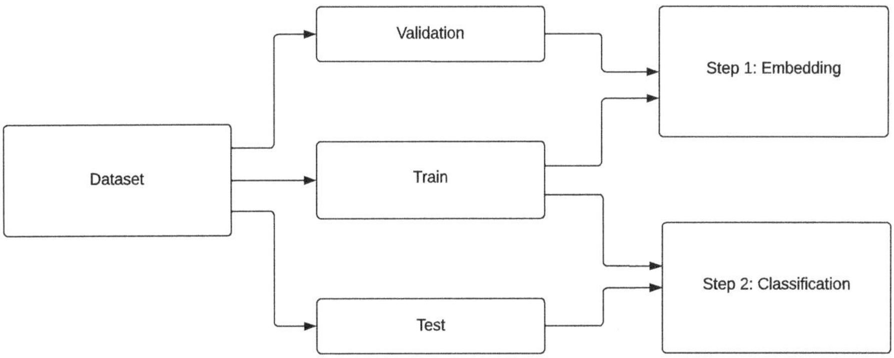

Figure 9: Design of evaluation framework.

4.4 Results

4.4.1 Embedding

Here, we show the results of the first step of our approach for the first three datasets used. In the first step, we trained a pre-trained language model and evaluated it on the validation data. Table 2 shows the training and validation results measured by cosine similarity accuracy, which is a measure of accuracy based on cosine similarity. Using a threshold of 0.5, we achieved an average cosine similarity accuracy of over 99% for all the three datasets. This indicates that the triplet loss methodology improves the representations such that the same entities are recognized as matches and different entities are recognized as non-matches and thus pushed away from one another. This resulted in vectors that were representative of the datasets used in the classification step.

Table 2: Triplet loss cosine similarity for the first three datasets.

Dataset		Number of records	Cosine similarity (%)
GeCo	Training	8,011	100
	Validation	2,670	99.9
	Total	10,681	–
Cora	Training	765	99.0
	Validation	255	98.8
	Total	1,020	–
Restaurant	Training	134	100
	Validation	45	100
	Total	175	–

4.4.2 Classification results (entity matching)

Here, we present the results of the second step of our framework. In the second step, we used the fine-tuned language model TriBERTa as a representation-learning approach and evaluated it on the test data. As depicted in Figures 10–12, the training and test results were measured using accuracy, recall, precision, and F1-score. Performance metrics are captured in Appendix 6.2.

TriBERTa outperformed all baseline methods by a margin of 3–19%. More specifically, TriBERTa improves the F1 measure by 5% when compared with RoBERTa + LR (which is the original SBERT model without fine-tuning for embedding plus logistic regression for entity matching) on average on all three datasets. The largest improvement is achieved in comparison with conventional TF-IDF + LR (which is TF-IDF as embedding plus logistic regression for entity matching), which is over 16% on average for all three datasets. Nevertheless, we observe a slight overfit in the restaurant data, with a drop of almost 5%. We believe that this is because the restaurant dataset was significantly reduced after we modified it to implement TriBERTa. The restaurant had the fewest duplicates, and thus the least amount of data to work with for the classification task. However, despite this deficiency, TriBERTa outperformed TF-IDF and RoBERTa.

Table 3 presents an assessment of the different models across both dirty and structured datasets. A significant metric that captures the eye is the average F1-score, which serves as a holistic indicator of a model's reliability across various datasets. The TriBERTa model recorded an average F1-score of 80.42%. What is remarkable about TriBERTa is not only its good performance on the dirty iTunes-Amazon dataset but also its steadfast consistency. It demonstrated a narrow oscillation in performance, with scores ranging from 72–91.81%. In contrast, the KAER model, which is a cross-encoder technique, attained an average F1-score of 84.82%. While at first glance this may seem commendable, it is crucial to observe the breadth of its performance oscillation. The KAER model exhibited scores that swung from a low of 54.90% to a peak of 98.99%. This wide oscillation suggests a pronounced sensitivity to dataset specifics, raising questions regarding its reliability across diverse real-world datasets. The broad oscillation in scores for models such as KAER might suggest unpredictable behavior when faced with unknown or new datasets.

It is worth noting that the limitation inherent to cross-encoders is their inability to yield embedding. Therefore, they are constrained only to entity matching. On the other hand, consistent performance, as displayed by TriBERTa, is invaluable in practical applications where data variability can challenge models. Another advantage of TriBERTa is its flexibility and wide range of applications. It can be used not only for entity matching, but also for clustering, data blocking, and other NLP tasks.

Table 3: F1 score of TriBERTa compared to cross-encoder approaches across different datasets [20].[8]

Models	Dirty data		Structured data		Average
	GoogleScholar	iTunes-Amazon	iTunes-Amazon	ACM	
TriBERTa + LR	82.90	72	75	91.81	80.42
Ditto	95.54	60.38	47.06	98.55	75.38
KAER	95.74	54.90	89.66	98.99	84.82

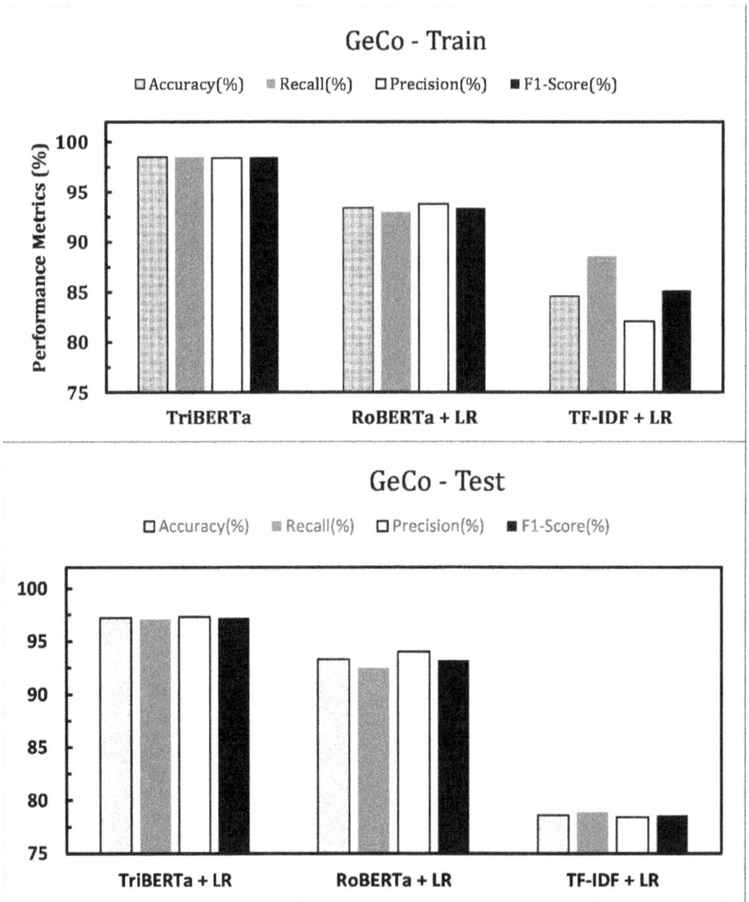

Figure 10: GeCo classification model performance.

8 Table data from [20].

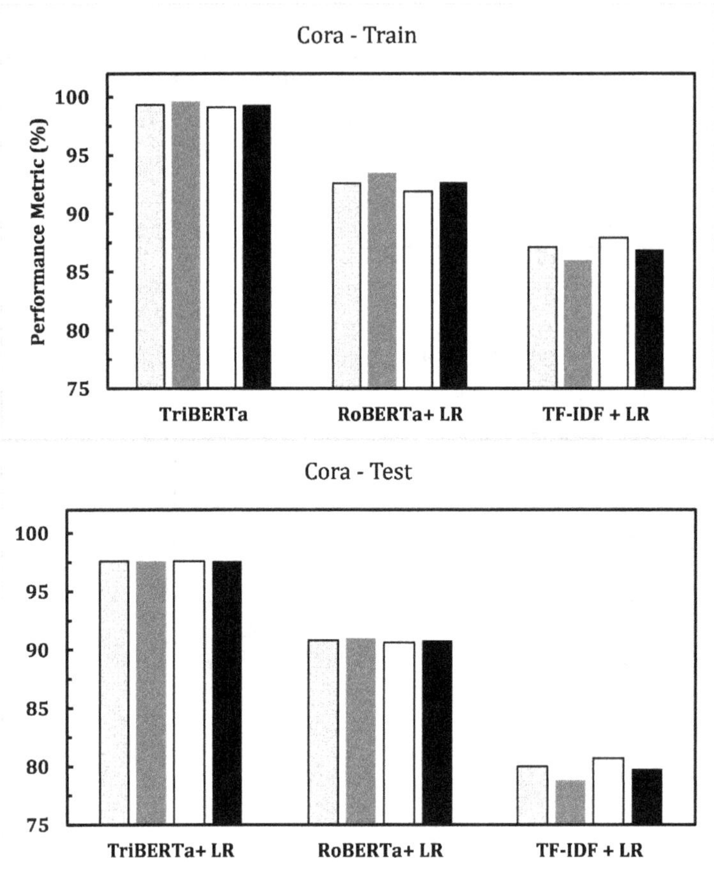

Figure 11: Cora classification model performance.

5 Conclusion and future work

In the vast landscape of data-driven research, the challenge often lies not in the sheer volume of data but in its effective interpretation and utilization.

Our work with TriBERTa underscores the transformative potential of representation learning to address this challenge, particularly within the realm of entity resolution. Through rigorous empirical evaluations, TriBERTa has demonstrated its capability to not only learn meaningful representations, but also to apply these representations effectively in the context of entity matching.

Our results, which span multiple datasets with varied characteristics, consistently highlight the superiority of TriBERTa's learned representations over traditional and state-of-the-art methods. Furthermore, the robustness exhibited by TriBERTa, especially

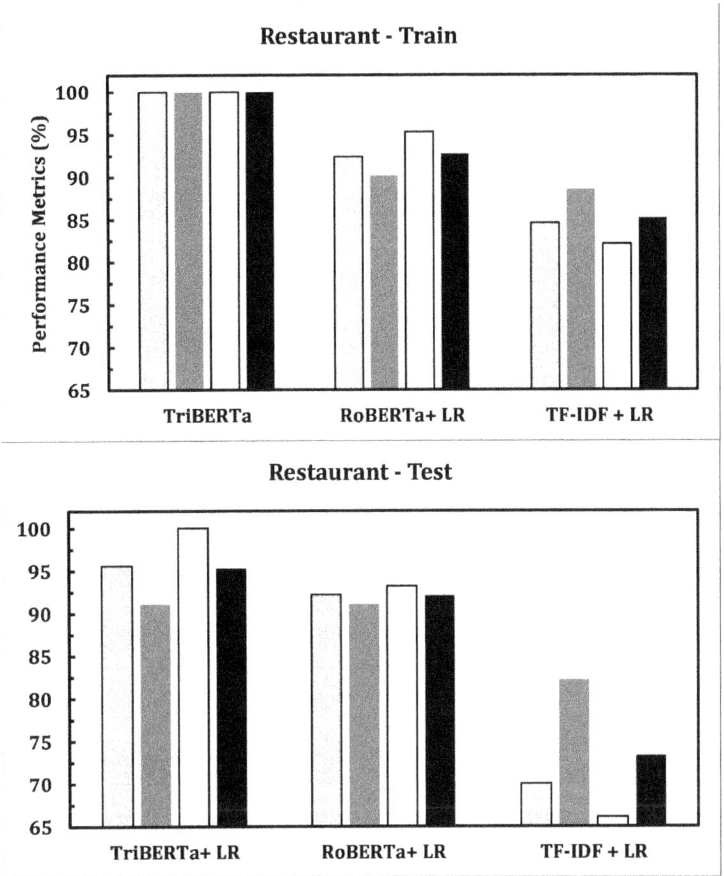

Figure 12: Restaurant classification model performance.

when compared with dedicated end-to-end entity-matching models, underscores its potential as a versatile tool in real-world scenarios characterized by data heterogeneity.

While our research is primarily evaluated on entity matching, the foundational principles of representation learning, as embodied by TriBERTa, hold promise for broader applications within the ER process. Future research could explore TriBERTa's efficacy in tasks, such as clustering, data blocking, and other NLP challenges.

In conclusion, TriBERTa stands as a testament to the power of representation learning, offering a fresh perspective on traditional ER tasks and setting the stage for future innovation in this domain.

6 Appendix

6.1 Models tested

Model name	Cosine similarity accuracy	Batch size	Epochs
distilBERT-base-uncased	0.77	64	10
RoBERTa-base	0.66	64	10
all-MiniLM-L12-v2	0.94	64	10
all-MiniLM-L6-v2	0.94	64	10
all-distilRoBERTa-v1	0.986	64	10
all-mpnet-base-v2	0.97	64	10
distiluse-base-multilingual-cased-v1	0.93	64	10
distiluse-base-multilingual-cased-v2	0.94	64	10
multi-qa-MiniLM-L6-cos-v1	0.9466	64	10
multi-qa-distilBERT-cos-v1	0.9666	64	10
multi-qa-mpnet-base-dot-v1	0.9666	64	10
paraphrase-ALBERT-small-v2	0.9266	64	10
paraphrase-multilingual-MiniLM-L12-v2	0.90	64	10
paraphrase-multilingual-mpnet-base-v2	0.98	32	10

6.2 Entity matching – classification – performance metrics

GeCo dataset	Model	Accuracy (%)	Recall (%)	Precision (%)	F1-score (%)
Train	TriBERTa	98.5	98.5	98.4	98.5
	RoBERTa + LR	93.4	93	93.8	93.4
	TF-IDF + LR	84.6	88.6	82.1	85.2
Test	TriBERTa + LR	97.2	97.1	97.3	97.2
	RoBERTa + LR	93.3	92.5	94	93.2
	TF-IDF + LR	78.6	78.9	78.4	78.6

Cora dataset	Model	Accuracy (%)	Recall (%)	Precision (%)	F1-score (%)
Train	TriBERTa	99.3	99.6	99.1	99.3
	RoBERTa + LR	92.6	93.5	91.9	92.7
	TF-IDF + LR	87.1	86	87.9	86.9
Test	TriBERTa + LR	97.6	97.6	97.6	97.6
	RoBERTa + LR	90.8	91	90.6	90.8
	TF-IDF + LR	80	78.8	80.7	79.8

Restaurant	Model	Accuracy (%)	Recall (%)	Precision (%)	F1-score (%)
Train	TriBERTa	100	100	100	100
	RoBERTa + LR	92.4	90.2	95.3	92.7
	TF-IDF + LR	84.6	88.6	82.1	85.2
Test	TriBERTa + LR	95.6	91.1	100	95.3
	RoBERTa + LR	92.2	91.1	93.2	92.1
	TF-IDF + LR	70	82.2	66.1	73.3

6.3 Evaluation metrics

Metrics	Symbol	Description (or formula)
True positive	TP	Number of records correctly classified as matches
True negative	TN	Number of records correctly classified as non-matches
False positive	FP	Number of records incorrectly classified as matches
False negative	FN	Number of records incorrectly classified as non-matches
Accuracy	Acc	(TP + TN)/(TP + TN + FN + FP)
Precision	P	TP/ (TP + FP)
Recall	R	TP/ (TP + FN)
F1-score	F1	2 * (P * R) / (P + R)
Cosine similarity	Similarity $(\boldsymbol{x}, \boldsymbol{y})$	$\cos(\theta) = \boldsymbol{x}.\boldsymbol{y} / \|\|\boldsymbol{x}\|\| * \|\|\boldsymbol{y}\|\|$

References

[1] V. Christophides, V. Efthymiou, T. Palpanas, G. Papadakis, and K. Stefanidis, "An Overview of End-to-End Entity Resolution for Big Data," *ACM Computing Surveys*, vol. 53, no. 6, Association for Computing Machinery, Feb. 01, 2021. doi: 10.1145/3418896.

[2] R. Peeters, and C. Bizer, "Supervised Contrastive Learning for Product Matching," *WWW 2022 – Companion Proceedings of the Web Conference 2022*, Association for Computing Machinery, Inc, Apr. 2022, pp. 248–251. doi: 10.1145/3487553.3524254.

[3] Z. Wang, B. Sisman, H. Wei, X. L. Dong, and S. Ji, "CorDEL: A Contrastive Deep Learning Approach for Entity Linkage," Sep. 2020, [Online]. Available: http://arxiv.org/abs/2009.07203.

[4] Paper With Code, "amazon-google dataset." Accessed: Sep. 12, 2023, [Online]. Available: https://paperswithcode.com/dataset/amazon-google

[5] Papers With Code, "Abt-Buy." Accessed: Sep. 12, 2023, [Online]. Available: https://paperswithcode.com/dataset/abt-buy.

[6] Papers With Code, "WDC LSPM Dataset." Accessed: Sep. 12, 2023, [Online]. Available: https://paperswithcode.com/dataset/wdc-products.

[7] J. Wang, T. Kraska, M. J. Franklin, and J. Feng, Crowder: Crowdsourcing entity resolution. arXiv preprint arXiv:1208.1927.

[8] H. Zhou, W. Huang, M. Li, and Y. Lai, "Relation-aware Entity Matching Using Sentence-BERT," *Computers, Materials and Continua*, vol. 71, no. 1, pp. 1581–1595, 2022, doi: 10.32604/cmc.2022.020695.

[9] S. Mudgal *et al.*, "Deep Learning for Entity Matching: A Design Space Exploration," *Proceedings of the ACM SIGMOD International Conference on Management of Data*, Association for Computing Machinery, May 2018, pp. 19–34. doi: 10.1145/3183713.3196926.

[10] R. Peeters, and C. Bizer, "Dual-Objective Fine-tuning of BERT for Entity Matching," *Proceedings of the VLDB Endowment*, VLDB Endowment, 2021, pp. 1913–1921. doi: 10.14778/3467861.3467878.

[11] L. Fang, L. Li, Y. Liu, V. I. Torvik, and B. Ludäscher, "KAER: A Knowledge Augmented Pre-trained Language Model for Entity Resolution," Jan. 2023, [Online]. Available: http://arxiv.org/abs/2301.04770

[12] R. Peeters, C. Bizer, and G. Glavaš, "Intermediate Training of BERT for Product Matching," [Online]. Available: https://github.com/Weyoun2211/productbert-intermediate.

[13] Y. Li, J. Li, Y. Suhara, A. Doan, and W. C. Tan, "Deep Entity Matching with Pre-trained Language Models," *Proceedings of the VLDB Endowment*, vol. 14, no. 1, pp. 50–60, Sep 2020, doi: 10.14778/3421424.3421431.

[14] T. Chen, S. Kornblith, M. Norouzi, and G. Hinton, "A Simple Framework for Contrastive Learning of Visual Representations," Feb. 2020, [Online]. Available: http://arxiv.org/abs/2002.05709.

[15] T. Gao, X. Yao, and D. Chen, "SimCSE: Simple Contrastive Learning of Sentence Embedding," Apr. 2021, [Online]. Available: http://arxiv.org/abs/2104.08821

[16] Z. Wu, Y. Xiong, S. Yu, and D. Lin, "Unsupervised Feature Learning via Non-Parametric Instance-level Discrimination," May 2018, [Online]. Available: http://arxiv.org/abs/1805.01978.

[17] O. J. Hénaff *et al.*, "Data-Efficient Image Recognition with Contrastive Predictive Coding," May 2019, [Online]. Available: http://arxiv.org/abs/1905.09272

[18] Y. Tian, D. Krishnan, and P. Isola, "Contrastive Multiview Coding," Jun. 2019, [Online]. Available: http://arxiv.org/abs/1906.05849.

[19] R. Gómez, "Understanding Ranking Loss, Contrastive Loss, Margin Loss, Triplet Loss, Hinge Loss and all those Confusing Names," *Raúl Gómez Blog*. Accessed: Sep. 12, 2023, [Online]. Available: https://gombru.github.io/2019/04/03/ranking_loss/.

[20] F. Schroff, D. Kalenichenko, and J. Philbin, "FaceNet: A Unified Embedding for Face Recognition and Clustering," Mar., 2015, doi: 10.1109/CVPR.2015.7298682.

[21] N. Reimers and I. Gurevych, "Sentence-BERT: Sentence Embeddings using Siamese BERT-Networks," Aug. 2019, [Online]. Available: http://arxiv.org/abs/1908.10084

[22] N. Reimers and I. Gurevych, "Cross-Encoders – Sentence-Transformers documentation," Accessed: Sep. 12, 2023, [Online]. Available: https://www.sbert.net/examples/applications/cross-encoder/README.html.

[23] M. Hernández, W. Cohen, and S. Tejada, "Duplicate Detection, Record Linkage, and Identity Uncertainty: Datasets," Accessed: Sep. 12, 2023, [Online]. Available: https://www.cs.utexas.edu/users/ml/riddle/data.html

[24] B. Foua, X. Xu and J. Talburt, Large Language Model-Based Representation Learning for Entity Resolution Using Contrastive Learning. CSCI, Las Vegas, 2023.

Xiaowei Xu*, Bi Foua, Xingqiao Wang, Vivek Gunasekaran,
Jonathan White, and John Talburt

TOAA: Train once, apply anywhere

Abstract: This paper presents a novel approach for entity matching using Generative Large Language Models (LLMs), such as OpenAI's GPT-3, Meta's Llama 2, and Data-Bricks' Dolly 2.0. The innovative approach of this study is based on training these models once (Train Once) on a singular dataset and then examining their adaptability and efficacy across a diverse array of datasets across domains (Apply Anywhere). The GPT-3 base model exhibited promising results in entity matching, achieving the highest average F-1 scores. This suggests a paradigm shift toward Train Once Apply Anywhere (TOAA) LLM-based approaches. However, traditional research protocols were followed in the initial phase of the training and testing models using domain-specific datasets, which resulted in limited success. The authors also discussed the limitations of their study and suggested directions for future research.

Keywords: generative large language models, entity matching, intelligent data matching, natural language processing

1 Introduction

Entity matching, the task of identifying different representations of the same real-world entity across different datasets or within a single dataset, is a fundamental component of various applications such as data integration, entity resolution, and data quality. With the proliferation of data from various sources, the ability to match entities accurately and efficiently across different domains has become increasingly important. However, this task is particularly challenging because of the heterogeneity

Acknowledgment: This research was supported in part by the US Census Bureau Cooperative Agreement CB21RMD0160002 for Record Linkage. This project used the facilities provided by the Arkansas High-Performance Computing Center, supported in part by grants from the National Science Foundation grants #0722625, #0959124, #0963249, #0918970, and a grant from the Arkansas Science and Technology Authority. This project was partially supported by the National Science Foundation under award no. OIA-1946391. Any opinions, findings, conclusions, or recommendations expressed in this material are those of the authors and do not necessarily reflect the views of funding organizations. The authors gratefully acknowledge the support.

***Corresponding author: Xiaowei Xu**, University of Arkansas at Little Rock, Little Rock, AR 72204, USA, e-mails: xwxu@ualr.edu, contact@OmniTrustAI.com
Bi Foua, Xingqiao Wang, Vivek Gunasekaran, John Talburt, University of Arkansas at Little Rock, Little Rock, AR 72204, USA
Jonathan White, U.S Census Bureau

https://doi.org/10.1515/9783111344126-018

of data, the presence of noisy and incomplete information, and the need for domain-specific adaptations. Traditional rule-based approaches for entity matching require frequent updates and modifications to keep up with changes in the incoming data pipeline environment. Although domain-specific machine learning-based solutions offer a more advanced alternative, they often demand fine-tuning of the model for each specific domain, which can be extremely time-consuming and sometimes impractical, particularly when labeled data or computational resources are limited. Additionally, it is worth noting that extending these domain-specific models to other domains is a challenging task that limits their application.

Generative Large Language Models have rapidly gained prominence, with models such as OpenAI's GPT-3, DataBricks' open-source Dolly 2.0, and Meta's Llama 2 leading the charge. Their rise is attributed not only to their proficiency in natural language generation but also to their remarkable ability to generalize across a plethora of tasks without exhaustive domain-specific training. This new transformative capability has significantly changed different areas adapted to natural language processing. This raises an interesting question: How well can these models be used for the complex task of entity matching?

In this study, we provide an answer by presenting a new approach to entity matching. Our approach uses the generalization ability of generative LLMs. We fine-tuned a pretrained LLM on a training dataset, applied the model directly to various domains without further adaptation, and evaluated its performance using metrics such as F-1 scores across all datasets. Our approach requires training only once, thereby significantly reducing the computational resources and time required for entity matching across different domains, and our major contributions are as follows:

- Introducing Train Once Apply Anywhere (TOAA): We propose a unique methodology for entity matching that leverages the generalization capabilities of Large Language Models. Training once and applying it across diverse domains can revolutionize entity-matching tasks, especially in resource-constrained and privacy-sensitive situations.
- Successfully applying the TOAA in an industry application: We successfully fine-tuned Llama2 using product domain data (Train Once) and applied it to the US Census Bureau's Entity Matching problem on Hard-to-Match cases (Apply anywhere).
- Cross-domain testing with single training: We demonstrate the feasibility and effectiveness of training a model on one dataset and then applying it to a myriad of different domain datasets. This pioneering effort marks a significant departure from the traditional methods and underscores the potential of LLMs.
- Sensitive domain training data: TOAA enables training on nonsensitive domain data and applies them to sensitive domains, such as personal information and health records. This prevents any privacy violation that might occur, owing to the use of sensitive data to train and validate a model.

- Comparative analysis with traditional protocols: We conduct a comprehensive analysis by training and testing LLMs on domain-specific datasets that adhere to traditional research protocols. This analysis not only offers a comparative perspective on the performance of LLMs such as OpenAI's GPT-3, DataBricks' Dolly 2.0, and Meta's Llama 2, but also highlights their superiority over other established methods in many tasks.
- In-depth analysis of performance across datasets: We explain why certain datasets yield better results, offering a nuanced understanding of the methodology's strengths and limitations.

This chapter is an extension of work originally presented in the paper titled 'Train Once, Match Everywhere: Harnessing Generative Language Models for Entity Matching,' [19] which was accepted for presentation at the 2023 International Conference on Computational Science and Computational Intelligence (CSCI) held in Las Vegas, NV, USA. The content here builds upon and expands the ideas and findings discussed in that initial paper, incorporating additional data, more comprehensive analyses, and developments that have occurred in the field since the conference.

The remainder of this paper is organized as follows. Section 2 discusses related work in the field of entity matching and the use of LLMs for various natural language processing (NLP) tasks. Section 3 describes the methodology of the proposed approach, including the model architecture, training procedure, and evaluation metrics. Section 4 presents the results of our experiments and compares them with those of the state-of-the-art methods. Finally, Section 5 concludes the implications of our findings, the limitations of our study, and directions for future work.

2 Related work

Entity matching, a pivotal task in data preprocessing, has experienced consistent methodological evolution over the years. However, most research efforts remain tethered to domain-specific training and testing, which often yield solutions with limited adaptability across diverse datasets. This recurring theme underscores the need for research and an innovative approach.

The initial methodologies were based on rule-based algorithms and heuristics. For instance, deterministic methods centered on semantic similarity have been proposed by different authors, such as Felligi et al. [1–5]. While groundbreaking for their times, these traditional methods can miss the context and often produce false-positive matches. The onset of machine learning introduced more dynamic approaches.

Bilenko and Moore employed SVM-based methodologies, tailored to specific domains, proving adept at handling domain-specific data [6]. The deep-learning revolution, evidenced by multiple papers, further entrenched the domain-specific nature of entity matching, with models often requiring extensive, domain-tailored training datasets [7–13]. Despite the promise of deep-learning methods, the necessity for task-specific fine-tuning remains persistent.

The emergent class of generative large language models, while transcending many traditional NLP tasks, is yet to be fully explored in domain-agnostic settings for entity matching. OpenAI's ChatGPT has emerged as a testament to advancements in the natural language processing domain. Its unparalleled capability to engage in coherent and contextually relevant conversations underscores the potential of generative language models in diverse applications, thereby establishing a benchmark for interactive and adaptive NLP tasks. Despite these strides, the research landscape is noticeably bereft of models trained once and broadly applicable across domains for entity matching.

Our research is a departure from this domain-centric norm, ushering in an era of adaptability and generalization. By proposing and validating a model that transcends domain boundaries, we challenge prevailing methodologies, and chart new territories, emphasizing a need for models that "Train Once, Apply Anywhere (TOAA)."

3 Methodology

Entity matching has historically necessitated rigorous domain-specific training to achieve satisfactory results. The rise of generative LLMs presents a special moment, offering the transformative potential to challenge traditional approaches. These models, celebrated for their ability to comprehend, generate, and adapt to diverse textual contexts, have introduced a novel avenue for entity matching.

3.1 Domain-specific approach

The domain-specific entity-matching approach followed a standard machine-learning approach. Using this approach, we evaluated the effectiveness of generative language models in domain-specific tasks. We aligned our experimentation with conventional research methods and training and testing LLMs, namely OpenAI's GPT-3 base-mode Ada, DataBricks' Dolly 2.0, and Meta's Llama 2, on domain-specific datasets. This provides a contextual comparison against established methodologies. Figure 1 shows that each model was trained on the same domain-specific dataset and was tested on data from the same domain. Twelve datasets were chosen from different domains. More details are provided in the following section. For every domain-specific task, we fine-tuned both models on their respective training splits to ensure that they grasped the intricacies of each domain. It was then tested using the test split data from the same domain. This standard approach enables us to compare the performance of LLM-based models against industry benchmarks.

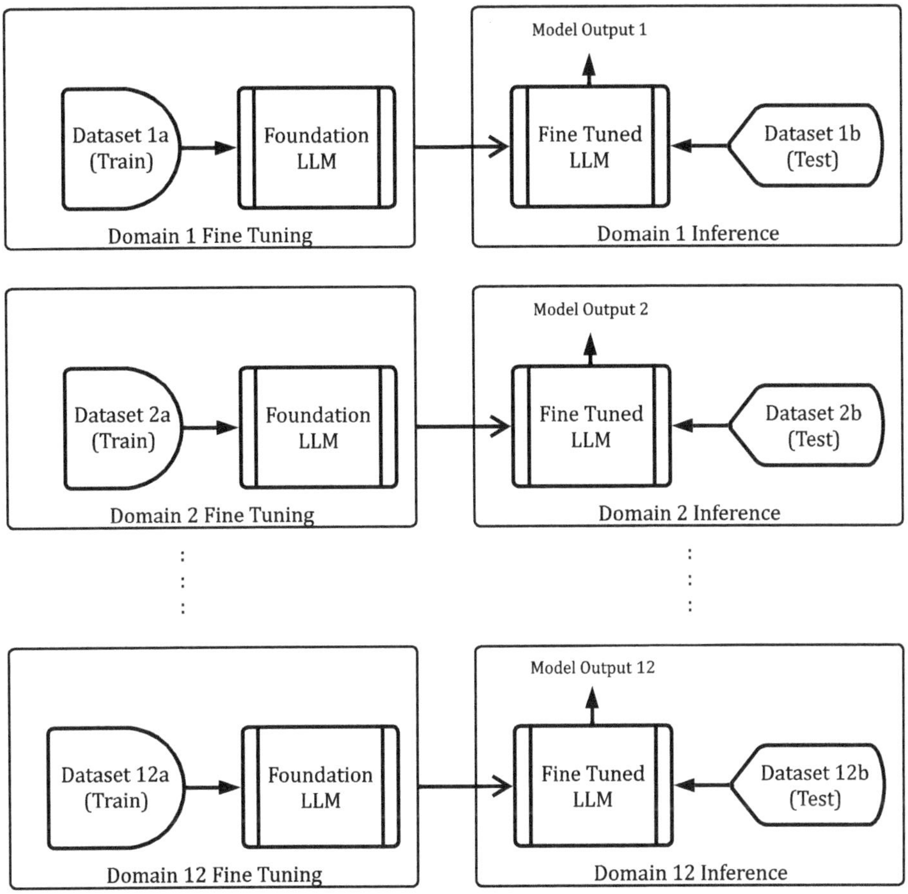

Figure 1: Domain-specific approach.

3.2 TOAA: Domain-agnostic approach

Train Once Apply Anywhere (TOAA) is our novel approach for performing domain-agnostic entity matching, which is straightforward yet comprehensive. After fine-tuning each model on a specific dataset (Train Once), we tested it across the remaining datasets (Apply Anywhere) as shown in the Figure 2. Because we worked with 12 datasets, every domain-specific fine-tuned model was tested on the remaining 11 datasets. This cross-evaluation approach enabled us to assess the generalization capability of our models across various domains. At the culmination of this rigorous testing phase, we computed the average performance across all datasets to gauge the overall efficacy of our domain-agnostic entity-matching technique.

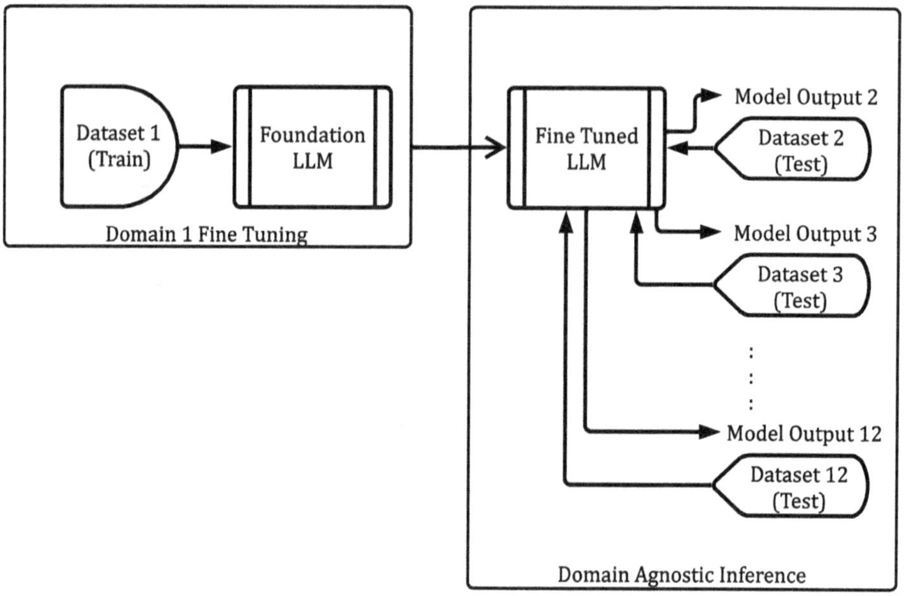

Figure 2: TOAA – Domain-agnostic approach.

3.3 Datasets

Central to our analysis were datasets spanning multiple domains, each with distinct challenges. We primarily sourced these datasets from the DeepMatcher repository [20], which encompasses structured, dirty, and textual data, as shown in Table 1. Structured data are organized into a specific format, such as fields or attributes. These structured data are derived from various domains such as software, music, and citations. The dirty data type comprises entities like the structured datasets, but is laced with inconsistencies. Dirty data poses a challenge to entity matching by introducing noise and variations. Textual data are tailored for entity-matching tasks in which textual descriptions are pivotal. These datasets provide a well-rounded platform for our evaluation.

Table 1 summarizes all datasets. A brief description of these columns is provided below.

- Identifier: This identifier denotes the dataset and the model trained using this dataset. For example, T1A2 refers to the model trained on dataset {1} Dirty_DBLP-ACM and Applied to dataset {2} Dirty_ DBLP-GoogleScholar.
- Number of tokens: A chunk of text that the model reads or generates.
- Positive ratio: Ratio of positive pairs in each dataset.
- No. of recording pairs: Count of pairs of records present in each dataset.
- Tokens per recording pair: Average number of tokens in each recording pair.

Table 1: Overview of datasets [14].

Identifier	Datasets	No. of tokens	Positive ratio	No. of recording pairs	Tokens per recording pairs
{1}	Ditry_DBLP-ACM	778,404	0.18	7,417	104.95
{2}	Ditry_DBLP-GoogleScholar	1,516,781	0.19	17,223	88.07
{3}	Dirty_iTunes-Amazon	60,370	0.24	321	188.07
{4}	Dirty_Walmart-Amazon	614,541	0.09	6,144	100.02
{5}	Structured_Amazon-Google	380,356	0.10	6,874	55.33
{6}	Structured_Beer	23,847	0.15	268	88.98
{7}	Structured_DBLP-ACM	741,031	0.18	7,417	99.91
{8}	Structured_DBLP-GoogleScholar	1,455,775	0.19	17,223	84.53
{9}	Structured_Fodors-Zagats	59,512	0.12	567	104.96
{10}	Structured_iTunes-Amazon	58,064	0.24	321	180.88
{11}	Structured_Walmart-Amazon	592,512	0.09	6,144	96.44
{12}	Textual_Abt-Buy	780,936	0.11	5,743	135.98

3.4 Model selection

Selecting an appropriate model was pivotal for the success of our experiments, ensuring both accuracy and adaptability in entity-matching tasks. Our selection process gravitated toward three distinct Large Language Models (LLMs), each offering unique advantages.

Ada from OpenAI. Ada is the base model in the OpenAI GPT-3 series. It is known for its simplicity, speed, and cost-effectiveness. Even as the base model, Ada provides a glimpse into the capabilities of OpenAI's more advanced offerings, such as DaVinci, Turbo, and GPT-4. This makes Ada not only an economical choice but also an insightful pick, giving an idea of the potential ceiling of performance we might expect from their high-end counterparts.

Dolly 2.0 from DataBricks. Dolly is noteworthy as the first truly open-source instruction-tuned model available for any use, whether academic or commercial. Dolly 2.0's open nature and adaptability, combined with the freedom of unrestricted application under its terms, make it invaluable for our domain-agnostic entity-matching experiments.

Llama 2 from Meta. Llama 2 is a collection of foundation language models, trained using trillions of tokens from publicly available open-source datasets, without resorting to proprietary and inaccessible datasets. This state-of-the-art model is freely available to the research community and outperforms many other models, such as GPT-3, on most benchmarks. Meta also recommends Llama 2 as a substitute for closed source models, for its performance on helpfulness and safety, based on Meta's human evaluation [18].

3.5 Implementation and fine-tuning

The core of our methodology lies in the implementation and subsequent fine-tuning of the selected models, ensuring that they are optimized for the challenges of entity matching across both domain-specific and domain-agnostic scenarios.

Foundation model setup: Ada is a closed-source model offered by OpenAI, and we accessed its API to integrate it into the experimental framework. In contrast, Dolly 2.0, and Llama 2, are open-source, offered with greater flexibility. We cloned the necessary repositories and integrated the model into our local environment by adjusting the parameters needed to suit the entity-matching tasks.

Fine-tuning strategy. Although Large Language Models such as Ada, Dolly and Llama are pretrained on vast corpora, their generic nature means that they can benefit from fine-tuning, aligning them more closely with the specifics of entity matching. Our strategy continues to train foundation models on the entity-matching task, using a domain-specific dataset. Given our goal of testing LLMs in both domain-specific and domain-agnostic scenarios, our fine-tuning strategy must be robust, yet adaptable.

Fine-tuning procedure for Ada (OpenAI model). We embarked on a systematic fine-tuning process for Ada, as instructed step-by-step by OpenAI [15]. We began by preparing our dataset in conversational chat format. We changed the structure of each dataset using the two inputs, records 1 and 2, as prompts and their outputs as completion. This helped to create a context for entity-matching queries. This dataset was then divided judiciously into training and test portions to ensure a balanced representation. Care was taken to adhere to the token limitations of the OpenAI model, ensuring that each example remained within the 4096-token constraint.

Fine-tuning procedure for Dolly (DataBricks' Model). Dolly 2.0, an open-source model, offers a different approach. We cloned Dolly's repositories and set them up in our environment, customizing their parameters to suit entity-matching tasks [16]. As Dolly comes with its own set of recommended practices for fine-tuning, we adhered to them to ensure that its performance was optimized for our experimental needs.

Fine-tuning procedure for Llama2 (Meta Model). The Llama2 is an open-source model that provides a safe and scalable approach. We implemented a tailored fine-tuning procedure using the guidance provided by the Meta for model optimization. To optimize the model's performance toward our experimental objectives while adhering to Meta's best practice recommendations, we initiated a fine-tuning process by adjusting Llama2's hyperparameters. This procedure aims to fully harness the capabilities of the Llama2 model and enhance the accuracy of the matching task.

Domain-specific model training. The objective of domain-specific model training is to adjust the model parameters to minimize the loss on a specific dataset. The loss function $\mathcal{L}$ can be defined as:

$$\mathcal{L}(\theta; A_{train}) = \frac{1}{N} \sum_{i=1}^{N} l(f(x_i; \theta), y_i)$$

$\mathcal{L}$ = Objective or Loss Function (difference between model's prediction and actual data)
θ = parameters of the Foundation Large Language Model
A_{train} = Domain A training dataset.
$\{(x_1, y_1), (x_2, y_2) \ldots, (x_N, y_N)\}$ represents the training dataset, where x_i are the inputs, and y_i are the corresponding labels from Domain A training datasets.
l is the loss function.

Domain-specific model applications. After domain-specific training, the model was evaluated by applying it to a test dataset from the same domain to assess its performance. The evaluation can be represented as

$$Performance\ (\theta; A'_{apply}) = Metric\ (f\ (x; \theta), y)$$

$$Domain\ (A_{train}) = Domain\ (A'_{apply})$$

$A'_{apply} = \{(x'_1, y'_1), (x'_2, y'_2) \ldots, (x'_M, y'_M)\}$ represents the testing or application of a dataset from the same domain.

Metric is a function that calculates the performance metric (accuracy, precision, recall and F1-score)

TOAA domain-agnostic model training. Previously trained domain-specific model $\mathcal{L}(\theta; A_{train})$ will be used for repurposing without additional training or fine-tuning.

$$\mathcal{L}(\theta; A_{train}) = \frac{1}{N} \sum_{i=1}^{N} l(f(x_i; \theta), y_i)$$

TOAA domain-agnostic model application. The model was evaluated by applying it to a testing dataset from a different domain to assess its performance. The evaluation can be represented as

$$Performance\left(\theta;\, B_{apply}\right) = Metric\ \left(f(x;\theta),\ y\right)$$

$$Domain\ (A_{train}) \neq Domain\ (B_{apply})$$

$B_{apply} = \{(x_1, y_1), (x_2, y_2) \ldots, (x_M, y_M)\}$ represents the testing or application of datasets from different domains.

Metric is a function that calculates the performance metric (accuracy, precision, recall and F1-score)

4 Experimental results

4.1 Domain-specific experiment

In our experiment, we first ventured into a domain-specific evaluation of three prominent large language models (LLMs): OpenAI's Ada, Databrick Dolly, and Meta's Llama, utilizing datasets leveraged in notable publications [7, 14, 17]. This approach aims to understand better how these models fare in the entity-matching domain, when tailored to specific datasets.

Exploration of entity matching using foundation LLMs (Ada, Dolly, and Llama) has demonstrated the distinct advantages and potential of large language models in this domain. We used the F-1 score to compare the models' performances. Observing the F-1 scores in Figure 3, it is evident that Ada not only outperformed Dolly on most domain-specific tasks across different datasets, but also surpassed all other models. The evaluation revealed the incredible performance of Ada (GPT-3) and Llama across diverse datasets. They achieved the highest average F-1 scores of over 92% and 90%, respectively, followed by KAER, which came as a distant second with an average F-1 score of approximately 86%. A detailed performance metrics table is available in Appendix A1.

In contrast, Dolly 2.0, despite being the first truly open-source model of its kind, demonstrated mixed performance across various datasets. Although it claimed the first rank in the DBLP-GoogleScholar dataset under the Dirty data category, its performance in other datasets, particularly in Structured iTunes-Amazon, seemed to lag. This type of performance distribution highlights the differences in capability and adaptability between Dolly and the other two LLMs (Ada and Llama), reflecting maturity and refinement of the products of OpenAI and Meta, with continued efforts to develop LLMs.

Despite disparities in performance, Dolly's open-source nature represents a significant milestone in the community. Dolly's unrestricted access offers a level of democratization in AI resources, opening doors to researchers and practitioners who may not have access to proprietary models or resources to utilize them. Moreover, Dolly's performance on certain datasets shows promising potential. With appropriate fine-tuning and domain-specific adjustments, it is plausible that Dolly's performance across various entity-matching tasks could be significantly improved.

Open-source LLMs such as Dolly and Llama foster a collaborative environment to enhance their performance, leading to rapid evolution and refinement. The DBLP-GoogleScholar dataset, where Dolly ranked first, showed its potential to excel when tailored to specific challenges. This demonstrates Dolly's capabilities and promises future advancement in entity-matching tasks. Llama even achieved a comparable performance with commercial Ada from OpenAI.

We also performed additional systematic domain-specific experiments using all the three foundation LLMs. They outperformed state-of-the-art models Robert, Ditto and KAER in all metrics. Refer to Appendix A2 for the detailed performance metrics.

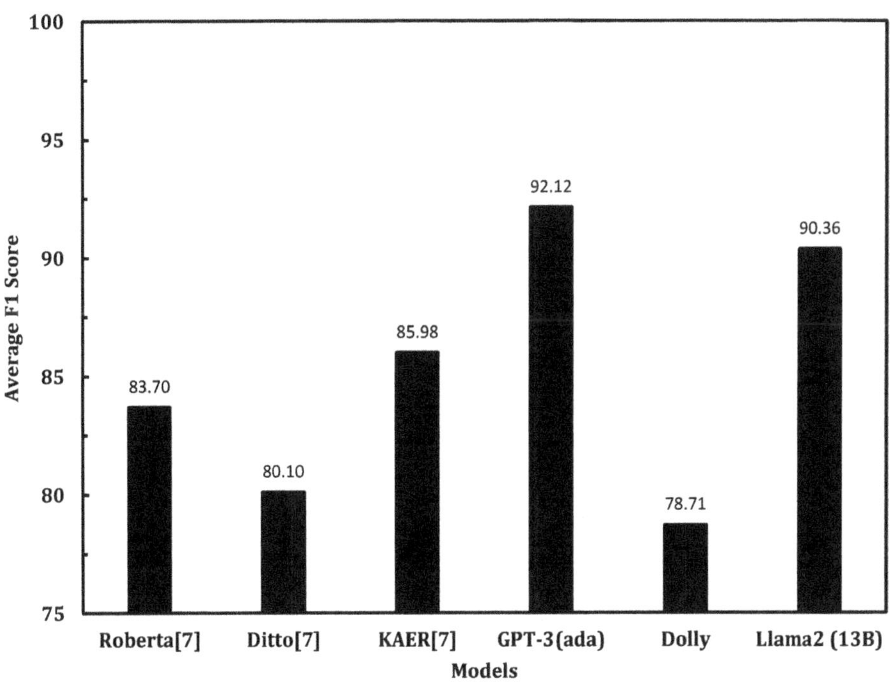

Figure 3: Domain-specific model performance.

4.2 TOAA: Domain-agnostic experiment

We employed a comprehensive multi-dataset evaluation methodology to evaluate the domain-agnostic capabilities of foundation LLMs, including Ada, Dolly, and Llama. The premise was to train each model once on each dataset and then test (apply them) the remaining 11 datasets, consequently calculating the average F1 score across all dataset applications. This approach allows for a robust assessment of the model's generalizability and adaptability across various data domains, without prior domain-specific training. Appendix B1, B2, B3, and B4 detail the Ada performance metrics. Appendices C1, C2, C3, and C4 detail Dolly's performance metrics.

Upon observation, we realized that some datasets had higher average F-1 scores across all 11 datasets than others did. One factor that may contribute to the observed performance differences is the number of tokens. We observed that the top five datasets, ranked in terms of average F-1 scores, all had 500 K or more tokens, as shown in Figure 4.

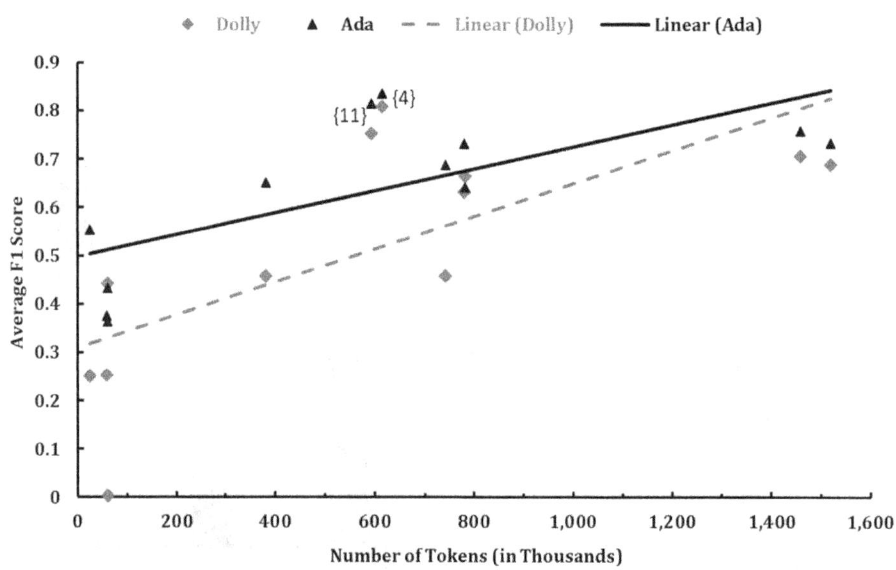

Figure 4: Relationship between the number of tokens and F-1 Score.

For instance, in Figure 5, Dirty_Walmart-Amazon and Structured_Walmart-Amazon emerged as the best performers, securing the 1st and 2nd rank, respectively, in terms of F1 score with Ada. In contrast, the Dirty_iTunes-Amazon dataset, having a significantly lower number of tokens, ranks the lowest in F1 score, indicating a possible challenge for Ada in dealing with datasets with such characteristics. Similarly, Structured_DBLP-GoogleScholar and Dirty_DBLP-GoogleScholar also exhibit commendable performance, ranking 3rd and 4th, respectively.

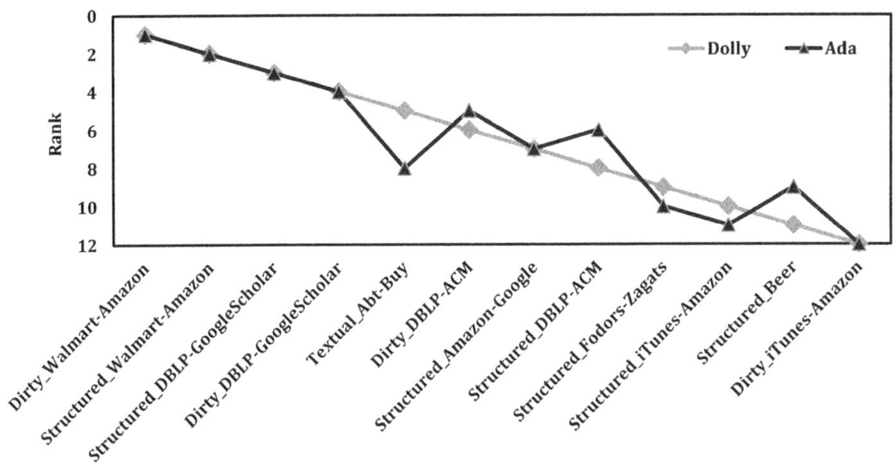

Figure 5: Rank by dataset.

The average F1 scores of the Dolly and Ada models demonstrated strong correlation with the number of tokens. Both models had a Pearson's correlation coefficient of 0.71.

4.3 TOAA: Domain-agnostic experiment – US Census Bureau

Domain-specific experiments showed some performance improvements over the state-of-the-art models. However, significant improvements were observed in domain-agnostic experiments. Therefore, in the next logical phase, we executed a domain-agnostic experiment on an entity-matching problem from the US Census Bureau.

Model selection and training of domain datasets are crucial for building truly domain-agnostic solutions. Based on previously conducted experiments, we chose open-source Llama2 as our foundation large-language model, and Dirty_Walmart-Amazon as the training dataset.

4.3.1 Model selection

At the time of the experiments, Llama2 outperformed open-source models on most benchmarks and was considered a substitute for closed-source models, owing to its helpfulness and safety features. Moreover, Meta provided detailed descriptions and documentation on the fine-tuning and safety of the models. This flexibility is essential for adapting the foundation model for safe and practical applications.

4.3.2 Training dataset selection

The Dirty_Walmart-Amazon dataset was selected for training Llama2 because it ranked first in both the Dolly and Ada domain agnostic experiments (refer to Figure 5). In the domain-specific Llama2 experiment, the F1 score was approximately 90%, well above the average F1 score (refer to Appendix A2).

4.3.3 Privacy concerns

The domain-agnostic approach addresses LLM adoption issues for organizations. One of the key factors inhibiting LLM-backed solutions is legal and compliance issues [21]. In the traditional Machine Learning or domain-specific LLM approach, training data must be sourced from the same domain, and more specifically, from the same system. However, this can lead to privacy, legal, and compliance issues. For instance, the US Census Bureau, Title 13 – Protection of Confidential Information, prohibits using Personally Identifiable Information (PII) for any model training exercise. Generating synthetic data and using them for training models may not be accurate because of inherent data quality issues, such as bias. Therefore, in domains with legal and compliance restrictions on the training data, we recommend a domain-agnostic (TOAA) approach.

4.3.4 Name-entity matching in US Census Bureau

The US Census Bureau handles millions of Name Entity (domain) records. One of their challenging processes is to match people's entity records. Current rule-based and machine learning methods fail to handle hard-to-match cases with serious data quality issues, including many missing values and noise, which must be handled manually. These hard-to-match cases are records without a social security number, missing or incorrect names (first, middle, last), date of birth, address, or many other key matching values. On average, 25% of the values in the hard-to-match records are missing, which prevents existing US census systems from processing these cases automatically.

The bureau provided us with two data files for this experiment, each with one million records. Sample records are provided in Appendices D1 and D2. The two files are:
- *Truth file*: Pseudo people record without any errors.
- *Hard-to-match file*: Pseudo people record excessive errors.

Using these two files, we generated one million matching pairs and one million nonmatching pairs. We mapped the Truth ID from the Truth file to the Hard-to-match file to generate one million match pairs. We also generated one million nonmatching pairs by randomly pairing the truth and hard-to-match records, whose truth ID do not match.

We then trained the Llama2 model using the Dirty_Walmart-Amazon {4} dataset. Once the model was trained, it was applied to the newly created 2 million record pairs, and the results are shown in Table 2.

Table 2: TOAA results for US Census Entity Matching.

Accuracy	Precision	Recall	F1 score
97.02	100.00	94.05	96.93

The above result suggests that Llama2 trained once on Dirty_Walmart-Amazon (product domain) data performed very well on the US Census (name entity domain) data.

The model performed excellently with a precision of 100%, indicating that the TOAA entity matching model did not produce any false positives. For the name-entity matching task in any organization, an incorrect match could have severe negative effects on downstream processes.

The recall score was 94%, which is slightly lower than that of the other metrics, but is still high. The key reason for this reduction is primarily the number of false negatives. Because an average of 25% of the data in the hard-to-match records was missing, the model failed to match it to its corresponding truth record. Table 3 explains why the model failed to match the truth records (first record) with the hard-to-match records (second record). Missing First name, City, State, Zip, DOB-day, and incorrect house number lead to very few tokens for comparison. Even for a human expert, with only the last name, Street name, DOB-Month, DOB-Year, occupation, and Salary, it is impossible to match these two entities with a high level of accuracy.

The TOAA-designed model demonstrated exceptional performance, achieving an accuracy and an F1 score of approximately 97%. These results indicate that the model successfully classified most of the inputs, even though it was not exposed to a new domain. High TOAA model performance suggests that the model is reliable and effective in any task-specific domain-agnostic environment.

5 Conclusion and future work

To explore the generalization capabilities of Large Language Models (LLMs) for entity matching, we ventured into uncharted territories, moving beyond domain-specific constraints to a broader, more inclusive domain-agnostic landscape. Our approach, rooted in training only once but applying across diverse domains, has provided illuminating insights.

Table 3: False negative example.

First	Last	House #	Street	City	State	Zip	DOB Day	DOB Month	DOB Year	Occupation	Salary
James	Sabbah	2184	Washington Drive	Winslow	NJ	8004	16	5	1949	Teacher Assistant	19,256
	Sabbah	2186	Washington Drive					5	1949	Teacher Assistant	19,256

Our primary LLM protagonists, Ada, Dolly, and Llama 2, show significant potential. Notably, Ada, OpenAI's base model, demonstrated that even a model designed for simpler tasks can provide competitive performance and, in some cases, superior performance. The success of Ada offers a tantalizing glimpse of the heights that more advanced models in the OpenAI lineup could reach in entity matching. Dolly, with its open-source lineage, stands as a testament to the possibility of unrestricted LLM usage, emphasizing that a community-driven approach can yield robust results, particularly in domain-agnostic scenarios. Llama2, an open-source model, outperformed most benchmark open-source and closed-source models. This has been favored for its safety and helpful features.

However, as with any other research, there are underlying nuances. For instance, the connection between the number of tokens and F-1 score partly aligns with the recognized scaling laws of neural language models, reaffirming that synergistic scaling of the model size, dataset size, and computation, often results in enhanced performance [22].

The domain of entity matching is at the cusp of a paradigm shift. Traditional methodologies, while reliable, may soon be overshadowed by the adaptability and expansive reach of LLMs. The findings from our study echo the sentiment that as technology, particularly in the realm of LLMs, continues to evolve, so will the methodologies, tools, and techniques in entity matching. The journey of "train once apply anywhere" might be nascent, but its trajectory promises a transformative impact on the future of entity matching.

6 Appendix

Appendix A1 Domain-specific experiment – comparison with existing entity matching models [7]

Models	Dirty		Structured				Textual
	{2}DBLP-GoogleScholar	{3}iTunes-Amazon	{8}DBLP-GoogleScholar	{10} iTunes-Amazon	{5} Amazon-Google	{7} DBLP-ACM	{12} Abt-Buy
Roberta[7]	**95.81**	**72.73**	95.32	61.76	73.12	98.77	88.41
Ditto[7]	95.54	60.38	**95.85**	47.06	74.44	98.55	88.89
KAER[7]	95.74	54.90	95.14	**89.66**	**74.5**	**98.99**	**92.61**
Llama2	95.77	96.30	95.87	94.34	58.31	98.54	93.36
GPT-3(ada)	95.74	91.23	96.21	96.43	73.71	98.87	92.68

(continued)

Models	Dirty		Structured				Textual
	{2}DBLP-GoogleScholar	{3}iTunes-Amazon	{8}DBLP-GoogleScholar	{10} iTunes-Amazon	{5} Amazon-Google	{7} DBLP-ACM	{12} Abt-Buy
Dolly	96.44	--	96.78	31.11	68.78	97.83	81.31
Δ***F1*** Ada vs best model	−0.07	+18.5	+0.36	+6.77	−1.14	−0.12	+0.07
Δ***F1*** Dolly vs. Best Model	+0.63	--	+0.57	−65.32	−6.07	−1.16	−11.37
Rank ada	3	1	2	1	2	2	1
Rank Dolly	1	--	1	5	5	5	5

- State-of-the-art models – Roberta, Ditto, KAER have results for seven datasets only. So, performances of Dolly, Ada and Llama2 corresponding to those seven datasets are recorded in this table.
- Dolly model for Dirty_iTunes-Amazon dataset F1 score cannot be calculated.

Appendix A2 Domain-specific – Llama2 – performance metrics

Dataset	Accuracy	Precision	Recall	F1
Dirty_Walmart-Amazon	0.980478	0.900524	0.891192	0.895833
Structured_DBLP-GoogleScholar	0.95472	0.843803	0.928972	0.884342
Dirty_DBLP-ACM	0.979377	0.91023	0.981982	0.944745
Structured_DBLP-ACM	0.976547	0.912393	0.961712	0.936404
Structured_iTunes-Amazon	0.917431	1	0.666667	0.8
Dirty_DBLP-GoogleScholar	0.952107	0.837723	0.921495	0.877615
Structured_Amazon-Google	0.893589	0.486772	0.786325	0.601307
Structured_Walmart-Amazon	0.979502	0.883249	0.901554	0.892308
Dirty_iTunes-Amazon	0.908257	0.904762	0.703704	0.791667
Structured_Beer	0.989011	0.933333	1	0.965517
Textual_Abt-Buy	0.964509	0.896552	0.757282	0.821053
Average	0.954139	0.864486	0.863717	0.855526

Appendix B1 Domain-agnostic – Ada – Accuracy

	A1	A2	A3	A4	A5	A6	A7	A8	A9	A10	A11	A12	Average
T1	0.99595633	0.96691048	0.88073394	0.85993167	0.87265591	0.97802198	0.99595633	0.96621386	1	0.88990826	0.88530991	0.81941545	0.92591784
T2	0.98301658	0.98415186	0.88073394	0.92484139	0.91670301	0.95604396	0.98261221	0.98484848	1	0.93577982	0.9238653	0.88830898	0.94674213
T3	0.19126567	0.19418321	0.95412844	0.10248902	0.10902747	0.28571429	0.19328751	0.19191919	0.27513228	0.96330275	0.10102489	0.12317328	0.307054
T4	0.95713708	0.92755138	0.89908257	0.97706198	0.90143916	0.98901099	0.94743227	0.93155695	0.97883598	0.91743119	0.97852611	0.96033403	0.94711664
T5	0.936514	0.946883	0.743119	0.928746	0.948975	0.989011	0.954307	0.95472	0.973545	0.752294	0.917521	0.859603	0.90877
T6	0.9074	0.851271	0.651376	0.733041	0.551243	0.978022	0.854023	0.782828	0.714286	0.541284	0.713031	0.635699	0.742792
T7	0.99474	0.968826	0.81651	0.91118	0.901439	0.967	0.995956	0.972309	0.984127	0.853211	0.909224	0.76618	0.920062
T8	0.986656	0.981017	0.944954	0.90776	0.920628	0.96703	0.98706	0.985893	1	0.944954	0.914104	0.892484	0.952712
T9	0.484432	0.330199	0.752294	0.463153	0.87222	0.450549	0.875455	0.719436	1	0.825688	0.747194	0.72286	0.686957
T10	0.192883	0.190352	0.963303	0.102489	0.116005	0.450549	0.19167	0.188959	0.396825	0.981651	0.113714	0.146138	0.336212
T11	0.938941	0.915361	0.862385	0.970717	0.901003	0.989011	0.961585	0.93922	0.989418	0.816514	0.976086	0.954593	0.93457
T12	0.82693085	0.86172065	0.34862385	0.80234261								0.98434238	0.76479207

T1: Model trained using Dataset 1
A1: Trained model applied to Dataset 1
Datasets 1 through 12 are mapped in Table 1

Appendix B2 Domain-agnostic – Ada – Recall

	A1	A2	A3	A4	A5	A6	A7	A8	A9	A10	A11	A12	Average
T1	0.98423423	0.88785047	0.74074074	0.47150259	0.41880342	0.92857143	0.98648649	0.8953271	1	0.62962963	0.36787565	0.56796117	0.73991524
T2	0.97972973	0.95514019	0.74074074	0.32124352	0.46581197	0.71428571	0.97972973	0.95607477	1	0.92592593	0.29533679	0.07281553	0.70056955
T3	0.99774775	1	0.96296296	1	0.9957265	1	0.99774775	1	1	1	1	0.99029126	0.99537302
T4	0.92792793	0.75420561	0.74074074	0.83937824	0.67521368	0.92857143	0.98198198	0.77943925	0.90909091	1	0.84974093	0.83009709	0.85136565
T5	0.806306	0.823364	0.111111	0.419689	0.700855	0.928571	0.957207	0.906542	0.772727	0.148148	0.554404	0.300971	0.619158
T6	0.975225	0.965421	0.962963	0.865285	0.884615	1	0.981982	0.984112	1	1	0.875648	0.752427	0.937307
T7	0.98649	0.866355	0.77778	0.33679	0.136752	0.929	0.986486	0.890654	0.863636	0.555556	0.207254	0.57767	0.676166
T8	0.977477	0.931776	0.851852	0.253886	0.440171	0.92857	0.981982	0.961682	1	0.925926	0.264249	0.281553	0.73326
T9	0.972973	0.957009	0.592593	0.611399	0.239316	0.928571	1	0.919626	1	0.296296	0.233161	0.34466	0.674634
T10	1	1	1	1	1	1	1	1	1	1	1	0.995146	0.999596
T11	0.790541	0.671028	0.814815	0.751295	0.692308	1	0.970721	0.841121	0.909091	1	0.839378	0.820388	0.841724
T12	1	0.96074766	1	0.89119171								0.9223301	0.95485389

Appendix B3 Domain-agnostic – Ada – Precision

	A1	A2	A3	A4	A5	A6	A7	A8	A9	A10	A11	A12	Average
T1	0.99318182	0.93137255	0.76923077	0.32971014	0.38582677	0.92857143	0.99095023	0.92115385	1	0.89473684	0.38586957	0.31283422	0.73695318
T2	0.92948718	0.95962441	0.76923077	0.72941176	0.62285714	1	0.92750533	0.96237065	1	0.83333333	0.74025974	0.39473684	0.82240143
T3	0.18140868	0.18781815	0.86666667	0.09498031	0.10241758	0.17721519	0.18178088	0.18739054	0.13836478	0.87096774	0.09484029	0.10839532	0.26602051
T4	0.84773663	0.840625	0.83333333	0.91011236	0.51298701	1	0.78136201	0.84157417	0.90909091	0.75	0.91620112	0.80660377	0.82913553
T5	0.834499	0.883651	0.428571	0.704348	0.777251	1	0.818882	0.858407	1	0.5	0.563158	0.331551	0.725027
T6	0.665131	0.558378	0.412698	0.242733	0.171216	0.875	0.552598	0.461235	0.289474	0.350649	0.230559	0.193267	0.416912
T7	0.98427	0.962617	0.6	0.54622	0.571429	0.867	0.99095	0.957789	1	0.789474	0.547945	0.247917	0.75544
T8	0.949672	0.96515	0.92	0.521277	0.668831	0.86667	0.947826	0.962582	1	0.862069	0.6	0.5	0.813673
T9	0.254867	0.212272	0.5	0.103237	0.327485	0.209677	0.590426	0.392188	1	1	0.108434	0.152034	0.404218
T10	0.181967	0.187096	0.870968	0.09498	0.103494	0.21875	0.181744	0.186834	0.161765	0.931034	0.096068	0.111413	0.277176
T11	0.858191	0.842723	0.6875	0.923567	0.511041	0.933333	0.840156	0.834106	1	0.574468	0.9	0.771689	0.806398
T12	0.50917431	0.57752809	0.2755102	0.30935252								0.93137255	0.52058753

Appendix B4 Domain-agnostic – Ada – F1 Score

	A1	A2	A3	A4	A5	A6	A7	A8	A9	A10	A11	A12	Average
T1	0.98868778	0.90909091	0.75471698	0.3880597	0.40163934	0.92857143	0.98871332	0.90805687	1	0.73913043	0.37665782	0.40344828	0.73223107
T2	0.95394737	0.95737705	0.75471698	0.44604317	0.53300733	0.83333333	0.95290252	0.95921238	1	0.87719298	0.42222222	0.12295082	0.73440885
T3	0.30699931	0.31624058	0.9122807	0.17348315	0.18573137	0.30107527	0.30753211	0.31563422	0.24309392	0.93103448	0.17324955	0.1954023	0.36347975
T4	0.88602151	0.79507389	0.78431373	0.87331536	0.58302583	0.96296296	0.87025948	0.80931587	0.90909091	0.85714286	0.88172043	0.81818182	0.83586872
T5	0.82016	0.852443	0.176471	0.525974	0.737079	0.962963	0.882658	0.881818	0.871795	0.228571	0.558747	0.315522	0.651183
T6	0.790868	0.707534	0.577778	0.379115	0.286902	0.933333	0.707218	0.628094	0.44898	0.519231	0.365011	0.30754	0.5543
T7	0.98538	0.911953	0.67742	0.41667	0.22069	0.897	0.988713	0.923002	0.926829	0.652174	0.300752	0.346939	0.687256
T8	0.963374	0.948169	0.884615	0.341463	0.530928	0.89655	0.964602	0.962132	1	0.892857	0.366906	0.360248	0.759321
T9	0.403927	0.347472	0.542373	0.176647	0.276543	0.342105	0.742475	0.549874	1	0.457143	0.148026	0.210996	0.433132
T10	0.307906	0.315216	0.931034	0.173483	0.187575	0.358974	0.307586	0.314845	0.278481	0.964286	0.175295	0.200391	0.376256
T11	0.822978	0.747138	0.745763	0.828571	0.588022	0.965517	0.900731	0.837599	0.952381	0.72973	0.868633	0.795294	0.815196
T12	0.67477204	0.72140351	0.432	0.45927904								0.92682927	0.64285677

Appendix C1 Domain-agnostic – Dolly – Accuracy

	A1	A2	A3	A4	A5	A6	A7	A8	A9	A10	A11	A12	Average
T1	0.9963607	0.93695576	0.83486239	0.91605661	0.90710859	0.93406593	0.99514759	0.92911877	0.99470899	0.82568807	0.92142509	0.89457203	0.92383921
T2	0.98706025	0.98258447	0.85321101	0.91068814	0.91016136	0.95604396	0.98544278	0.98310693	1	0.87155963	0.90873597	0.89665971	0.93710452
T3	0.82110839	0.81317721	0.75229358	0.90531967	0.89790576	0.84444444	0.81976509	0.81278379	0.88359788	0.75229358	0.90527344	0.89220304	0.84168049
T4	0.97250303	0.92685475	0.87155963	0.97364568	0.92193633	0.97802198	0.97007683	0.92650644	0.98941799	0.88073394	0.97559785	0.96033403	0.94559904
T5	0.83704003	0.73058168	0.32110092	0.26110298	0.92280855	0.27472527	0.837849	0.75566005	0.57142857	0.44036697	0.28062469	0.51356994	0.5622382
T6	0.22857143	0.16936672	0.24271845	0.16935484	0.14945055	0.23529412	0.11052632	0.17465475	0.23333333	0.32692308	0.17563739	0.12254902	0.194865
T7	0.99231702	0.93086033	0.7706422	0.91117618	0.90100305	0.89010989	0.99555196	0.93207941	0.96825397	0.7706422	0.91312836	0.89300626	0.9057309
T8	0.98827335	0.98188784	0.76146789	0.90580771	0.91801134	0.96703297	0.98706025	0.98241031	0.98412698	0.82568807	0.90824793	0.90292276	0.92607812
T9	0.82369592	0.80094044	0.74311927	0.67252318	0.85870039	0.58241758	0.89324707	0.84970394	1	0.69724771	0.58321132	0.76304802	0.77232124
T10	0.22604125	0.22535702	0.7706422	0.21278673	0.11077191	0.24175824	0.18762636	0.22030651	0.12169312	0.71559633	0.1283553	0.30423419	0.2887641
T11	0.93327942	0.88140021	0.87155963	0.96876525	0.91234191	0.96703297	0.9704812	0.92459073	0.98941799	0.83486239	0.97852611	0.94050104	0.93106324
T12	0.9175091	0.90856844	0.46788991	0.86041972	0.80418666	0.8021978	0.90820865	0.90978753	0.96825397	0.3853211	0.81795998	0.98068894	0.81091598

Appendix C2 Domain-agnostic – Dolly – Recall

	A1	A2	A3	A4	A5	A6	A7	A8	A9	A10	A11	A12	Average
T1	0.98423423	0.67943925	0.55555556	0.29015544	0.16239316	0.71428571	0.97747748	0.63364486	0.95454545	0.40740741	0.30051813	0.02427184	0.55699404
T2	0.97747748	0.94953271	0.59259259	0.29015544	0.25641026	0.78571429	0.97972973	0.95420561	1	0.74074074	0.25906736	0.05339806	0.65325202
T3	0.00000	0.00188	0.00000	0.00000	0.00000	0.00000	0.00000	0.00000	0.00000	0.00000	0.00000	0.00000	0.00015635
T4	0.94369369	0.66915888	0.59259259	0.84974093	0.41452991	1	0.94594595	0.67009346	0.95454545	0.51851852	0.86010363	0.77669903	0.76630184
T5	1	0.99813084	1	0.97927461	0.93162393	1	1	0.99719626	1	1	0.97409326	0.88834951	0.98072237
T6	0.83783784	0.8556701	0.96153846	1	0.94444444	1	0.67857143	0.79452055	0.45454545	1	0.94827586	1	0.87295035
T7	0.96621622	0.63551402	0.07407407	0.06735751	0.07264957	0.28571429	0.98423423	0.64766355	0.72727273	0.07407407	0.07772021	0.00485437	0.38477874
T8	0.98198198	0.95981308	0.74074074	0.30051813	0.30769231	1	0.98423423	0.97850467	1	0.92592593	0.3626943	0.15048544	0.72438257
T9	0.43693694	0.37383178	0.37037037	0.45595855	0.40598291	0.92857143	0.86486486	0.31028037	1	0.48148148	0.57512953	0.27669903	0.54000894
T10	0.97297297	0.95514019	0.22222222	0.83419689	0.9957265	0.92857143	0.99099099	0.95607477	1	0.22222222	0.94818653	0.72330097	0.81246714
T11	0.67117117	0.4046729	0.66666667	0.68911917	0.3034188	0.85714286	0.95045045	0.64672897	0.90909091	0.88888889	0.83419689	0.48058252	0.69184418
T12	1	0.92897196	1	0.87046632	0.72649573	1	1	0.92803738	1	1	0.89637306	0.90291262	0.93777142

Appendix C3 Domain-agnostic – Dolly – Precision

	A1	A2	A3	A4	A5	A6	A7	A8	A9	A10	A11	A12	Average
T1	0.99544419	0.97453083	0.71428571	0.61538462	0.69090909	0.83333333	0.99541284	0.97835498	1	0.78571429	0.69047619	0.83333333	0.84226495
T2	0.95175439	0.9566855	0.76190476	0.54901961	0.65217391	0.91666667	0.94155844	0.95509822	1	0.74074074	0.53191489	0.78571429	0.81193595
T3		0.28571		0.00000			0.00000	0.00000			0.00000		0.05714286
T4	0.90692641	0.91560102	0.84210526	0.86772487	0.69784173	0.875	0.89361702	0.91221374	0.95454545	1	0.87830688	0.84210526	0.88216564
T5	0.52420307	0.4087256	0.26732673	0.11124191	0.57519789	0.175	0.525444	0.43250912	0.21359223	0.30681818	0.11345806	0.16758242	0.31842492
T6	0.11567164	0.13112164	0.24509804	0.14876033	0.11724138	0.13333333	0.07210626	0.10526316	0.22727273	0.27835052	0.16034985	0.07969152	0.15118837
T7	0.99076212	0.98981077	1	0.86666667	0.62962963	1	0.99092971	0.9815864	1	1	1	1	0.95411544
T8	0.95404814	0.94393382	0.51282051	0.5	0.73469388	0.82352941	0.94588745	0.93066667	0.88	0.5952381	0.51851852	0.73809524	0.75645264
T9	0.51052632	0.45819015	0.47619048	0.13455657	0.33928571	0.26	0.65306122	0.72647702	1	0.40625	0.12570781	0.15745856	0.43730865
T10	0.18508997	0.18849133	0.6	0.0924225	0.10259797	0.16049383	0.1799591	0.18760315	0.11702128	0.375	0.09341501	0.10470836	0.19890021
T11	0.94006309	0.90775681	0.7826087	0.97080292	0.65137615	0.92307692	0.89217759	0.92637216	1	0.61538462	0.93063584	0.93396226	0.87285142
T12	0.68518519	0.68884269	0.31764706	0.39160839	0.30630631	0.4375	0.66169896	0.69246862	0.78571429	0.28723404	0.32889734	0.91625616	0.54161325

Appendix C4 Domain-agnostic – Dolly – F1 Score

	A1	A2	A3	A4	A5	A6	A7	A8	A9	A10	A11	A12	Average
T1	0.98980747	0.80066079	0.625	0.3943662	0.26297578	0.76923077	0.98636364	0.76914351	0.97674419	0.53658537	0.41877256	0.04716981	0.63140167
T2	0.96444444	0.95309568	0.66666667	0.37966102	0.36809816	0.84615385	0.9602649	0.95465171	1	0.74074074	0.34843206	0.1	0.6901841
T3		0.00373											0.00372787
T4	0.92494481	0.77321814	0.69565217	0.85863874	0.52010724	0.93333333	0.9190372	0.77262931	0.95454545	0.68292683	0.86910995	0.80808081	0.809352
T5	0.68783888	0.57996199	0.421875	0.19978858	0.71125612	0.29787234	0.688906	0.60333616	0.352	0.46956522	0.20324324	0.28197227	0.45813466
T6	0.20327869	0.22739726	0.390625	0.25899281	0.20858896	0.23529412	0.13036021	0.18589744	0.3030303	0.43548387	0.27431421	0.14761905	0.25007349
T7	0.97833523	0.77404667	0.13793103	0.125	0.1302682	0.44444444	0.98757062	0.78040541	0.84210526	0.13793103	0.14423077	0.00966184	0.45766088
T8	0.96781354	0.95180723	0.60606061	0.37540453	0.43373494	0.90322581	0.96467991	0.95398633	0.93617021	0.72463768	0.42682927	0.25	0.7078625
T9	0.47087379	0.41173443	0.41666667	0.20779221	0.36964981	0.40625	0.74418605	0.43483955	1	0.44067797	0.2063197	0.20070423	0.44247453
T10	0.31101512	0.31484904	0.32432432	0.16640827	0.18602794	0.27368421	0.30460367	0.31365936	0.20952381	0.27906977	0.17007435	0.18293432	0.25301452
T11	0.78318003	0.55979315	0.72	0.80606061	0.41399417	0.88888889	0.92039258	0.7616951	0.95238095	0.72727273	0.87978142	0.63461538	0.75400458
T12	0.81318681	0.79108635	0.48214286	0.54019293	0.43092522	0.60869565	0.79641256	0.79313099	0.88	0.44628099	0.48122392	0.90953545	0.66440114

Appendix D1 US Census Truth – Sample data

SSN	First	Last	House number	Street address	City	State	Zip	DOB-day	DOB-month	DOB-year	Phone	Occupation	Salary	TruthRowNum
213-21-3564	Walter	Johnson	2725	Cherry Blvd.	Baltimore	MD	21212	29	7	1929	(410) 475-4717	Chief Executive	$136,555.00	2
236-10-5542	Connie	Dorey	2244	North Ave.	Elgood	WV	24740	21	3	1925	(304) 436-7969	Food Servers, Nonrestaurant	$16,983.00	3
251-67-1413	James	Kurtz	1209	Church Ave.	Aiken	SC	29801	8	10	1944	(803) 507-7329	Customer Service Representative	$27,193.00	4
551-12-9616	Monica	Butler	1412	Ridge Dr.	Brooklyn	NY	11215	30	1	1929	(925) 385-6646	Real Estate Sales Agent	$43,414.00	5
601-28-5870	James	Plair	438	Church Blvd.	Mesa	AZ	85201	3	6	1976	(520) 796-6417	Educational, Vocational, and School Counselor	$45,285.00	6
680-30-3400	Dennis	Mcbride	2102	Hickory Dr., Apt. 31	Berwyn	PA	19312	6	12	1934	(702) 723-6879	Teacher Assistant	$19,261.00	7
163-18-1061	Geraldine	Nutting	1316	Thirteenth Blvd.	Altoona	PA	16601	11	2	1949	(814) 327-5531	Food Preparation and Serving Related Occupation	$17,907.00	8
167-72-3807	Marcella	Dewberry	2411	Railroad Blvd.	Manheim	PA	17545	8	6	1945	(717) 664-8442	Nursing Aide, Orderly, and Attendant	$19,664.00	9
563-52-1384	Carl	Moxley	811	Fifth St	Stanford	CA	94305	6	3	1935	(650) 345-2428	Cook, Institution and Cafeteria	$19,805.00	10
534-72-7183	Michael	Garver	359	Ridge Street	Tacoma	WA	98405	4	4	1907	(253) 404-7440	Retail Salesperson	$21,448.00	11

Appendix D2 US Census – Hard-to-match sample data

First name	Last name	Street num	Street address	City	State	Zip	DOB-day	DOB-month	DOB-year	Occupation	Salary	TruthFileRowNum
Evelyn		1634	Main Ave., Apt. 38		NJ	7054	30	9	1912	Food Preparation and Serving Related Occupation	$16,945.00	517747
Heaher	Riddle	2751	UNLake Street		WA	98604	25	1		Receptionist and Information Clerk	$21,849.00	647426
	Doss	234	River Avenue	Sterling Heights	MI	48310	17		1964	ZFirst-Line Supervisor/Manager of Mechanics, Installers, and RepairerIO	$51,144.00	686569
Helen	Martinez		Central Blvd.		AZ	85015		7	1940	Laborer and Freight, Stock, and Material Movers, Hand	$21,210.00	286365
Barbara	Garner	2987	Maple Dr	Carrollton	TX	75007		9	1994	Child, Family, and School Social Worker	$37,096.00	48744
Golda	Ott	1845	Sycamore St.	Phoenix City	AL		3	6		Retail Salesperson	$20,685.00	856815

Kevin		1727	Chestnut Ave	Rockaway	NJ	7866	11	1	1988		$18,864.00	185567
PChristopherF	MBaltzellG	2962	Lake St	Austin	TX	78749	13	11		Team Assembler	$25,211.00	271229
	GWalker	914		Saint Joseph	MN	56374	1	9	1985		$27,809.00	100570
MDonald	Clark	1090	Church St.	Alto		71269	5	4	1950	Office Clerk, General	$23,180.00	257353

Some of the key missing features of the hard-to-match file are – Missing SSN for all records, Missing last names, Incomplete address, and Incomplete or missing data of birth

References

[1] P. I. Fellegi, and A. B. Sunter. "A theory for record linkage," *Journal of the American Statistical Association* vol. 64.328, pp. 1183–1210, 1969.

[2] M. A. Hernández, and S. J. Stolfo, "The Merge/Purge Problem for Large Databases," *Association for Computing Machinery (ACM)*, 1995, pp. 127–138, doi: 10.1145/223784.223807.

[3] W. W. Cohen, "Integration of heterogeneous databases without common domains using queries based on textual similarity". In *Proceedings of the 1998 ACM SIGMOD international conference on Management of data*, pp. 201–212, June, 1998.

[4] B. Kilss, and W. Alvey, "Record Linkage Techniques-1985," *Proceedings of the Workshop on Exact Matching Methodologies Co-Sponsored with the Washington Statistical Society and the Federal Committee on Statistical Methodology.*

[5] M. Sreejam, and P. K. Wilson, "A Review on Rule Based Method for Entity Resolution," [Online]. Available: www.iosrjournals.org.

[6] M. Bilenko, and R. J. Mooney, "Adaptive duplicate detection using learnable string similarity measures." In *Proceedings of the ninth ACM SIGKDD international conference on Knowledge discovery and data mining*, pp. 39–48, August, 2003.

[7] L. Fang, L. Li, Y. Liu, V. I. Torvik, and B. Ludäscher, "KAER: A Knowledge Augmented Pre-trained Language Model for Entity Resolution," Jan. 2023, [Online]. Available: http://arxiv.org/abs/2301.04770.

[8] Y. Li, J. Li, Y. Suhara, A. Doan, and W. C. Tan, "Deep Entity Matching with Pre-trained Language Models," *Proceedings of the VLDB Endowment*, vol. 14, no. 1, pp. 50–60, Sep. 2020, doi: 10.14778/3421424.3421431.

[9] R. Peeters, C. Bizer0, and G. Glavaš, "Intermediate Training of BERT for Product Matching," [Online]. Available: https://github.com/Weyoun2211/productbert-intermediate.

[10] R. Peeters, and C. Bizer, "Dual-Objective Fine-tuning of BERT for Entity Matching," *Proceedings of the VLDB Endowment*, VLDB Endowment, pp. 1913–1921, 2021. doi: 10.14778/3467861.3467878.

[11] H. Zhou, W. Huang, M. Li, and Y. Lai, "Relation-aware Entity Matching Using Sentence-BERT," *Computers, Materials and Continua*, vol. 71, no. 1, pp. 1581–1595, 2022, doi: 10.32604/cmc.2022.020695.

[12] Z. Wang, B. Sisman, H. Wei, X. L. Dong, and S. Ji, "CorDEL: A Contrastive Deep Learning Approach for Entity Linkage," Sep. 2020, [Online]. Available: http://arxiv.org/abs/2009.07203.

[13] R. Peeters, and C. Bizer, "Supervised Contrastive Learning for Product Matching," *WWW 2022 – Companion Proceedings of the Web Conference 2022*, Association for Computing Machinery, Inc, Apr. 2022, pp. 248–251, doi: 10.1145/3487553.3524254.

[14] S. Mudgal *et al.*, "Deep Learning for Entity Matching: A Design Space Exploration," *Proceedings of the ACM SIGMOD International Conference on Management of Data*, Association for Computing Machinery, May 2018, pp. 19–34, doi: 10.1145/3183713.3196926.

[15] OpenAI, "Fine-tuning," Accessed: Sep. 16, 2023. [Online]. Available: https://platform.openai.com/docs/guides/fine-tuning/common-use-cases.

[16] Databricks, "Dolly", Accessed: Sep. 16, 2023. [Online]. Available:: https://github.com/databrickslabs/dolly.

[17] M. Akbarian Rastaghi, E. Kamalloo, and D. Rafiei, "Probing the Robustness of Pre-trained Language Models for Entity Matching," *International Conference on Information and Knowledge Management, Proceedings*, Association for Computing Machinery, Oct. 2022, pp. 3786–3790, doi: 10.1145/3511808.3557673.

[18] Meta, *Llama 2: Open Foundation and Fine-Tuned Chat Models.* n.d.

[19] B. T. Foua, X. Wang, J. Talburt, and X. Xu, (2023). "Train Once, Match Everywhere: Harnessing Generative Language Models for Entity Matching." In *Proceedings of the 2023 International Conference on Computational Science and Computational Intelligence (CSCI)*, Las Vegas, NV, USA., 2023.
[20] S. Mudgal (2018). "Datasets for DeepMatcher Paper," Retrieved from: https://github.com/anhaidgroup/deepmatcher:https://github.com/anhaidgroup/deepmatcher/blob/master/Datasets.md.
[21] M. Loukides, 2023 *Generative AI in the Enterprise*. O'Reilly, https://ae.oreilly.com/Generative_AI_in_the_Enterprise. Accessed 04/23/2024.
[22] J. Kaplan, S. McCandlish, T. Henighan, T. B. Brown, B. Chess, R. Child . . . D. Amodei, "Scaling Laws for Neural Language Models," *OpenAI. ArXiv*, 2020, Retrieved from: https://arxiv.org/pdf/2001.08361.pdf.

Index

https://doi.org/10.1515/9783111344126-019